RISK-BASED CONTAMINATED LAND INVESTIGATION AND ASSESSMENT

RISK-BASED CONTAMINATED LAND INVESTIGATION AND ASSESSMENT

JUDITH PETTS

Loughborough University

TOM CAIRNEY

W. A. Fairhurst and Partners

MIKE SMITH

M. A. Smith Environmental Consultancy

JOHN WILEY & SONS

Chichester · New York · Weinheim · Brisbane · Toronto · Singapore

Copyright © 1997 by John Wiley & Sons Ltd,
        Baffins Lane, Chichester,
        West Sussex PO19 1UD, England
        National   01243 779777
        International (+44) 1243 779777
        e-mail (for orders and customer service enquiries): cs-books @wiley.co.uk
        Visit our Home Page on http://www.wiley.co.uk
                or http://www.wiley.com

Other Wiley Editorial Offices

John Wiley & Sons, Inc., 605 Third Avenue,
New York, NY 10158-0012, USA

WILEY-VCH Verlag GmbH, Pappelallee 3,
D-69469 Weinheim, Germany

Jacaranda Wiley Ltd, 33 Park Road, Milton,
Queensland 4064, Australia

John Wiley & Sons (Canada) Ltd, 22 Worcester Road,
Rexdale, Ontario M9W 1L1, Canada

John Wiley & Sons (Asia) Pte Ltd, 2 Clementi Loop #02-01,
Jin Xing Distripark, Singapore 0512

Library of Congress Cataloging-in-Publication Data

Petts, Judith.
    Risk-based contaminated land investigation and assessment.
        p.   cm.
    Includes bibliographical references (p.   ) and index.
    ISBN 0-471-96608-8
    1. Hazardous wastes—Risk assessment.   2. Hazardous waste sites—Management.   I. Cairney,
    T. (Thomas)   II. Smith, Mike.
    TD1050.R57P48   1997
    363.738′4—dc21   97-7957
                                                                                                CIP

British Library Cataloguing in Publication Data

A catalogue record for this book is available from the British Library

ISBN 0-471-96608-8

Typeset in 10/12pt Times from the author's disks by Vision Typesetting, Manchester
Printed and bound in Great Britain by Biddles Ltd, Guildford and King's Lynn

This book is printed on acid-free paper responsibly manufactured from sustainable forestation, for
which at least two trees are planted for each one used for paper production.

Contents

Preface

Recognition of contaminated land as a potential environmental risk problem has come late. In less than 20 years there has been a move from a focus merely on the recycling of derelict urban land to a recognition that the risks which may be associated with currently operating industries, vacant industrial sites, and with land which has already been developed with no regard to contamination must be addressed. The concept of a systematic approach to dealing with contaminated sites has underpinned UK guidance for over a decade. However, it is only in the 1990s that there has been general recognition that dealing with such problems would benefit from a risk management approach, not least to ensure that limited resources are directed appropriately to the most significant risks.

Not surprisingly, such a "young" subject has been suffering from a dearth of texts. The idea for this book came from our direct observation of the deficiencies of site investigation and assessment in the context of managing risks. A series of short courses for engineers, environmental consultants and regulators, developed and run at Loughborough University, has provided additional background regarding the problems which people are facing in understanding the nature and role of risk assessment, and the information requirements to allow robust judgements about risks to be made. The two most fundamental problems we have observed have been the failure to recognise site investigation as a component of (not merely an adjunct to) risk assessment, and an over-reliance on measuring soil concentrations to compare with the (limited) available soil guidelines. Site investigation and assessment has often been an overly mechanistic activity undertaken with little regard to the nature and range of potential risks presented by sites or to the importance of high-quality data of the right type to allow judgements about risks to be made. There has also been evidence either of a sense of mystique or of disciplinary protectionism surrounding risk assessment. Some people familiar with quantified assessments, such as those which have formed the basis of the remediation of sites on the National Priorities List in the United States, seem unsure either about the implications of such assessments in risk management, or about the broader potential of risk assessment in decision making.

This book is intended to support students and practitioners in their understanding and practice of contaminated land risk assessment. To keep the book within reasonable bounds, we have limited the discussion to the information required to understand, and make judgements about, the nature and extent of risks presented by sites. Whilst setting this in the context of identifying and implementing appropriate remediation measures, these two, equally large, topics are not addressed here. We have assumed some familiarity with environmental processes and with the general contaminated land guidance, referring readers to appropriate texts for the necessary information. The other limitation is that ecological risk assessment is not specifically addressed. There is currently less attention to this issue than to human health and in many countries there is negligible experience. The subject is referred to but not covered in detail, as it has not been possible to apply significant practical experience to the discussion.

This is an integrated and multidisciplinary subject. The book provides for an understanding of the regulatory and social context in which contaminated land risks are assessed, the interdisciplinary requirements, as well as the methodologies and their limitations for investigation and assessment within the risk management context. Whilst referring to UK experience and situations, we have tried to provide readers with international comparisons. The issues which underpin effective contaminated land risk assessment are universal. Readers should note that for ease of reference the UK is referred to in discussion of legislation and regulatory agencies, although Scotland and Northern Ireland have separate systems of control, which, while similar in principle to that in England and Wales, differ in their regulatory basis.

Chapters 1–3 provide background to contaminated land risk management, developing approaches, the risk assessment framework, and to site investigation as a component. They "define" risk assessment as a process of making judgements about the nature and likelihood of adverse effects being realised. We support pragmatic approaches – qualitative, semi-quantitative and quantitative – to the making of such judgements. Ultimately, the definition of risk assessment is less important than an understanding of its objectives and what can be achieved.

Chapters 4–7 discuss site investigation as a component of the risk assessment process. They address the means of optimising the gathering of systematic information which is representative of the contaminant conditions and environmental pathways by which contaminants may move, and relevant to the understanding of the characteristics (sensitivity and exposure) of the targets (humans, water, flora and fauna, buildings) which may be at risk. The primary messages of these chapters are the importance of gathering information relevant to the source–pathway–target chains of concern and the value of multistage investigations which allow for resources to be directed to the most critical areas of potential risk.

Chapters 8–10 discuss the assessment of the collected information and data. Most sites have been, and will continue to be, assessed using generic guidelines and standards which are compared to measured concentrations of contaminants. No site-specific estimate of the risk to actual targets is produced, the generic guidelines acting as a surrogate. However, the effective use of such guidelines requires a full understanding of their basis and of their relevance to the source–pathway–target chains of concern. Chapter 10 addresses quantified, site-specific risk assessment including the derivation of remediation values. This is a large subject, and the objective here has been to provide readers with an understanding of the general approaches and the significant issues, particularly uncertainty, which form the basis of current developments in practice.

The final chapter links to the use of risk assessment within decision making. It discusses key issues which surround its use, not least risk communication and how interested parties can be involved in the process.

At the time of writing there has been significant development in the collation of new UK guidance on risk assessment. It is hoped that this will all be published by the time that this book is available. Most particularly, new guideline values for soil relevant to human health derived by a risk assessment approach will be published. This book is intended to complement the new guidance and to contribute underpinning knowledge relevant to its use by people with different backgrounds and responsibilities. Contaminated land risk management practice will face many challenges in the coming years. This text is presented as a contribution to consideration of best practice, and to assist in the education of young professionals who will take up the challenge.

The support of colleagues and of John Wiley & Sons in what has been an unintentionally extended period of writing is gratefully acknowledged. The help of Rachel Lindley with the final, laborious, wordprocessing has been invaluable. The support of friends and colleagues involved in the development of the risk-based approach has been important.

Judith Petts, Tom Cairney and Mike Smith
January 1997

Abbreviations

COUNTRIES

UK	United Kingdom
USA	United States of America

ORGANISATIONS

ACE	Association of Consulting Engineers, UK
ACS	American Chemical Society
AEC	Association of Environmental Consultancies, UK
AGS	Association of Geotechnical Specialists, UK
ASTM	American Society for Testing and Materials
ANZECC	Australian and New Zealand Environment and Conservation Council
BSI	British Standards Institution
CCME	Council of Canadian Ministers of the Environment
CEC	Commission of the European Communities
CIRIA	Construction Industry Research and Information Association, UK
COC	Committee on Carcinogenicity of Chemicals in Food, Consumer Products and the Environment, UK
GLC	Greater London Council
HSE	Health and Safety Executive
IACR	International Agency for Research on Cancer
ICE	Institution of Civil Engineers, UK
ICRCL	Interdepartmental Committee for the Redevelopment of Contaminated Land, UK
IEA	Institute of Environmental Assessment, UK
ISO	International Standards Organisation
IWM	Institute of Wastes Management, UK

LPC	Loss Prevention Council, UK
MAFF	Ministry of Agriculture, Fisheries and Food, UK
MVROM	Ministerie van Volkhuisvesting, Ruimteljke Ordening en Mileubeheer, the Netherlands
NAS	National Academy of Sciences, USA
NATO	North Atlantic Treaty Organisation
NNI	Netherlands Normalisatie Institute
NRA	National Rivers Authority, UK
OECD	Organisation for Economic Cooperation and Development
OMEE	Ontario Ministry of Environment and Energy
RCEP	Royal Commission on Environmental Pollution, UK
USEPA	United States Environmental Protection Agency
WDA	Welsh Development Agency
WHO	World Health Organisation

LEGISLATION

CDM	Construction (Design and Management) Regulations 1994 (UK)
CERCLA	Comprehensive Environmental Response, Compensation and Liability Act, 1980 (USA)
COSHH	Control of Substances Hazardous to Health Regulations, 1995 (UK)
HSW Act	Health and Safety at Work Act, 1974 (UK)
SARA	Superfund Amendment and Reauthorisation Act, 1986 (USA)

TERMS

ACQUIRE	Aquatic Information Retrieval Database
ADI	Acceptable Daily Intake
AERIS	Aid for Redevelopment of Industrial Sites
AOD	Above Ordnance Datum
CDI	Chronic Daily Intake
CEC	Cation Exchange Capacity
CLEA	Contaminated Land Exposure Assessment (Model) UK
CRO	Contamination Related Objective
DPM	Defense Priority Model
DSS	Decision Support System
EAI	Estimated Average Intake
EDTA	Ethylenediaminetrata acetate
ESES	Environmental Sampling Expert System
GC/MS	Gas Chromatography/Mass Spectrometry
GWL	Groundwater Level
HALO	Hazard Assessment of Landfill Operations
HI	Hazard Index

ICPS	Inductively Coupled Plasma Spectrometry
ICPS-AES	Inductively Coupled Plasma-Atomic Emission Spectrometry
LEL	Lower Explosive Limit
LNAPLs	Light Non-Aqueous Phase Liquids
MACs	Maximum Allowable Concentrations
MDI	Mean Daily Intake
MEI	Maximally or Most Exposed Individual
MF	Modifying Factor
MTR	Maximum Tolerable Risk
NAPLs	Non-Aqueous Phase Liquids
NOEL	No Observed Effect Level
OESs	Occupational Exposure Standards
OPRA	Operator and Pollution Risk Appraisal
PAHs	Polycyclic Aromatic Hydrocarbons
PCBs	Polychlorinated Biphenyls
PCP	Pentachlorophenol
QA	Quality Assurance
QC	Quality Control
RBCA	Risk-Based Corrective Action
RfC	Reference Concentration
RfD	Reference Dose
RMEI	Reasonably Maximally or Most Exposed Individual
SF	Slope Factor
TDI	Tolerable Daily Intake
TEM	Toluene Extractable Matter
TLC	Thin Layer Chromatographic Method
UCL	Upper Confidence Limit
UEL	Upper Explosive Limit
UF	Uncertainty Factor
VOCs	Volatile Organic Compounds

MEASUREMENTS

EC_{50}	Effect-specific concentration lethal to 50% of exposed organisms
ft^2	square feet
ha	hectare
kPa	kilo Pascals
K_d	soil solid/soil liquid partition coefficient
kg	kilogram
l	litre
LC_{50}	concentration lethal to 50% of those exposed
LD_{50}	dose lethal to 50% of those exposed
m	metre
m^2	square metre

m^3	cubic metre
mm	millimetre
mg/day^{-1}	milligrams of chemical exposure per unit body weight of exposed target per day
mg/kg	milligram(s) per kilogram weight of sampled medium
mg/litre	milligram(s) per litre of total liquid volume
mg/m^3	milligram(s) per cubic metre of total volume
pH	negative log of the hydrogen ion activity
ppm	parts per million
μg/kg	microgram(s) per kilogram weight of sampled medium
μg/m^3	microgram(s) per cubic metre of total volume
μg/l	microgram(s) per litre of total fluid volume

1

Contaminated Land Risk Management: Policies and Issues

1.1 INTRODUCTION

Contaminated land, as a recognised environmental risk problem, is a recent phenomenon, coming relatively late to public attention and political concern compared to issues such as air and water pollution. Contaminated land is primarily a post-1800s problem in terms of cause, but a post-1970s phenomenon in terms of effect: if effect is equated with recognition and action to manage risks.

Lack of definitional clarity throughout the environmental and engineering literature serves to illustrate the diversity of awareness, concerns and priorities in relation to contaminated land as an environmental risk problem. "Soil pollution", "land pollution", "polluted sites", "waste disposal sites", "abandoned hazardous waste sites", "derelict land" and "marginal land" are all terms found in the literature. Some terms reflect specific interests and experience, not least in relation to abandoned hazardous waste disposal sites in the United States of America (USA). Some terms denote political interests which have focused on the barrier of contamination to the reuse of land as opposed to the protection of soil as a natural resource, not least in the United Kingdom (UK).

As with air and water pollution, contaminated land is the product of minerals extraction, industrial synthesis of chemicals, manufacturing growth, the generation of wastes (gaseous, aqueous and solid) and lack of control over their disposal. Direct emissions to air and water and their resulting visible environmental problems provide convincing proof of direct cause and effect. However, in relation to intentional and non-intentional emissions to land, the risks presented in terms of damage to the soil ecosystem, loss of soil resources, water pollution, damage to building materials, public health impact or loss of fauna and flora have been less readily defined and identifiable. Land, and more specifically its key component soil, acts as a sink, a filter and a bioreactor for contaminants (Alloway, 1990). It is a complex medium and different soil fractions and constituents have variable degrees of reactivity to any contaminants (Cairney, 1993). Most of the soil on the earth's surface is contaminated to a degree. The problem is to determine the point at which a contaminated site is determined to present an unacceptable risk. The latter is primarily a matter of socio-political imperative and priority.

In most developed countries, political awareness of the problem of land contamination has arisen from two primary sources:

- sites presenting an immediate risk which have come to public attention;
- the need for development of, and demand for, land for residential and urban use.

However, it would be wrong to consider contaminated land as solely a problem of long-industrialised countries. In many developing countries, with significant economic growth, rapidly developing industries and poor waste disposal infrastructure, land contamination is a growing, although as yet largely unmanaged (Butler, 1996), problem.

As the diversity of sources of land contamination and the potential scale of the problem in different countries have become apparent, there has been recognition of the need to adopt a structured and objective approach to identification, investigation and assessment. In this context there have been two primary stimuli promoting a *risk management* approach to the contaminated land problem:

(i) recognition that spending has to be prioritised because no country can afford to eliminate all of the environmental effects of human activity; and
(ii) the need to optimise the extent to which decisions are made on the basis of an objective consideration of the problem and possible actions, not least when the decisions will always have to be made with incomplete information (Ruckelshaus, 1983).

This book supports a risk management framework for the investigation, assessment and remediation of contaminated land. Figure 1.1 depicts the risk management framework. It includes:

(i) the process of *assessing risks*: i.e. the identification of potentially contaminated sites; an analysis of the hazards or harm that might arise from the exposure of vulnerable targets to the contaminants; an estimate of the likelihood that the hazards will occur; and an evaluation of the acceptability of this risk; and
(ii) the act of *reducing risks*: the selection, implementation and monitoring of appropriate remediation strategies, defined as any action to remedy a problem, including clean-up.

This book concentrates on the process of assessing risks. It promotes a structured, rational and pragmatic approach to the adoption of risk assessment approaches – qualitative, semi-quantitative and quantitative – of relevance to environmental, engineering and regulatory professionals involved in con-

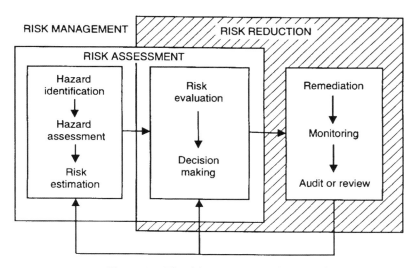

Figure 1.1. The risk management framework

taminated land management. It focuses on identification, investigation, assessment and evaluation so as to inform effective risk reduction decisions.

It is not the intention to duplicate the large amount of official guidance, not least the guidance in the UK being developed at the time of writing. Rather it is hoped to provide further background, and to identify divergent views and approaches, so as to optimise best practice. The book builds on practical experience of common failings in site investigation and assessment. This chapter defines the nature and scope of the contamination problem and introduces key issues surrounding the management of risks. The risk assessment process is discussed in more detail in Chapter 2.

1.2 CONTAMINATED LAND DEFINITIONS

1.2.1 Contamination versus pollution

In the UK, the definition of contaminated land has been a controversial issue, as much to do with political, financial and social interests as with concerns over the potential risks (defined in section 1.3) of contaminants. The roots of a definition can be found in the official discussions of *derelict* land which emerged in the 1960s: i.e. "land so damaged by industrial or other development that it is incapable of beneficial use without treatment" (Ministry of Housing and Local Government, 1963). This definition provides no indication of the potential risks except in as much as the use of the land could be restricted. The British Standards Institution (BSI) adopted the basis of this view in its draft standard relating to the identification, investigation and assessment of contaminated land (BSI, 1988). In

1990 the Department of the Environment was criticised (House of Commons, 1990) for maintaining the view of contaminated land as a subset of derelict land and so potentially underestimating a genuine environmental problem.

Scientific distinction exists between *contamination* as the mere presence of a foreign substance, possibly harmless, and *pollution* where causation of harm is inherent. The Royal Commission on Environmental Pollution (RCEP) supported this scientific distinction, stressing that contamination is therefore a necessary, but not sufficient, condition for pollution (RCEP, 1984). In an earlier, but enduring, definition of pollution – "the introduction by man into the environment of substances or energy liable to cause hazards to human health, harm to living resources and ecological systems, damage to structures or amenity, or interference with legitimate uses of the environment" (Holdgate, 1979) – the potential to cause harm can be seen to be an initiator of action to investigate contamination.

Herein lies the crux of decisions related to contaminated land. Once a site or a group of sites has been identified as potentially containing contaminants that may cause harm, social expectations may require that action be taken and that, in the absence of evidence to the contrary, the land must be treated as polluted. Regulatory and economic provision has to be made for these expectations to be met. Concern that social expectations will equate contamination with pollution was one of the main reasons for the widespread institutional opposition to the proposals to create "Registers of Contaminative Land Uses" (Department of the Environment, 1991) in the UK.

The NATO Committee on Challenges of Modern Society (NATO/CCMS) (Smith, 1985), echoing Holdgate (1979), defined contaminated land as:

> land that contains substances that when present in sufficient quantities or concentrations, are likely to cause harm directly or indirectly, to man, to the environment, or on occasions to other targets.

The UK government adopted this definition in its report on sustainable development (Anon, 1994). However, it did not recognise the NATO/CCMS broad view of "targets" that may be harmed, including not only people and ecosystems, water and air, but also physical structures and hence property and financial interests.

1.2.2   Harm

Harm is defined in UK legislation (Environmental Protection Act 1990) to include harm to humans, living organisms, offences to any of man's senses and harm to property. Harm may include physical damage, acute or chronic health damage or damage to financial investments.

In 1997, England and Wales implemented the provisions of the Environment Act 1995 which relate to contaminated land. Under the Act land is defined as contaminated only if it is causing, or has the potential to cause, "significant harm" or "pollution of controlled waters". However, this is only in respect of specific powers of the local authorities and the Environment Agency to take enforcement action to require the remediation of sites. In other words, this can be considered a limited definition for defined legal purposes rather than a general definition of contaminated land.

In specifying pollution of controlled waters the legislation has rectified the low priority which has historically been given to the effects of land contamination on groundwater; partly reflecting a lower national perception of the importance of groundwater as a resource, as on average only 35% of public water supply demand in England and Wales is met from groundwaters. The awareness of the importance of groundwater increased with the formation of the National Rivers Authority in 1989 and the production of the first groundwater protection policy (NRA, 1992).

Most legislative systems stress long-term (chronic) as well as short-term (acute) human health impacts. In the USA, the Environmental Protection Agency (USEPA) is authorised to undertake emergency removal programmes for sites which might catch fire or explode or which in some way might result in acute human exposure. However, by far the largest amount of money has been spent on the remediation of sites where chronic health effects might result. This policy was founded in the belief that long-term exposure to low doses of industrial contaminants (particularly synthetic chemicals) could cause cancer and other grave illnesses. However, despite the identification of thousands of sites across the USA where chemicals have to some degree migrated into the surrounding environment, there is no peer-supported epidemiological evidence of inactive waste sites causing chronic illnesses (Wildavsky, 1995). In considering human targets, these might include occupants of a site, investigators working on a site, recreational users or visitors, and residents living in the surrounding area.

The World Health Organisation defines human health as a "a state of complete physical, mental and social well-being and not merely the absence of disease and infirmity". Thus, if we are concerned about harm to humans, direct health impacts in terms of disease and physical disorders should not be our only concern, even though they may be more easily assessed and criteria for determining acceptable degrees of harm more readily determined (discussed in Chapter 2). Consideration of the impact of contaminated land may be extended to include detrimental effects on the amenity of an area which prevents its continued enjoyment, and psychological effects including fear and stress.

A concentration on human health impacts has, until the 1990s, led to a lower priority being attached to the assessment and management of ecotoxicological risks (toxic effects on non-human organisms, populations and communities), even though non-human organisms may be sensitive to certain contaminants at levels below those set as the safe threshold for human health.

The concept of harm, if truly reflective of targets which might be at risk from soil contamination, should include damage to property. The UK Environment

Act 1995 definition of significant harm in this respect refers to damage or failure such that a building ceases to be capable of being used for the purpose for which it was intended.

1.2.3 Definition by source

Two sources of contamination are relevant:

- point;
- diffuse.

The first usually results from the direct introduction of contaminants from a point source, such as accidental spillages or leakages of chemicals at a manufacturing or storage site, or disposal of wastes to land. The second results from indirect transfer of contaminants to soils and usually affects much larger areas. Diffuse contamination includes lead fall-out from car exhaust fumes, deposition of sulphur dioxide and other acidic chemicals emitted to atmosphere in industrial areas, and the spreading of nitrogen and phosphate fertilisers, pesticides and sewage sludge for agricultural use.

Point sources are easier to control by regulatory action (usually end-of-pipe, i.e. at the discharge or disposal point) than are diffuse sources. In a review of environmental risk reduction for the US government, the Environmental Protection Agency's Science Advisory Board viewed the contemporary problems of diffuse sources of pollution as one of the challenges for the 1990s (USEPA, 1990a).

In the UK, the institutional and regulatory compartmentalisation of the problem of contaminated land has led to an artificial definition which, for example, left all pollution sources relating to diffuse sources and to underground mining activities outside of the suggested listing of contaminative land uses (Department of the Environment, 1991). Land has been defined as "any ground, soil or earth, houses or other buildings, the airspace above it, but excluding all mines and minerals beneath the land" (House of Commons, 1990). Concerns over the statutory exemption for discharges from abandoned mines led to proposals to end this long-standing anomaly (now incorporated in the Environment Act 1995). Incidents such as the pollution of the Carnon Valley and Falmouth Bay, Cornwall, arising from discharges of heavy metals (primarily zinc, cadmium and iron) from the Wheal Jane mine served to refocus attention. The discharge was probably one of the most visual pollution incidents ever recorded in the UK (NRA, 1994b).

Enhanced concentrations of potentially harmful substances occur naturally in the form of highly mineralised soils and radionuclides. The gas radon-222 has attracted particular concern because of its accumulation in residential properties. Resulting from the decay of radium-226 which occurs in small concentrations in rocks and soils, its highest concentration in commonly occurring rocks is in granite (Wild, 1993), although it can be found in widely diverse geological settings (Phillips, Denman & Barker, 1997).

The question arises as to whether such occurrences should be regarded as "contamination", taking into account the fact that the definition of pollution quoted above (Holdgate, 1979) and employed in some European Union legislation requires the contaminant/pollutant to be "introduced by man". The preference is to refer to "elevated naturally occurring concentrations of potentially harmful substances". The distinction is important, because while we can seek to avoid adding further contamination to land, we have to live with what nature gives us although recognising the potential hazards. For convenience the term "natural contamination" is used below, but readers should remain aware that this goes beyond the definition of contamination which underpins the discussion in the rest of the book.

The tendency to leave "natural" contamination out of regulatory definition is likely to reflect:

● the large areas which would be covered;
● the inherent political and psychological difficulty of accepting that areas where people have lived for centuries may be harmful to health;
● public perceptions, concerns and priorities relating to industrial pollution and the view that industry is the primary polluting source;
● the difficulties and cost of dealing with "natural" contamination, which usually relate to the installation of physical barriers between the contaminant and the target;
● the relative ease with which regulatory (and hence governmental) action can be directed to non-natural sources.

Even soils not influenced by cultivation practices and aerial deposition can display higher than normal levels of natural elements. In particular, the natural levels of inorganic elements, e.g. heavy metals, vary widely. Table 1.1 compares the "trigger concentrations" which have been used in the UK as a basis for considering soil contamination (discussed in Chapters 2 and 8) with the reported "normal" background range of some elements in the soils. The table highlights the problem of determining contamination levels relative to background levels.

1.2.4 "Soil" versus "land"

In some countries (e.g. the Netherlands) contaminated land is seen primarily as a *soil* issue, with soil a significant and essential resource that requires specific protection. Soil is:

● the control over element and energy cycles within the ecosystem;
● the essential substrate for terrestrial life;
● an essential biochemical filter or buffer between atmosphere and lithosphere (the earth's rock crust);
● a reservoir for water;
● essential for food and timber production.

Table 1.1. Comparison between normal heavy metal concentrations in soils with UK contaminant thresholds (mg/kg)

Metal	"Normal" range in soils	High levels from industrial influence	Threshold trigger concentration[3]
Arsenic (As)	0.1–50[1]	<9000[5]	10–40
Cadmium (Cd)	0.01–2.4[1]	168[6]	3–15
Chromium (Cr)	5–1500[1]	27–90[6]	600–1000
Copper (Cu)	2–250[1]	<900[6]	130[4]
Lead (Pb)	2–300[1]	250–37 200[7]	500–2000
Mercury (Hg)	0.01–0.3[1]		1–20
Molybdenum (Mo)	<1–5[2]		
Nickel (Ni)	2–1000[1]		70[4]
Selenium (Se)	<1–2[1]		3–6
Zinc (Zn)	10–300[1]	250–37 200[7]	300[4]

[1] Wild, 1993.
[2] Bullock & Gregory, 1991.
[3] ICRCL, 1987. Data represented threshold values at which significance of contamination relative to end-use should be considered. All data referred to total concentrations. Except for Cu, Zn and Ni, the lower figure referred to gardens and allotments, the upper figure to parks and open space.
[4] Data referred to potential phytotoxic hazards and any land uses where plants are to be grown.
[5] Bullock & Gregory, 1991 – soil from mining contamination, Cornwall.
[6] Bullock & Gregory, 1991 – soil near metal smelters in the lower Swansea valley.
[7] Morgan & Simms, 1988.

The 1972 European Soil Charter recognised soil as a vital and limited resource which is easily destroyed and which is a major support for human life and welfare (Council of Europe, 1972). The Charter was strengthened in 1992 with the addition of a further recommendation on soil protection (Council of Europe, 1992). In a country with the physical and hydrogeological characteristics of the Netherlands, the importance of the soil as a resource is immediately and directly apparent to every individual. The same could not be said to be true of the UK. Indeed, in a 1993 survey of public concerns about the environment on behalf of the Department of the Environment, neither soil pollution nor any form of related issue featured in the "top ten" list of concerns (Department of the Environment, 1994d).

The UK's 1994 Strategy on Sustainable Development was the first policy document that specifically identified soil as a sustainable resource (Anon, 1994). A study by the Royal Commission on Environmental Pollution (RCEP, 1996) has called for a much higher priority to be given to soil protection.

The Dutch criteria for clean-up have stressed the need to protect the *multifunctionality* of the soil, i.e. providing for any use in the future and specifically the most sensitive use. This contrasts with the UK approach which is that land should be "suitable for its current use" (Department of the Environment, 1994a). However, an alternative policy to that of multifunctionality has been suggested in the Netherlands, recognising the excessive cost of the latter, but still regarding soil as an essentially non-renewable resource (Palsma & Diependaal, 1993). The policy document suggested a form of land-use zoning which would permit lesser standards of remediation for some types of development, beginning to bring the Dutch approach more in line with others in Europe.

1.3 HAZARDS AND RISKS

1.3.1 The source–pathway–target framework

Much of the preceding discussion has served to introduce the key concept which is essential to effective understanding and management of the risks of contaminated land, i.e. the definition of the *source–pathway–target* framework. Chapter 2 discusses the framework more specifically in the context of risk assessment.

The *source* refers to the identified potentially contaminated site(s) and is defined as a function of the nature of the contaminant(s) that may be present and of the harm that they may present. The *pathway* is the route (direct and indirect) in the environment by which the contaminant(s) may be transferred to the *targets* of concern. The *target* (sometimes also referred to as the *receptor* or in ecological assessments the *endpoint*) is the point at which damage may occur if the contaminant is present at a level sufficient to cause harm. It should be apparent that in order for there to be a *risk* (i.e. the likelihood of harm being realised), and therefore for a degree of intervention or management to be required, the source–pathway–target framework (chain or linkages) must be complete. Thus, even where contaminants are present on a site above a background concentration or a guideline value of acceptability, there will not be a risk if there is neither a pathway by which the contaminant could reach a sensitive target, nor any targets on or in the vicinity of the site. This is a fundamental point.

The source–pathway–target framework is depicted conceptually in Figure 1.2, with the contaminated site (an abandoned waste disposal site) linked by direct and indirect exposure pathways to five potential targets (water, air, soil, humans, flora and fauna).

1.3.2 Sources

Table 1.2 presents an example list of contaminative land uses which may represent sources of potential risk.

Table 1.2 does not provide for a clear definition of sources. Any contaminative use on the list will only represent a significant source which warrants further consideration if the activities at the site involved contaminants of potential concern which could have been (or could be) released to the environment. The form of hazardous substances will vary on different sites depending on the specific processes used, and may include gaseous, solid and/or liquid phases. Table 1.3 lists some general categories of hazardous contaminants associated with sites.

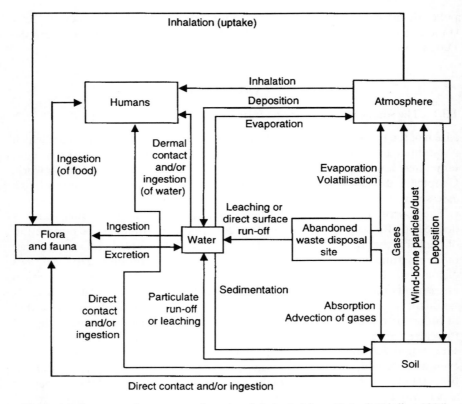

Figure 1.2. Source–pathway–target framework (adapted from Petts & Eduljee, 1994)

A large number of potentially harmful substances may be present on any one site, although in most cases their concentrations will be low. Contaminants will most often be located close to the point at which chemicals were processed, stored or used, and this has specific implications for the conduct of effective studies of site history (see Chapter 4). Where contaminants are mobile contamination may spread to adjoining ground and/or groundwater. This has implications for the understanding of the nature of the contaminants present and of the surrounding geology and hydrogeology.

1.3.3 Pathways

The pathway is the environmental route by which contaminants reach targets, i.e. it is the link between the source and a target which could be exposed. Within the pathway chemical release mechanisms (e.g. leaching), transport mechanisms (e.g. groundwater flow), transfer mechanisms (e.g. sorption), and transformation

Table 1.2. Potentially contaminative land uses

Agricultural/horticultural activities	Mineral works and mineral-processing
Airports	works
Animal-processing works	Munitions production and testing
Asbestos works	Oil refineries
Brickworks	Paint manufacturers
Burial of diseased livestock	Paper and printing works
Car breakers	Petroleum storage and distribution
Cemeteries	Power stations
Chemical works	Radioactive material storage or disposal
Coalmines and coal-preparation plants	Railways, railway goods yards
Docks and dockyards	Road haulage maintenance
Dry-cleaning establishments	Rubber processing
Electrical equipment manufacturers	Scrapyards
Engineering	Service stations
Food-processing industry, particularly pet	Sewage farms and works
foods or animal feedstocks	Sheep and cattle dip stations
Gasworks and other coal carbonisation	Shipbuilding and breaking
works	Tanneries and associated trades
Glassmaking and ceramics manufacture	Tyre manufacture
Hospitals	Wood preservative manufacture, storage
Iron and steel works	and use
Laboratories	
Landfills and other waste disposal sites	
Metalliferous mines	
Metal smelters, foundries and metal	
finishing	

mechanisms (e.g. biodegradation) will be important in determining the concentration of a contaminant at the point at which a target is exposed (e.g. residents drinking water from a well). Some environmental pathways and the fate of contaminants along these pathways are indicated in Figure 1.2.

It is important to appreciate the linkages between the presence or release of a contaminant and an alleged effect. The chemical form and physical properties of the contaminants determine behaviour in the environment. For potential human health effects a chemical must be toxic, mobile and in a bioavailable form. In the environment the chemical may change characteristics by combination and its concentration be modified by dispersion, dilution, degradation, adsorption etc. A potential pathway, perhaps through rock or soil, may not be complete or may be so extended (spatially and temporally) that at the point of contact with a target the chemical may no longer present a risk.

Significant contamination of environmental media is a prerequisite for human exposure and the latter is a necessary precondition for potential health effects. Chemicals received at the boundary of the body must penetrate three principal barriers – the lungs (inhalation), the skin (dermal contact) and the gastro-intestinal tract (ingestion). At each of these resistance is offered to the

Table 1.3. Hazard categories associated with contaminated sites

General category	Examples
Toxic gases	Carbon dioxide (CO_2), carbon monoxide (CO), hydrogen cyanide (HCN), chlorine, phosphine (PH_3), hydrogen sulphide (H_2S), sulphur dioxide (SO_2)
Flammable and explosive gases	Ethene, butane, methane (CH_4), hydrogen, carbon monoxide (CO), ethene, hydrogen cyanide (HCN), phosphine (PH_3), hydrogen sulphide (H_2S)
Combustible materials	Fuel oils, solvents, process feedstocks, intermediates and products: paper, grain, sawdust; spent oxides, clinker
Corrosive substances	Acids and alkalis, reactive feedstocks, intermediates and products
Zootoxic (toxic to animals) metals	Cadmium (Cd), lead (Pb), mercury (Hg), arsenic (As), beryllium (Be)
Phytotoxic (toxic to plants) metals	Copper (Cu), zinc (Zn), nickel (Ni), boron (B)
Reactive inorganic salts	Sulphate, cyanide (CN), ammonium, sulphide
Aliphatic hydrocarbons	Low molecular weight hydrocarbons, mineral oils
Aromatic hydrocarbons	Benzene, toluene, xylene, phenol
Polycyclic aromatic hydrocarbons	Naphthalene, pyrene, fluoranthene, anthracene
Substituted aliphatic compounds	Trichloroethane, tetrachlororethylene, brominated compounds
Substituted aromatic compounds	Pentachlorophenol, polychlorinated biphenyls (PCBs), polychlorinated dibenzodioxins (PCDDs), polychlorinated dibenzofurans (PCDFs)
Biological agents	Anthrax, polio, tetanus, Weil's
Radioactive substances	Radon, radium, cesium[137], actinides etc.

transfer of the chemical: the eventual intake is often less than 100% of the chemical presented. For the majority of chemicals even after intake, an adverse health effect can only occur if the concentration in the body exceeds a minimum level.

The full understanding of the pathways of exposure is an essential part of the risk assessment framework for contaminated sites. This places particular pressure on the site investigation stage, where a tendency to concentrate on gathering soil and water samples has been to the detriment of the collection of data on the nature of the surrounding environmental media, topography, geology, hydrogeology, land use and targets. The complexity of contaminant fate screening, analysis and modelling (USEPA, 1988a; 1991b) has led to a tendency to make conservative or cautious assumptions, i.e. that the contaminant will reach a target in a concentration equal to the highest concentration measured at the site. Such assumptions are justifiable only if they are properly discussed in the assessment (see Chapters 2 and 8) and the implications (often financial) can be fully understood in the risk management decision.

1.3.4 Targets

A sensitive target must be present if a risk is to be realised. *Risk* is defined as a minimum of a two-dimensional concept involving (i) the possibility of an adverse outcome, and (ii) a degree of uncertainty over the timing, magnitude or occurrence of that outcome (Covello & Merkhofer, 1993).

Table 1.4 lists five types of targets which may be of concern and some effects of concern. These effects may be of two types:

(i) *direct or first-order effects*, where the direct contact with the chemical and the target has an adverse effect (e.g. increase in ambient air concentrations, or increase in chemical concentration in receiving waters), or

(ii) *indirect or second-order effects*, in which the target is in contact with environmental media (e.g. air or water) or materials which have been contaminated by chemicals in other media (e.g. health effects through ingestion of food grown on contaminated soil).

As identified earlier, human health effects attract a great deal of attention, although evidence of actual adverse effects from contaminated sites worldwide is low (Grisham, 1986; Marsh & Caplan, 1987; Andelman & Underhill, 1987; British Medical Association, 1991; RCEP, 1996). Although recent work in the USA determined that both defects and other disorders are associated with living near sites, a causal link has not been proved and the location of sites in poor communities and industrial areas could explain disease incidence (Anon, 1996a). Public concern has often targeted chemicals which have a generally high profile (e.g. lead and cadmium), as well as specific chemicals such as mercury and polychlorinated biphenyls (PCBs), which have acquired a degree of notoriety as a result of highly publicised incidents such as Minamata Bay (discharge of mercury into the bay from a fertiliser plant resulting in serious ill-effects in humans) and the "Yusho" incident (PCB contamination of rice oil) (British Medical Association, 1991).

The effects of stress on ill-health also have to be considered. Residents on, or in the vicinity of, a contaminated site who experience an elevation in stress levels may be subject to an increased vulnerability to a range of disorders which are not specifically related to exposure to any chemical (Petts & Eduljee, 1994; Petts, 1994a). Psychological effects potentially connected to negative perceptions of health effects have been reported in a study of a closed industrial waste landfill in France (Deloraine *et al.*, 1995). Other socio-economic effects (such as blight on land values) have formed an important consideration in regulatory decision-making relating to contaminated sites, although they have rarely been measured objectively (Petts, 1994a).

Suter (1993) suggests five criteria which the definition of targets of concern will generally satisfy. Although he suggests these entirely within the context of ecological risk assessment, a closer examination suggests that they are relevant to the broader contaminated land management context:

Table 1.4. Targets and impacts of potential concern

Target	Impact
Environmental quality	Soil contamination, including possible long-term effects on soil's ability to support biodiversity and produce food
	Air pollution, including odour, lowering of air quality
	Surface-water pollution, including visual pollution, lowering of water quality
	Groundwater pollution
Flora and fauna	Death
	Reduced growth – population and individuals
	Stress
	Reproductive effects
	Toxic effects – acute and chronic
	Teratogenicity (damage to foetus leading to birth defects)
	Loss of habitat
Humans	Death
	Toxic effects – acute and chronic
	Carcinogenicity (development of malignant tumours and neoplasms)
	Teratogenicity (damage to foetus leading to birth defects)
	Mutagenicity (damage that results in changes in DNA structure in genes)
	Sensitisation (e.g. induction of dermatitis following skin contact)
	Adverse health effects as a result of stress
Materials and buildings	Explosion and fire damage
	Deterioration of materials by chemical attack
	Corrosion of materials and services
	Collapse due to subsidence
Socio-economic	Loss of land value
	Additional costs to develop
	Costs of long-term monitoring
	Delays in selling/developing land
	Stress in affected populations
	Perceived loss of amenity
	Disruption to personal plans
	Fear (particularly of potential health impacts)

- *Societal relevance* – i.e. the target is understood and valued by society and decision makers.
- *Biological relevance* – i.e. the importance of the target within the biological/ecological hierarchy.
- *Susceptibility to the source* – i.e. the potential for exposure and the responsiveness to exposure.
- *Unambiguous operational definition* – i.e. the potential for the target effects to be tested and modelled.
- *Accessibility to analysis and prediction* – i.e. the availability of data and appropriate models.

The first criterion is controversial. Suter suggests that satisfaction of this criterion leads to a concentration of concern about "cancer and birds" rather than "skin rashes and orbatid mites", i.e. what society dreads and values will provide the focus for attention in contaminated land risk management. Furthermore, this focus will change with time: partly through experience of different types of risks; partly as regulatory action begins to deal with certain types of risks, providing for attention to move to those previously given a lower priority; and partly as scientific knowledge improves in relation to recognition of the vulnerability of different components of the ecosystem.

1.4   ISSUES IN MANAGING RISK

1.4.1   The influence of problem sites

Contaminated site risk management internationally and nationally has been significantly affected by the identification of "problem sites". Love Canal, USA, Lekkerkerk, the Netherlands, and Loscoe, UK, provide key examples. These cases have five particularly important features which can be seen to be replicated in reviews (e.g. USEPA, 1986; 1992a) of other "problem" sites and which should provide important lessons for our current approach to the risk management of potentially contaminated sites:

(i) They relate to sites where the reason for the contamination was the known disposal of waste materials at a time when the consequences of that disposal were neither fully recognised nor understood.
(ii) They represented neither examples of necessarily "bad" practice at the time the activities were undertaken nor of illegal practice.
(iii) The disposal practices assumed that there would be a degree of ongoing institutional control (e.g. land-use control in the case of Love Canal) which would mitigate any pollution problems, but which in the event was not forthcoming.
(iv) There was a delay between public reports of effects and official linkage of these effects with contamination. The source–pathway–target chain was not immediately recognisable to officials and the process of assessing and evaluating the risk added to public concern and the loss of credibility in the risk assessors and in management.
(v) They each led to significant regulatory intervention.

Love Canal, USA

Love Canal was a 6.5 hectare (ha) chemical-disposal site located about 1.5 km from the industrial area of the city of Niagara Falls. The area was used as a disposal site for waste chemicals as early as the 1920s. In 1953 the Hooker

Chemical Company had filled the concrete-lined disposal "canal", capped it with a clay cap and sold it to the Niagara Falls Board of Education (for a nominal sum of $1) for use as the site of an elementary school.

During the 1950s and 1960s the area grew rapidly as a residential suburb. Streets and utilities were cut through the clay cap and builders used soil from the site as fill for homes (Levine, 1981; Tarr & Jacobsen, 1987). There was no real public issue during this period, even though exposed chemicals from the disposal site caused some physical injuries (Levine, 1981; Brown, 1979).

During heavy rains and snow-melt in the winter of 1977–78 groundwater flooded basements and some buried drums broke through the canal cap. Analysis of the wastes from the basement of one home revealed organic chemicals, including benzene, polychlorinated biphenyls (PCBs) and other halogenated hydrocarbons. Reports of health defects, including headaches, respiratory discomfort, skin ailments and rashes, breast cancer, birth defects and miscarriages, began to circulate.

State and federal agencies began to investigate the site and to assess the risks. The state report released in the summer of 1978 seemed to confirm that a risk existed, at least to those residents who lived closest to the disposal site (New York Department of Health, 1978). A state of emergency was declared, the school was closed and pregnant women and children under two years of age living in close proximity to the canal were evacuated. In order to prevent further leaching of chemicals, remediation work was commenced.

Love Canal was to become a national media story. The state health department's attempts to limit the nature of the risk appeared callous and irresponsible in the face of the media stories of hardship, poor health and deformed babies. Finally, Congress approved an emergency appropriation allowing President Carter to spend $20 million on the relocation of residents and a loan and grant to New York State for the purchase of Love Canal homes.

The direct outcome of Love Canal was the Comprehensive Environmental Response, Compensation and Liability Act 1980 (CERCLA): the so-called "Superfund" legislation amended by the Superfund Amendment and Reauthorisation Act (SARA) 1986 and subject to further reauthorisation in 1995. CERCLA authorised the federal government to respond directly to releases or threatened releases of hazardous substances that may endanger human health, welfare or the environment. It established a federal trust fund – "Superfund" – to provide for prompt governmental remedial action. It also established a liability regime under which the US government can recover its expenses from "potentially responsible parties". Cleaning up abandoned hazardous waste sites has become one of the most costly undertakings of US environmental policy, with projected costs reaching hundreds of billions of dollars into the new millennium (Wildavsky, 1995).

Lekkerkerk, the Netherlands

In 1978, severe soil contamination was discovered under a new housing development at Lekkerkerk, a small village northeast of Rotterdam on the River Lek.
The housing had been built over the period 1972–75 on land reclaimed using
household refuse and industrial waste (USEPA, 1992a).

Residents had been in their new homes for up to three years when a workman
working on some underground service pipes was overcome by fumes. Prior to
this there had been a number of complaints about burst water pipes, noxious
odours inside the houses and toxicity symptoms in garden plants, but these had
not been connected with possible contamination problems. Excavations un-
covered a significant amount of chemical wastes, both loose and in drums. Over
1600 containers were found, some containing residual chemicals. Groundwater
was found to be polluted with heavy metals and organic pollutants, including
aromatic hydrocarbons, alcohols, ketones and esters. Plastic drinking-water
pipes were found to have deteriorated. Evacuation of residents commenced in
the summer of 1980. The policy laid down by the Minister for Public Health and
Environmental Hygiene was that all waste and all polluted soil had to be
removed and the materials processed in the Netherlands. Furthermore, 250
houses had to be abandoned temporarily and approximately 156 000 tonnes of
contaminated fill removed and transported by barge to Rotterdam for destruction by high-temperature incineration. Polluted water was treated in a
physicochemical purification plant.

The significant effect of Lekkerkerk was to galvanise the government to
introduce the Soil Remediation (Interim) Act 1983 which formed the basis for the
development of Dutch legislation and the risk control of contaminated sites with
proactive identification of sites, funding of clean-up, development of remediation
technologies (particularly *in situ* technologies) and the development of clean-up
criteria based on the multifunctionality approach.

Loscoe, UK

On 24 March 1986 at 6.30 a.m. a bungalow in Loscoe, Derbyshire was destroyed
by a methane explosion (Anon, 1986). The three occupants of the property
survived, although they were injured. Subsequent investigations revealed the
source of the methane to be an adjacent closed landfill site. A quantity of gas
migrating from the landfill through the sandstone strata had collected below
ground level in a void. As a result of an unusual and sudden drop in atmospheric
barometric pressure, gas was released from the void and an explosion was
triggered by the burner or pilot light of the central heating boiler in the bungalow.

Residents had been reporting dead vegetation and grass in their gardens,
smells and white fungus in the soil since closure of the landfill two years
previously. Investigations by local authority officials in response to the com-

plaints, by the local gas board, and also by the National Coal Board and the Mines and Quarries Inspectorate, had provided some indication that the vegetation damage and the gas might be connected with nearby mine workings; there appeared to be no connection made with the landfill (Anon, 1986).

In December 1987, as a direct result of Loscoe, Her Majesty's Inspectorate of Pollution sent a letter to all waste disposal authorities informing them of the hazards of landfill gas, the need for control and asking for all active sites and those closed in the last 10 years to be notified to the Inspectorate. In 1988, local planning authorities were required to consult their waste disposal authority on all development proposals on, or in the vicinity (within 250 m) of, active and closed landfills, and in 1989 the Department of the Environment produced Waste Management Paper No. 27 (Department of the Environment, 1989) providing guidelines on the management of landfill gas.

1.4.2   The scale of the contamination problem

As indicated by the above case studies, the identification of contaminated sites is the key to effective risk management, whether at the national, or local, government level or by corporate and institutional interests. There are two divergent approaches to identification:

(i) *reactive* – where sites are considered as the need arises, usually when some form of development is proposed or when there are public complaints about the condition of a site; and
(ii) *proactive* – where a formal system is introduced to survey an area to identify all sites (vacant and occupied) which may be contaminated.

Proactive mechanisms require:

(i) priorities to be identified in terms of land uses which may present significant contamination;
(ii) systems to be devised by which sites can be identified on a consistent, objective basis;
(iii) the development and management of systems for holding information on land where contamination is known, or suspected, to be present; and
(iv) prioritisation tools to identify those listed sites which should be investigated further.

The UK has witnessed over 15 years of activity to develop methodologies for identifying sites on a national basis, commencing with research projects for the Department of the Environment in 1980 and 1986 and work by the Welsh Office to develop a regional-level register of vacant sites in Wales (Welsh Office, 1988). Proposals for Registers of Contaminative Land Uses (Department of the Environment, 1991) foundered on the premise that a list of potentially contaminated sites which was made public would lead to blighting of land values until the actual state of the land could be proved. The UK government remained of the opinion that the normal process of development and redevelopment of land provided the best means of tackling much past contamination, although the powers of authorities to identify and act on land contamination have been re-emphasised through the Environment Act 1995.

The UK's primarily reactive approach has differed from that in other countries, including Norway (Jordfald, 1991), the Netherlands (Holtkamp & Gravesteyn, 1993), Austria (Kasamas, 1991), Canada (Foote, 1993), Denmark (Welinder, 1993), Germany (Franzius & Grimski, 1995), Australia (McFarland, 1992) and the USA.

Knowledge of the number of sites in different countries which may be contaminated land is relatively poor. Although it is possible to collate information on numbers of sites, it is not possible to make direct comparisons between countries because of the varied definitions of contaminated land and the basis of the figures (e.g. national surveys versus local surveys versus estimates derived from other surrogate sources). Table 1.5 provides some data. A distinguishing feature of the lists derived in different countries has been the fact that they have grown considerably over the years, and now all include significantly larger numbers of sites than originally estimated.

In the UK, significant areas of contamination relate to sites where the manufacture of coal gas was undertaken, with about 1000 sites still owned by British Gas plc, many of which are available for redevelopment. The sale and redevelopment of former military sites provides a significant example of contaminated sites requiring remediation to optimise the opportunities for change of use. As in many countries, former landfills represent the most numerous category of contaminated sites, with problems of identification caused by a lack of records on the location of sites where waste was deposited prior to the 1970s.

1.4.3 The liability dimension

Contaminated land presents a potential financial liability (risk) to a number of different parties:

- government and public authorities;
- corporate landowners;
- financial institutions;
- individual landowners;
- consultants and contractors who undertake investigation, assessment and remediation.

The scale of the contaminated land problem has been discussed in terms of the liabilities or costs of clean-up. As with figures on numbers of sites, the estimates produced throughout the literature have to be treated with caution. Nevertheless, the sums involved are significant. By 1994 Dutch government

Table 1.5. International estimates of numbers of contaminated sites

Country	Number of sites	Notes
Austria	1807	1995 figure of sites on register of potentially contaminated sites, assessed from 24 155 sites reported by provincial governments; 94% inactive landfills (Kasamas, 1995)
Belgium	2000	1995 figure of sites on inventory of potentially contaminated sites (Van Dyke, 1995)
Canada	5000	"High-risk sites" 1993 up to 10 000 possible; national inventory being compiled (Foote, 1993)
Denmark	2200	1989 figure of number of "problem" sites (Welinder, 1993)
Germany	143 252	1994 figure of suspected contaminated sites (Franzius & Grimski, 1995)
The Netherlands	120 000	1991 figure (Holtkamp & Gravesteyn, 1993); 25 000 may present a serious risk to the environment; over 60% are former industrial sites
Norway	2500	Sites registered in 1991 by surveys by State Pollution Control Authority; 50% landfills; register of landfills, mining sites, industrial activities which generated hazardous wastes (Jordfald, 1991)
UK	100 000 +	Estimate based on derelict land survey and other local and national sources
	5000–20 000	Estimate of number of sites which may present significant harm (House of Commons Select Committee on the Environment, 1996)
USA	1355	On the National Priorities List (1994)
	38 000	EPA database to be screened (Kovalick & Kingscott, 1995)

funding for remediation was to have reached Dfl365 million ($207 million approximately), compared to Dfl242 million ($138 million approximately) anticipated as coming from private sources (Harris, Herbert & Smith, 1997). In the former German Democratic Republic substantial public funds have been made available for remedial purposes, DM35 billion ($22 billion approximately) for the period 1994–2004 (Franzius & Grimski, 1995). Under the Danish Chemical Waste Site Act 1983, Dkr400 million ($68 million approximately) was committed for the ten-year period to 1993 (Harris, Herbert & Smith, 1997).

The relative role of government versus private sector liability and of the insurance and banking industry to "cover" contaminated land liabilities is a significant risk management issue in the 1990s. The "polluter pays" principle is advocated as the primary policy approach across Europe and North America. For example, the Dutch system identifies a hierarchy of liability: first the polluter, then the owner and in the last resort the government. The UK Environment Act 1995 places primary liability for remediation on the person who "caused or knowingly permitted" the contamination (i.e. the polluter), or if that person cannot be found the current owner or occupier. However, where the latter (perhaps an individual householder) is an innocent party, then costs of remediation fall on the public purse.

The size of liabilities which parties may face, not only where responsibilities for clean-up can be apportioned but also for the significant number of "orphan sites" where no responsible party is available, is placing considerable demands on the risk management process. Priorities for expenditure have to be agreed based on an objective and structured assessment of the risks. Business and investment companies argue that any extension of liabilities for remediation could act as a disincentive to investment in potentially polluting processes and to the redevelopment of contaminated land (Advisory Committee on Business and the Environment, 1993).

For industry, managing the liability risk is not only about managing or minimising the costs of remediation and of civil or criminal legal action, but also the costs of loss of business, of land value, and of societal and institutional confidence in the event of a contamination incident. Since the mid-1980s there has been growing recognition of the need to expand insurance cover to provide for potential liabilities. However, at the same time cover has become difficult to obtain for some of these risks as insurance companies have been faced with pollution liabilities. In a number of European countries the reluctance of individual insurance companies to bear the risks of environmental pollution has resulted in national "pooling" arrangements, e.g. GARPOL in France, the Italian *Pool Inquinamento*, the MAS-Pool in the Netherlands (Berliner & Spuehler, 1990). In 1991 the Association of British Insurers advised its members to change the wording of their policies to limit cover to "sudden and accidental events", although a few companies are providing environmental liability cover.

The modern, environmentally alert corporation now realises that insurance has to be the "last resort" and that risk management is primarily about pollution prevention and risk reduction (Shrivastava, 1993). Corporate risk management of contaminated land requires a three-pronged approach:

(i) identification of existing land holdings which may be potentially contaminated and investigation and assessment of priority sites to determine the need for remediation;
(ii) the implementation of environmental management systems (BSI, 1993a; CEC, 1993b) to ensure effective minimisation of future contamination problems, including ongoing auditing of activities to verify environmental performance; and
(iii) auditing of all potential acquisitions for contamination liabilities.

The rising corporate awareness is partly in response to the threat of retrospective strict liability, i.e. liability of a polluter without the need for the regulator or plaintiff to establish some degree of fault for contamination that occurred as a result of past activities. The US "Superfund" works on this basis and in 1993 the European Commission opened a debate on the subject of strict liability with a Green Paper on Remedying Environmental Damage (CEC, 1993b). The Green Paper examined the usefulness of civil liability as a financial tool for making those responsible for causing damage pay compensation for the costs of remediation – i.e. making the polluter pay. The Green Paper did not refer to contamination or pollution but rather to the notion of "environmental impairment". Returning to mirror our theme at the beginning of this chapter, the Green Paper raises the thorny question of determining the point at which contamination causes pollution and that at which pollution causes actual damage.

Finally, while emphasising business risks, liabilities may also be significant for site investigation and assessment consultants and contractors if they provide inappropriate and/or inaccurate advice or if the investigation itself results in environmental pollution.

1.4.4 Evaluating contamination risks

There are two overlapping approaches to the assessment of significance of risk (i.e. whether a site is contaminated to an extent that requires action):

(i) the *generic approach* in which guidelines or standards are developed which can be applied to all sites; and
(ii) the use of *site-specific criteria* that are developed on a site-by-site basis taking into account the actual hazards and exposure that prevail on each site.

Guidelines are numerical values issued by an authoritative body and intended to assist in the assessment of risk using appropriate professional judgement. Standards are limits made binding through legislation and must be applied in all cases where they are applicable.

Generic guidelines or standards can relate specifically to soil, water or flora and fauna as key targets for protection, i.e. as dedicated guidelines. Non-dedicated guidelines or standards derived for some other purpose can be used as surrogates for contamination decisions where the target and exposure pathway is appropriate; for example, air quality standards might be used where the route of target exposure is inhalation or acceptable or tolerable daily intakes (ADIs and TDIs) used where ingestion is the primary route.

The generic and site-specific risk assessment approaches have sometimes been portrayed as alternatives, with the site-specific approach equated with formalised, quantified risk assessments as used in the USA and the generic approach with a non-quantified assessment, often not even regarded as a risk assessment. The latter view has to an extent prevailed in the UK, without doubt to the detriment of the adoption of structured and professional approaches to contaminated land assessment. A reliance on the generic approach which indicates when a site is contaminated and when it is not has produced an almost obsessive search for figures to compare with criteria often without regard to their derivation or relevance. There has been a tendency to refer to guidelines developed in other countries relative to very different physical, geological and hydrogeological characteristics and social and political priorities.

Most countries with an active policy to deal with contaminated sites use generic criteria (Siegrist, 1989), and many countries have in recent years been revising these values either to broaden the basis of protection to include ecotoxicological as well as human health risks, or to extend the range of contaminants covered, or to present figures founded on a more risk-based model. In the UK, the generic approach was typified by the "action" and "threshold" trigger levels developed under the auspices of the Interdepartmental Committee for the Redevelopment of Contaminated Land (ICRCL, 1987). Revised guidelines derived using the Contaminated Land Exposure Assessment (CLEA) model (Ferguson & Denner, 1993a; 1994; Department of the Environment, 1997a) focus on pathways which could lead to direct risks to human health. In the Netherlands new target intervention values (Van den Berg, Denneman & Roels, 1993) have been derived to provide for ecotoxicological as well as human risk protection (discussed in Chapter 8). Australian and New Zealand soil criteria provide investigation levels developed using a typical residential setting, lifetime exposure factors and a young child as the "critical receptor" (ANZECC, 1992).

The generic approach has the advantage of convenience and ease of application, relatively modest demands on data and expertise, and it may encourage consistency of use, particularly across regulatory authorities.

However, a site-specific assessment may be required in circumstances where:

(i) relevant generic criteria are not available for potentially significant contaminants; or
(ii) available generic criteria would not be sufficiently protective, e.g. because of complex chemical mixtures, or proximity of highly sensitive human or ecological targets; or
(iii) available generic criteria would be overly protective as a result of the conservative assumptions used in their derivation; or
(iv) the contaminant concentration lies above the threshold value; or
(v) local background levels are high compared to generic guidelines; or
(vi) the site is already causing considerable local concern; or
(vii) owners or potential purchasers of an operating site wish to understand their liabilities.

The derivation of guidelines and also of acceptable criteria for specific sites represents a balancing of scientific knowledge (always limited) with social, economic and political priorities. The framework of guidance or regulation in which risk management decisions are taken has to balance the need to safeguard human health and the environment with the promotion of appropriate remediation. If guidelines for clean-up are set at levels which would make remediation prohibitively costly, redevelopment of many older industrial sites would not proceed, with consequent impacts on job creation and demands on greenfield sites (Ferguson & Denner, 1994).

Table 1.6 lists a range of clean-up criteria developed for use on a number of US Superfund sites in relation to the same contaminant – arsenic. The guidelines range from <1 milligram per kilogram (mg/kg) to 300 mg/kg depending on the type (i.e. generic versus site specific) and their basis (i.e. the protection required). Such a large range is not surprising, reflecting different source–pathway–target frameworks, divergent scientific approaches including exposure and dose-response assumptions, as well as social and political priorities and pressures depending on the nature of intended use of the sites and concerns over risks.

1.4.5   Risk management issues – conclusions

This chapter has provided an introduction to the primary issues relating to the risk management of contaminated land in order to set the context for the following chapters. At the time of writing, UK experience and practice are evolving from a position of reactive, limited assessments of sites based on restricted criteria and guidelines with relatively little professional attention to the source–pathway–target framework, to one where the risk-based approach is playing an increasingly important role.

A key question remains about the potential extent and impact of risk assessments in decisions on contaminated land. The highly structured and quantified assessments of the US Superfund have taken their toll in terms of some of the credibility of risk assessment as a decision-making tool. A Presidential Commission on Risk Assessment and Management reported in 1996 that an overemphasis on quantification was failing to provide for societal priorities to be taken into account (NAS, 1996). The reputed costs of Superfund and its relatively limited practical impact in terms of actual risk reduction have been affecting European views and fears of liabilities. A lack of monitoring of site remediations generally means that the impact of risk reduction measures has not been proven.

There is a new corporate, institutional and societal "mood" which supports the adoption of more proactive, structured, objective and transparent assessments of the risks presented by sites. The impact of the public demand for more direct involvement in risk decision-making generally will be felt in relation to contaminated land. Yet a pragmatic balance has to be maintained between comprehensive and robust risk decisions and the cost of risk reduction. The rest of this book sets out a framework for achieving this, by looking at site investigation and assessment as part of risk assessment in the context of the risk management process.

Table 1.6. Listing of remediation guidelines for arsenic reported in the Record of Decision for US Superfund Sites ($n = 16$) (Sheppard *et al.*, 1992)

Level mg/kg	Type	Derivation
< 1	Generic	Human health, cancer
1.1	Site-specific	Cancer risk, residential exposure
2	Site-specific	Human health
5	Site-specific	Human health risk assessment
7	Site-specific	Human health risk assessment
14	Site-specific	Background
14	Site-specific	Background
15	Site-specific	Groundwater, drinking-water criteria
20	Generic	Jurisdictional guidelines (applied to seven sites)
20	Site-specific	Human health risk assessment
65	Site-specific	Human health risk assessment
92	Site-specific	Groundwater guidelines
94	Site-specific	Risk assessment industrial land use
100	Site-specific	Human health
200	Generic	Guidelines set by independent agency
300	Site-specific	Human health, commercial land use, no leaching to groundwater

Note: for comparison the UK threshold trigger level for arsenic (ICRCL, 1987) was 10 for domestic gardens and allotments and 40 for parks, playing fields and open spaces.

2

Contaminated Land Risk Assessment: An Overview

2.1 INTRODUCTION

Chapter 1 introduced the risk management framework, consisting of two over-lapping activities: risk assessment and risk reduction. This book concentrates on *risk assessment*, addressing the requirements to ensure the robustness of the process but also to optimise risk reduction decisions. Good site investigation as a key component of risk assessment is an important focus of the book. This chapter provides an overview of the risk assessment process.

In its most basic form the application of risk assessment to human decision-making can be traced back thousands of years (see historical reviews in Covello & Mumpower, 1985; Paustenbach, 1989). The roots of the more formalised process that we know today are in the occupational and user chemical assessments of the 1930s, the nuclear and major hazard safety assessments commencing in the 1960s–1970s, and the cancer health assessments of the 1970s–1980s. These diverse roots have provided for fragmentation in the way in which risk assessment is perceived and discussed, and of terminology and objectives.

The primary influence on the discussion and application of risk assessment to contaminated land has been North American. In the context of contaminated land, risk assessment has been applied for the longest period and most directly (i.e. as part of regulatory control) in the USA. In 1976, the United States Environmental Protection Agency (USEPA) adopted the first policy for the use of risk assessment in relation to toxic chemicals that were potentially human carcinogens (USEPA, 1976). However, it was the publication of a landmark report on human health risk assessment by the National Academy of Sciences (NAS, 1983) that served to focus discussion and provided for a generic system applicable across a range of environmental health risk problems.

In the investigation of Superfund sites, risk assessment is a formalised method-ology used to determine the potential for human health impacts and to evaluate the benefits to be gained from remedial actions. USEPA has looked to extend its use of risk assessment to the characterisation of adverse ecological effects

(USEPA, 1992b). In 1996, Canada extended its guidance to include ecological risk assessment (CCME, 1996a). The bias which has been apparent towards human health risk assessment has in part arisen from a mistaken belief that protection of human health automatically protects non-human organisms and a failure to recognise that the latter may be more sensitive to exposure than are humans (Suter, 1993).

In the UK, risk assessment was used predominantly as a tool for managing engineering safety risks until the mid-1980s. Not until 1995 was official guidance published on risk assessment for environmental protection (Department of the Environment, 1995a). Guidance relating to risk assessment for contaminated sites (WDA, 1993; Harris, Herbert & Smith, 1995; Department of the Environment, 1997b; 1997c) has chosen to reflect the broader systems and engineering basis of UK experience (Royal Society, 1992) compared with the public health experience of the USA.

As presented in Chapter 1 (Figure 1.1) risk management represents the entire process of assessing risks, taking decisions on those risks and actions to reduce them as appropriate. Risk assessment has often been discussed in the technical literature as a process separate from the decision about control and risk reduction; presented as being predominantly scientific while risk management is primarily legal, political and administrative (Royal Society, 1992). The Department of the Environment guidance (1995a) has also not stressed risk assessment as a part of risk management but rather as an input to it, as depicted in Figure 2.1. However, this book does not find such distinctions helpful. As suggested by the UK's Interdepartmental Liaison Group on Risk Assessment (HSE, 1996, p. 3):

> a risk assessment is invariably a mixture of science and policy with one of the two predominating and with often no consensus where the boundary between the two lies.

Promotion of a risk management approach to dealing with contaminated sites recognises three main advantages (WDA, 1993):

(i) adoption of a systematic and objective assessment method;
(ii) provision of a rational, consistent, transparent and defensible basis for discussion of a proposed course of action; and
(iii) provision of an assessment of uncertainties.

Optimisation of these advantages requires that all of the activities relating to contaminated site identification, investigation and assessment are conducted with regard to the requirements of the risk-based approach. Risk management objectives and information requirements must drive all of these activities.

US experience has shown that risk assessments are often overly conservative evaluations, primarily due to the selection of assumptions to compensate for data limitations and uncertainties. There is a view that improved risk assessments must be reliant on accurate initial site characterisation (Anderson, Chros-

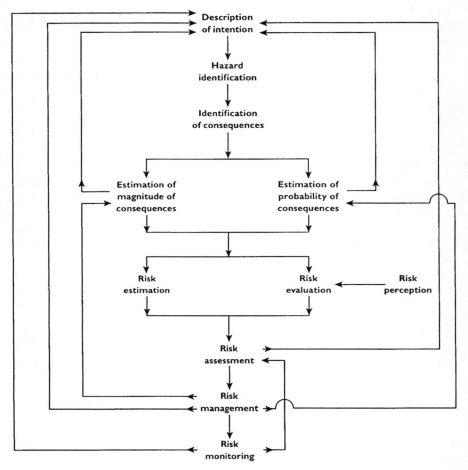

Figure 2.1. Risk assessment framework (Department of the Environment, 1995).
Reproduced with the permission of Her Majesty's Stationery Office

towski & Vreeland, 1990). This book concurs with this view: site characterisation, primarily by preliminary and detailed site investigations (Chapter 4), must support and inform the risk assessment process. Site investigation must be viewed as a part of the risk assessment process, not merely as a means of describing site conditions prior to an assessment of the risks.

The book also supports the view that the application of risk assessment in relation to a specific site must be with full regard to the uncertainties inherent in the assessment. Risk assessment is not a rigid set of procedures which are followed on every occasion. Rather it is a process which requires professional and expert application only to an extent which is sufficient and relevant to the situation.

2.2 RISK ASSESSMENT AND RISK MANAGEMENT DEFINITIONS

In Chapter 1 concepts of hazard and risk were introduced. A *hazard* is an event or situation (including a contaminant source) which has the potential to cause harm to targets of concern (human, ecological, physical, financial, psychological). *Risk* is a combination of the probability and frequency of a harm being realised with an estimate of its magnitude or scale.

Risk assessment

The term risk assessment is often applied loosely by both experts and laypersons, and can be used to describe a whole range of procedures, from a simple statement about possible hazards and risks, to formalised, quantitative risk estimates. In this book we stress risk assessment as a process of making judgements about the nature of potential adverse effects and the chance that they will be realised. The judgement may be expressed in a qualitative, semi-quantified or quantitative manner.

In the context of contaminated land, risk assessment can be applied to:

- the assessment of a group of sites to determine priorities for action or further investigation, i.e. site screening or prioritisation;
- the assessment of the risk presented by a single site;
- the derivation of clean-up criteria or action values for a specific site;
- the derivation of generic guidelines relative to the specific media and targets;
- the demonstration that generic criteria provide sufficient protection for a specific site;
- the balancing of risks and benefits;
- the consideration of long-term legal and financial liabilities for current and future landowners.

Risk assessment is "a systematic process for identifying and analysing the risks inherent in a system or situation and their significance in an appropriate context" (adapted from Royal Society, 1992). It seeks to identify, characterise and measure each of the three components of the source–pathway–target framework (see Chapter 1). Risk assessment also includes the element of risk evaluation, i.e. judging the significance of the assessed risk to determine whether it is acceptable or tolerable and the need for risk reduction or control measures. In some senses risk assessment can be regarded as a "diagnostic tool" for risk reduction (Asante-Duah, 1996), although, as already indicated, this book supports risk assessment as a "process" rather than merely as a scientific tool (discussed further in Chapter 11).

Returning to our definitions of harm and risk considered in Chapter 1, risk assessment is an appropriate decision process relevant to the consideration of human health risks, ecological and other environmental risks, financial risks and physical risks to property. However, it is primarily in relation to human health risks that its application has developed. Indeed, during the USEPA consideration of ecological risk assessment, it was noted, that while there are many similarities between the paradigm for human health risk assessment and the processes used for ecological risk assessment, there are also major differences (USEPA, 1992b).

A distinguishing feature of ecological risk assessment is that it is not limited to how a particular species (or the most sensitive individual) is affected, as in human health risk assessment, but also addresses the complexity of factors affecting the health and vitality of whole ecosystems. The human health risk assessment is directed to identifying factors affecting susceptibility, i.e. describing not only how the population in general may respond, but also how individuals at high risk may be affected. The human health assessor is interested in all impacts, from subtle biochemical changes to severe toxic responses. The ecological assessor, while following the same procedural pathway, focuses on defining the impacts of exposures in terms of growth, maintenance and reproduction. Thus, the high-risk group is studied as part of a sensitive life-cycle evaluation (Calabrese & Baldwin, 1993; Suter, 1993; Maughan, 1993). Stress-response may be a more appropriate term than dose-response because of the importance of non-chemical stresses such as habitat alteration (USEPA, 1992b). For contaminated sites the ecological risk assessment may be evaluating whether adverse effects have already occurred, as well as predicting their potential occurrence perhaps resulting from development activity. Human health risk assessment is more frequently a predictive activity.

Risk management

Risk management is the process whereby decisions are made to accept a known or assessed risk and/or the implementation of actions to reduce the consequences or probability of its occurrence (Royal Society, 1992). It is dependent on the risk assessment as input to the decision. Risk management also includes risk reduction, comprising the decision and then the process of implementation of required actions (e.g. remediation of a site) and the monitoring and auditing of the effectiveness of those actions.

2.3   RISK ASSESSMENT

2.3.1   The risk assessment process – an overview

The risk assessment process includes the following stages:

   (i) Hazard identification.
  (ii) Hazard assessment.
 (iii) Risk estimation.
 (iv) Risk evaluation.

In simple terms, the four stages equate with the following questions:

(i) What is the possible problem?
(ii) How big a problem might it be?
(iii) What will be the effect?
(iv) Does it matter?

It is easy and convenient to consider risk assessment in terms of these four stages, and there is a tendency in such a definition to suggest that they are linear steps. However, in practice the most effective risk assessment is an iterative process, with the stages merging and blurring. The assessor must always be prepared to revisit previous assumptions, to return to collect further data, or to stop at a particular point if the potential risks are considered not to be significant.

The US Superfund public health evaluation includes five steps: (i) selection of indicator chemicals; (ii) determination of human exposures; (iii) estimation of human intakes; (iv) evaluation of toxicity; and (v) characterisation of the risk. The latter can be matched broadly with the definitions used in the UK as follows:

UK	USA
Hazard identification	Indicator chemical selection
Hazard assessment	Determination of human exposures
Risk estimation	Estimation of intakes and evaluation of toxicity
Risk evaluation	Characterisation of the risk

The risk assessment stages require information from the site investigation and assessment stages in the following manner:

- Hazard identification = • Site identification and site reconnaissance (preliminary investigation)
- Hazard assessment = • Exploratory through to detailed site investigation
- Risk estimation = • Detailed site investigation and site assessment
- Risk evaluation = • Site assessment

Nevertheless, again it is important to stress that this is not a rigid categorisation, merely an indicator of activities relevant to different elements of the process. There is often overlap: for example, some limited exploratory investigation may assist in the hazard identification.

2.3.2 Hazard identification

Hazard identification is the systematic identification of the hazards that may be associated with a site or a group of sites, considering both the existing or proposed use of the site and its environmental setting. Table 2.1 identifies some contaminants which may be associated with contaminated sites and places them in the context of the source–pathway–target chain.

The objectives of the hazard identification stage are to: (i) produce a qualitative understanding of the potential for the site to present a risk; (ii) highlight those sources of risk which will require detailed assessment; and (iii) enable the assessor to discount those sources as not requiring further assessment when the source–pathway–target chain (or linkages) may not be complete or plausible. This stage allows the assessor to develop a conceptual model of the source–pathway–target linkages. At this stage the objectives and interests of the risk manager can also be identified, e.g. regulatory requirements for groundwater protection or proposed development of the site for residential use, so that specific elements of risk can be considered in the light of these objectives.

The hazard identification will:

- determine the source of contamination;
- identify the specific chemicals/constituents of potential concern;
- identify the environmental media which could be affected;
- delineate potential contaminant migration pathways;
- identify potential targets, including their relative potential sensitivity to contaminants given their inherent characteristics and nature of exposure;
- construct a conceptual model for the site to assist in the focusing of further assessment.

The main stages of site investigation which contribute to the hazard identification stage are the desk study, site reconnaissance and possibly some initial limited exploratory work (see Chapter 4). Hazard identification aims to ensure that:

- all of the potential hazards likely to be present on the site and all of the targets which could be potentially affected are identified;
- the site investigation is adequately designed so that it is properly focused on the hazards of concern and can be conducted safely and effectively;
- potential acute risks are identified;
- the need for any immediate or interim remedial works (such as fencing of the site to prevent access) can be identified.

Covello & Merkhofer (1993) criticise the USEPA risk assessment definition in including hazard identification as the first step. They advocate that this should be a totally separate step *always* undertaken before the risk assessment process can commence, and that including it within the risk assessment definition has tended to downplay its importance. They relate hazard identification to the identification of hazard sources and the conditions under which they could produce adverse consequences, and prefer the term *release assessment* as the first stage of the risk assessment when the potential of the sources actually to introduce harmful substances into the surrounding environment is quantified.

Experience in the UK suggests that Covello & Merkhofer's concerns are justified. Failures to undertake good site characterisation and historical studies, often through pressures of time, can have a major adverse impact on the later stages of investigation and assessment, to the extent that either key potential hazards are totally missed, or too detailed an investigation is undertaken, so wasting money. Indeed, an important objective of the hazard identification stage is to screen out negligible risks so that the risk assessment can focus on the critical source–pathway–target scenarios.

The USEPA framework report for ecological risk assessment refers to *problem formulation* (USEPA, 1992a). This is viewed as a planning and scoping process, the end result of which is to identify the environmental values to be protected, the data needed in order to conduct an assessment, and the relevant analyses to be used.

2.3.3 Hazard assessment

The hazard assessment stage allows for the conceptual model, derived from the hazard identification, to be refined. It assesses the degree of hazard associated with a site or a group of sites. Hazard assessment aims to understand and describe the plausible and critical pathways by which substances could reach targets, the fate of the substances in the environmental media through which they are transported or move and the characteristics of the targets at risk. Figure 2.2 (Harris, Herbert & Smith, 1995) provides examples of possible exposure pathways related to human exposure in particular which may need to be considered at the hazard assessment stage.

Hazard assessment is dependent on good site investigation data (from exploratory and detailed investigation) to provide observed contaminant concentrations (Chapters 5–7). The information and data collected during the investigations must be relevant to the source–pathway–target scenarios of concern; representative of the contaminant conditions; provide for the understanding of the likelihood of target exposure; and be relevant to any generic guidelines which will be used.

The hazard assessment stage could stop with the comparison of measured soil and/or water and/or air concentrations with existing generic guidelines and standards which are applicable to the pathways and targets of concern (Chapter 8). This provides for plausible exposure scenarios for different targets to be tested against guidelines indicating concentrations which are judged not to present a risk.

For example, if the primary identified target at risk on a site is a child who may ingest soil, then either a soil guideline value which is known to protect against this target-specific risk for the contaminant of concern would be used or, as a surrogate, an acceptable or tolerable daily intake for that contaminant might be

Table 2.1. Source–pathway–target chain examples*

Chemical	Characteristics	Sources	Hazards	Pathways	Targets/risk
Arsenic (As)	Widely distributed in environment. Green colour; spoil heaps deposits mostly white. Organic forms less toxic than inorganic forms. Toxicity affected by concentrations of other metals, especially iron. In soils predominantly in an adsorbed form. May become more mobile in alkaline soils. Arsenate dominates in aerobic soils; arsine and elemental arsenic in reduced conditions.	By-product of copper and lead smelting; wood preservatives; timber treatment; agricultural chemicals; electroplating; sewage sludge; coal burning.	Toxic. Soluble inorganic compounds of As(III) considered principal toxic species. Water pollution. Proven systemic carcinogen.	Inhalation; direct soil ingestion, dermal contact. Accumulates in roots of vegetables etc. rather than in leaf. Humans are normally exposed through food, air, water.	Primarily humans – acute effects with ingestion; irritation of gastrointestinal tract. Inhalation – irritation of bronchial tubes. Potential cancer of lung, liver, bladder through ingestion.
Cadmium (Cd)	Natural occurring but non-essential heavy metal. Coloured deposits white, yellow, orange. Low pH increases toxicity with enhanced mobility in soil. Soil temperature, texture, moisture and redox potential all affect toxicity	Metal mine waste; metal smelters; incineration; coal burning; sewage disposal; manufacture of batteries, fertilisers and pesticides; ceramics; pigments and glass manufacture; foundries; electroplating.	Toxic. Phytotoxic. Water pollution. Covanogen by inhalation.	Direct soil ingestion; inhalation; plant up-take; food. Humans are normally exposed through water, food and air.	Humans – chronic long-term exposure may lead to kidney damage, lung damage. Plants.

Chromium (Cr)	Hard, brittle grey metal. Salts have strong and varied colours. Hexavalent compounds are most relevant. All are soluble in water and/or acids.	Metallurgy industries; fly ash; sewage sludge; timber preservatives; pigments tanning; plating; mining/smelting; chemical industries; natural occurrences.	Hexavalent compounds carcinogenic. Corrosive effect on tissue. Phytotoxic. Water pollution.	Direct contact. Air-particulate inhalation. Water; plant uptake; indirect to humans by ingestion.	Humans – ulcers, dermatitis, bronchial carcinoma. Plants.
Copper (Cu)	Malleable, ductile, reddish colour; non-combustible except as a powder. Commonly occurs as sulphates, sulphides and carbonates in the soil. Many organic and inorganic compounds, some soluble in water.	Smelting; waste from electroplating, chemical and textile industries. Waste from manufacture of pesticides and pigments and antifouling paints; shipbuilding industries; scrap yards; sewage sludge; wood preservatives.	Toxic. Irritant. Phytotoxic (especially at low soil pH and low organic matter). Corrosive to rubber.	Air-inhalation; ingestion; dermal contact; plant uptake.	Humans – chronic toxicity rare. Plants. Materials (rubber) – corrosion.
Cyanides (CNs)	Complex (e.g. ferri/ferrocyanide; blue or blue/grey may cause staining of soils etc. Thiocyanate – red staining of soils and watercourses; HCN odour of sweet almonds. Spent oxides, musty odour. Simple salts (e.g. potassium or sodium cyanide) present greatest risk to humans.	Iron and steel manufacture. Spent oxides from town gas manufacture; electroplating effluent; non-ferrous metal production; photography and pigment manufacture.	Simple cyanides toxic; phytotoxic; groundwater pollution; toxic to fish. Cyanide salts very soluble.	Plant uptake; water; humans – absorbed from all entry routes.	Plants, fish, humans – chronic exposure, nervous system disturbances; dermatitis. Thiocyanate low acute toxicity.

Table 2.1. (*cont*)

Chemical	Characteristics	Sources	Hazards	Pathways	Targets/risk
Lead (Pb)	Heavy, ductile, soft grey, solid, insoluble in water.	Natural occurrences. Mining and smelting; batteries; scrap metal; petrol additives; pigment; paints; glass manufacture; fluorspar; mine waste.	Toxic. Water contamination.	Air inhalation; ingestion; water supplies.	Human – pica children who ingest dust, consumption of water, ingestion of contaminated crops – behavioural disorders, central nervous system, blood and kidneys affected.
Mercury (Hg)	Silvery, heavy, insoluble in water, highly volatile; forms inorganic and organomercury compounds and amalgams with other metals.	Mining and smelting; electrical apparatus; paints; catalysts; dental products; plastics; wood preservatives; burning of coal, gas, wood and oil; iron and steel works; foundries; electroplating; glass manufacture.	Toxic, phytotoxic.	Ingestion; skin absorption. Air inhalation, water.	Humans – skin contact causes burns and blistering; absorption results in digestive and nervous symptoms. Chronic exposure to vapour produces neuropsychiatric symptoms. Plants.
Nickel (Ni)	Malleable, silvery metal; inflammable as a dust or powder.	Refining of impure nickel oxide; wastes from metal-finishing processes including electroplating, alloy and stainless steel manufacture; enamel and battery production; sewage sludge; shipbuilding; scrap yards.	Toxic; fire (dust or powder). Phytotoxic especially in acid soils.	Dermal contact; air inhalation; ingestion.	Humans – carcinogenic effects associated with occupational exposure and children, dermatitis, rhinitis, and chronic pulmonary irritation. Plants and indirect to humans via ingestion.

Zinc (Zn)	Shining white metal with bluish-grey lustre. Most simple salts are water soluble.	Smelting of ore; wastes from metal finishing pigment; plastics and cosmetics manufacture; sewage sludge; shipbuilding industries; scrap yards.	Fire and explosion from dust; phytotoxic	Plant uptake; dermal contact with $ZnCl_2$ and ZnO fumes.	Plants – toxic synergistic effect with Cu and Ni especially at low pH. Low risk to humans.
Sulphur and compounds (S)	Sulphur compounds frequently white unless pigmented by cation – e.g. copper sulphate is blue. Sulphides in anaerobic conditions generally black. At pH < 4 hydrogen sulphide (H_2S) is liberated, giving odour of bad eggs.	Metal ores; waste from pigment manufacture; ceramics; spent oxide from gas manufacture (contain up to 60% free sulphur and up to 3% sulphate).	Corrosive, hydrogen sulphide at low pH phytotoxic, toxic (according to metal salt). Sulphur flammable.	Water.	Plants and construction materials. Humans from products combustion (e.g. SO_2).
Phenols	Class of aromatic organic compounds with characteristic antiseptic odour and acrid burning taste. Simpler compounds, soluble in water.	Coal carbonisation; wastes from gasworks (in coal tars) ammoniacal liquors; pharmaceutical dyes; rubber; solvents; paper; paints/wood preservatives manufacture; iron and steel.	Toxic, corrosive; phytotoxic; water contamination.	Air inhalation; ingestion; dermal contact (readily absorbed through skin); plant uptake; water; food.	Humans – tissue damage at >1%; ingestion causes intense burning, nausea; nervous system damage; materials – plastic water piping and rubber attacked. May affect concrete; toxic to fish; taints fish flesh. Plants.

*Summary example information only

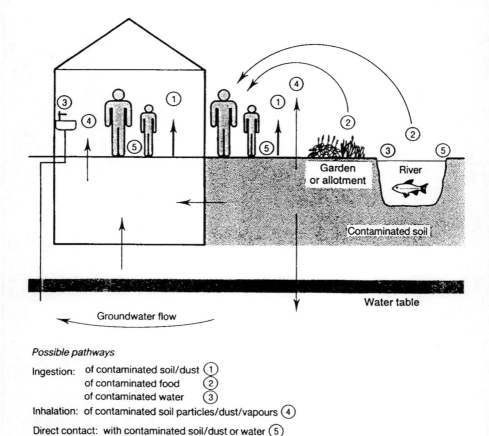

Possible pathways

Ingestion: of contaminated soil/dust ①
               of contaminated food   ②
               of contaminated water   ③

Inhalation: of contaminated soil particles/dust/vapours ④

Direct contact: with contaminated soil/dust or water ⑤

Figure 2.2. Examples of human exposure pathways (Harris, Herbert & Smith, 1995). Reproduced by permission of CIRIA

used. If the primary identified target at risk is groundwater which could be polluted by hydrocarbons, then a water quality standard relevant to that contaminant could be used to assess the risk.

Thus, hazard assessment does not produce a quantified estimate of the site-specific risks; rather it provides for judgement about risk through the comparison of site-specific data with appropriate national or international standards or guidelines, which in their derivation have included assumptions about acceptable exposures and intakes. The hazard assessment stage can also provide for an understanding of the contribution of a site to the risk to specific targets compared with the background exposure of those targets.

However, caution has to be exercised because:

(i) the available generic guidelines and standards may not include all of the substances of interest at the site being investigated;
(ii) the guidelines or standards may not have been derived for the source–pathway–target scenario of interest: for example, the UK ICRCL threshold trigger concentrations (ICRCL, 1987) for copper, nickel and zinc in soils related to their potential phytotoxic hazards rather than human toxicity effects; and
(iii) an observed concentration that is below a guideline or standard does not necessarily represent no hazard.

The hazard assessment will indicate either that:

(i) observed levels of contamination are unlikely to present a risk to specific targets and therefore no further action is required; or that
(ii) further investigation and/or assessment (perhaps involving site-specific risk estimation) is required so that the significance of observed levels of contamination can be judged; or that
(iii) the level of contamination is such that there can be no doubt that remedial action is required.

The hazard assessment stage can be a *qualitative* (i.e. primarily narrative or descriptive) risk assessment (see Chapter 8). The hazard assessment stage can also be used to compare the potential risks from a number of sites. Termed *hazard ranking* or *hazard prioritisation*, it is common for some type of scoring system to be devised which puts all information about sites, including known site contamination characteristics, onto a common scale. This type of assessment may be *semi-quantitative* (i.e. semi-numerical). The objective is to allow those sites which have the potential to present a significant risk to be identified from a group of sites so that further investigation or assessment can be appropriately targeted. Chapter 9 discusses hazard ranking and provides examples of systems developed in different countries.

In this book, we use the term "qualitative" risk assessment to describe judgements about risks made using generic guidelines as a surrogate for a risk estimate. We reserve the term quantitative risk assessment for situations where a site-specific estimate of risk to defined targets is derived. As generic guidelines have started to be derived using quantitative risk assessments, there has been a suggestion that these enable risk estimates to be derived for a site, because they have in-built judgements and assumptions about exposure and the dose-response of likely targets. However, we believe that it is important to retain a questioning view of the relevance of generic guidelines to specific sites, to ensure that if used they are adequately protective of targets at that site and cover all contaminants of potential concern.

2.3.4   Risk estimation

Risk estimation is the process of estimating the probability that an unwanted "event" or outcome will occur under defined conditions (Harris, Herbert & Smith, 1995). Both quantitative and semi-quantitative estimates may be the outcome.

If the hazard assessment indicates that the frequency and level of exposure are likely to be high and effects significant, then a site-specific estimate of the risk may be required. Other situations in which such an estimate may be required include where there is public concern about a site and only a full and objective assessment of the site-specific risks will be acceptable, or where local background concentrations or contaminants are high relative to generic standards, prompting consideration of the contribution of the site to local environmental burdens.

The full quantification of risk can be expensive, time consuming and problematic, and is unlikely to be justified except where dealing with complex, high-risk and/or contentious problems. Certainly, the quantification of risk allows for site-specific action and remediation targets to be derived. Furthermore, the insight into the behaviour of contaminants gained as a result of a detailed assessment may be beneficial in the selection and evaluation of alternative remedial strategies. In forcing the assessor to characterise fully the source–pathway–target framework, a quantitative assessment avoids the "list mentality" where only the scenarios covered by generic guidance are evaluated.

Risk estimation involves:

- exposure assessment; and
- effects assessment.

Exposure assessment

Exposure assessment is the process of measuring or estimating the intensity, frequency and duration of human or other exposures to the hazards of concern. The exposure assessment identifies the rate of contaminant movement and potential movement; the characteristics of the host medium (soils, rock, groundwater etc.); the extent to which environmental factors such as dispersion, dilution, degradation, adsorption etc. could modify contaminant concentrations at points along the pathway; and the characteristics of the exposure route that determine how much of the contaminant is taken in by the target. Some of this information will be relevant at the hazard assessment stage, but during the latter will be largely considered in a descriptive rather than a quantitative manner. Exposure assessment is often one of the most difficult components of the risk assessment, as the assessment often depends on factors which are hard to estimate and for which there are few data (Covello & Merkhofer, 1993). The exposure assessment places demands on site investigation which are often not fully appreciated. At this stage in the risk assessment process detailed site investigation data and information may have to be supplemented by further work.

The primary routes of human exposure which are considered in contaminated land assessments are inhalation of dusts and vapours; ingestion of contaminated food and water; direct ingestion of dust and soil (both intentional, as in the "pica" child who habitually eats non-food substances, and non-intentional); and skin contact with contaminated soil and dust. A significant area of uncertainty in the risk assessment of human health impacts is the strong influence of individual personal habits on exposure. For example, how much soil is ingested? This is a deceptively simple question which in itself has prompted a large and often contradictory literature (Ferguson, 1996). The assessment at this stage has to identify which groups may be affected given their likely exposure frequency and their sensitivity to exposure.

Exposure assessment is typically based on environmental transport and fate models which model the *likely* movement and dispersion in the environment using site investigation data, in the absence of monitoring and time-dependent data which can show how the contaminant is *actually* moving and behaving in the environment. The latter type of data is obviously preferred for risk assessment purposes, but is rarely available except in the case of sites already known to present a contamination problem, and even then it is rare that monitoring (e.g. of concentrations and contaminant dispersion patterns in air, or of pollution plume movement in the unsaturated zone) will have been proceeding long enough to provide comprehensive and representative data. However, contaminant transport models have a range of limitations. In general they often lack verification, i.e. the assumptions inherent within them have not been tested against actual data and behaviour. The utility of models depends more on the credibility of the data and the user than on the proven capability of the model to make correct predictions (Barnthouse, 1992; Keenan, Finley & Price, 1994). Chapter 10 discusses the data demands of environmental transport and fate models and identifies sources of such models.

Effects assessment

The results of the contaminant transport and fate analysis allow the assessor to evaluate targets contacting the contaminants. This *effects assessment* is the process of describing and quantifying the relationship between exposure to the contaminant(s) and the adverse health or environmental effects which result. The assessment requires a full characterisation of the targets (gender, age, general health status, species composition, physical properties of the building fabric etc.). In addition to the characterisation there is a need to determine how, and with what frequency and duration, contact may occur. For example, compare dermal exposure to contaminated water through daily showering with possible unintentional ingestion of soil, or with inhalation of vapours inside a property.

Traditionally the risk of exposure (i.e. contact with a chemical or physical agent) is estimated at the point of maximum exposure. Further conservatism can be built into the assessment by referring to the most sensitive individual or species. Human health risk assessments have often assumed the presence of a "most" or "maximally" exposed individual (MEI) or the reasonably maximally exposed individual (RMEI) based on 95 percentile exposure values. This hypothetical individual depicts a worst-case scenario with respect to the location, lifestyle and habits of the exposed population (discussed in Chapter 10). However, the conservatism or caution inherent in the consideration of the MEI while providing for the uncertainties in assessments to be addressed has often been misinterpreted as actual or plausible health risks. US practice has been moving away from this (USEPA, 1992c) in recognition of the impact on decisions – i.e. overcautious and expensive remedial requirements. One means of dealing with conservatism is the use of Monte Carlo analysis (discussed in Chapter 10).

The basis of the risk estimation is to determine the dose-response relationship between a pollutant and a target (whether humans, ecological systems or buildings). The general assumption is that there is a relationship between the concentration of a pollutant and the probability of an effect occurring and the magnitude of the impact. However, for many pollutants the nature of this relationship remains uncertain. As a result of the ethical considerations inherent in direct human studies, the basic assumption underlying all risk assessments is that adverse human health effects can be inferred from adverse health effects observed in experimental animals.

A wide variety of effects can be produced by exposures. The nature of an effect can be minor and temporary (such as a skin rash or irritation) or severe and permanent (such as irreversible organ damage or death). Either effect is a risk and its importance is often determined by legislation. For example, the Environment Act 1995 in England and Wales is limited to consideration of significant harm, which is defined as death or serious irreversible effects. However, the risks of minor injury may also be significant, and understanding the risks of more minor injuries is important in the management of sites and the protection of workers on sites.

Within the international literature and practice there are two divergent approaches to estimating human health risk: the threshold and no-threshold approach. It is common practice to group chemicals that produce effects into two categories: non-carcinogens and carcinogens. The estimation of risk is then simplified to these two broad categories of chemicals. For health effects arising from genotoxic carcinogens, i.e. those which affect DNA, the US approach has been to assume that there is theoretically no level of exposure to such a chemical that does not pose a small, but finite, probability of generating a carcinogenic response: i.e. the no-threshold approach. For non-genotoxics and non-carcinogens there is the assumption of a threshold dose. This threshold is termed the *reference dose* (RfD) in the USA (see Barnes & Dourson, 1988 for a discussion of use) and *tolerable daily intake* (TDI) in the UK. In the UK and Europe the concept of the threshold dose is applied to all carcinogens, diverging from US practice (McDonald, 1996). The UK Committee on Carcinogenicity of Chemicals in Food, Consumer Products and the Environment (COC) has not supported the US's no-threshold approach to carcinogens, primarily because of model difficulties and uncertainties. This divergence of approach between the USA and Europe in the treatment of carcinogens requires recognition by the risk assessor who refers to the international literature. In the procedural risk assessment guidance (Department of the Environment, 1997b) the use of the threshold-based tolerable soil daily intake is preferred, mainly as a means of simplifying risk estimation.

Chapter 10 discusses the different models for extrapolating from the high doses typical in laboratory experiments to the low doses typical in environmental exposures. The risk estimate output is either a hazard index or an estimate of the increased lifetime risk (particularly of cancer). In ecological risk assessments the estimation of risk to targets involves consideration of the credibility of effect rather than the probability. The most familiar example of the latter is the weather forecaster's stated probability of rainfall, often expressed as the chance of rain occurring. The ecological assessor must estimate the chance of occurrence of an unreplicated event using statistical models (Suter, 1993). As with human health risk assessment, it is often necessary to derive appropriate standards against which to assess ecological impacts from ecotoxicological tests conducted on aquatic and terrestrial species (see Chapter 10).

2.3.5 Risk evaluation

Risk evaluation involves consideration of:

- the qualitative or quantitative statements about risk derived from the risk estimation process;
- other site-specific factors which may affect the risks – e.g. sea-level rise; propensity to flooding; construction activity such as piling;
- the uncertainties in the risk estimates;
- the costs and benefits of taking action to control or reduce unacceptable risks;
- the social pressures for action;
- the significance and acceptability of the risks in relation to current and future land use.

Risk evaluation involves consideration of the results of the hazard assessment or the risk estimation, depending on the decision which has been taken as to the appropriate point to stop the risk assessment. The output may be a qualitative statement, or a semi-quantitative ranking of the significance of contamination, or a quantitative estimate of risk. The risk evaluation stage requires the identification of appropriate standards or criteria which provide an indicator of acceptability of the risk in human or ecotoxicological terms. The risk evaluation also requires an understanding of the uncertainties inherent in the risk assessment output (see below).

Table 2.2 provides a list of some sources of information which may be relevant to the identification of standards or criteria against which human health hazards may be evaluated. As introduced in Chapter 1 and discussed further in Chapter 8, the most important requirement for the effective use of generic standards and criteria is a clear understanding of their source, derivation, intended use and current status.

Numerical risk targets can be defined to determine the upper and lower boundaries of acceptability. In the USA, formal protocols have been developed to assist decisions on the acceptability of human health risks for both carcinogens and non-carcinogens. For carcinogenic risks, where the no-dose threshold approach is used, an upper boundary on acceptable life-time risk in the range 10^{-4} to 10^{-6} chance of mortality from cancer is considered acceptable. However, remediation of all hazardous waste sites up to 1996 has generally achieved risk levels of only about 10^{-2} to 10^{-4}, and evidence suggests that no clean-ups have achieved the goal of 10^{-4} or less (see discussion in Kocher & Hoffman, 1996). For non-carcinogens the hazard index must be less than 1 to represent an acceptable risk. In the UK there are no comparable protocols, although reference can be made to criteria used by the Health and Safety Executive (HSE) to define an acceptable risk of receiving a "dangerous dose" from major chemical accidents (HSE, 1989). In the Netherlands similar acceptable risk criteria have been defined in relation to the risk of death, and to the risk of deleterious ecosystem impact.

The output from the risk evaluation stage includes statements on the magnitude of the risks and their effects; the uncertainties inherent in the assessment and the steps which have been taken to deal with them; consideration of how changes in assumptions may change the estimates; and a view as to the need for action and where risk reduction and control activities should be directed. A decision to take action requires that both the magnitude and potential consequences of the risks are taken into account. For example, some risks may be of a very low probability but the consequences could be significant (e.g. the risks of an explosion involving landfill gas which results in the death of individuals). Conversely, a high risk in terms of probability may be tolerated in the light of other environmental considerations and the cost of action to remediate a problem (e.g. the risks of death of trees from phytotoxic effects) which are part of a landscaping scheme. Chapter 11 provides further discussion on the use of risk assessment output in decision-making and its effective presentation to assist the decision process.

Table 2.2. Examples of sources of information relevant to criteria for human health risks

Medium	Type and source	Comment
Soil	UK guideline values (Department of the Environment, 1997e)	Guideline values for chronic health risks related to different land uses. Limited number of contaminants.
Soil	Trigger concentration values (ICRCL, 1987)	As above – to be superseded by above.
Soil gas	Guideline concentration values for methane and carbon dioxide (Department of the Environment, 1989; 1992a)	Guideline values, relating to measured concentrations on and in the vicinity of landfills, and in ground beneath buildings.
Air	Occupational exposure standards (OESs) and maximum exposure levels (MELs) (Health and Safety Executive annual publication, EH40)	Legally enforceable limits related to occupational exposure via the inhalation pathway only – not strictly applicable to the population at large or ambient conditions, but application of safety factors to provide for general exposure can provide indicative data. Indicative data on potential skin contact hazards are given in documentation.
	Air quality standards (AQSs) and WHO air quality guidelines	Guideline values which will not have adverse effects on human health and in the case of odorous compounds will not create a nuisance of indirect health significance.
Water	Drinking water standards (the water supply) (Water Quality Regulations 1989)	Legally enforceable limits on the concentrations of substances permitted in water intended for human consumption.
	Environmental quality standards and objectives for drinking water implementing EC Directives for List I and List II substances	
Substance specific	Tolerable daily intakes (World Health Organisation; Department of the Environment, 1997c)	The average daily intake of a contaminant, expressed in mg/day^{-1}, that can be ingested over a lifetime without appreciable health risk.
	Limits on contaminants in food (Ministry of Agriculture, Fisheries and Food)	Statutory limits set for lead and arsenic. Guideline limits for specific substances (e.g. PCBs) in food are published in Food Surveillance Papers.
	Reference doses (RfDs) (USEPA, 1989a, and USEPA databases: Integrated Risk Information System (IRIS), Health Effects Assessment Summary Tables (HEAST))	An estimate of a daily exposure to humans which is likely to be without an appreciable risk of deleterious effects during a lifetime.

2.4 RISK MANAGEMENT

2.4.1 The risk management process

The NAS four-step process separated risk assessment from risk management functions where they were to be undertaken by a single agency, taking the view that the fundamentally scientific process of risk assessment should not be forced to conform with the management and decision-making preferences and values of the authority. This separation of risk assessment and risk management has not been favoured in some UK guidance (e.g. WDA, 1993; Harris, Herbert & Smith, 1995). Regarding risk assessment as a *component* of risk management acknowledges that the objectives of the latter influence the former; what has been termed the "top-down" rather than "bottom-up" approach (Hertzman, Ostry & Teschke, 1989). The top-down approach recognises the time-consuming nature of risk assessment when it is separated from management objectives and priorities, and also the tendency to produce large quantities of data which may be at best confusing and at worst irrelevant.

The relative role of site-specific risk assessments and remediation decisions based on the use of generic criteria, as introduced in Ontario, Canada (OMEE, 1996), provides an example of the concerns which have become evident about an over-reliance on quantified risk assessments. The Ontario approach stresses a preference for the use of risk-based criteria. The guidance states that:

> Proponents choosing to use Site Specific Risk Assessment (SSRA) should be aware that the process of risk assessment and risk management are necessarily lengthier than that of utilizing the generic criteria, and that site clean-ups based on them may result in increased restrictions for future site uses. It is also noted that, since the generic criteria have both ecological and human health components built into them, the use of SSRA in the guideline (with the exception of the specific situations where the guideline requires that an Ecological Risk Assessment (ERA) must be conducted) should also consider both human health and ecological protection aspects.

An effective risk management process is dependent on full and open liaison and discussion between those responsible for assessing a contaminated site, the relevant statutory and regulatory authorities, and any other interested parties. The latter could include conservation and ecological interests, the local community, local residents and financial (e.g. banks, insurance) interests (see Chapter 11). At the beginning of the site investigation and assessment stage it is essential to understand the objectives which different parties may have with regard to a site, although experience indicates that this is rarely done.

2.4.2 Risk control

Where the risk assessment indicates that the risks associated with a site are unacceptably high, consideration has to be given to risk control and reduction. A large range of options may be relevant, from comprehensive remedial action to remove or detoxify contaminants, through to isolation of parts of a site or restrictions on access, to ongoing monitoring of the site to ensure that contaminant levels are not increasing (particularly common in relation to gassing landfills).

The selection of the most appropriate risk reduction option requires a careful balancing between the costs of the different options, technical constraints, administrative factors including regulatory requirements, and the reduction in risk which is required and can be achieved (Harris, Herbert & Smith, 1995). The selection of an appropriate remedial scheme will also require consideration of various policy and social factors, such as the acceptability to a local community of certain actions, for example the removal and transport off-site of large quantities of soils compared with the installation of a barrier system which will leave the contaminants on a site.

Derivation of site-specific remediation values

Risk control measures, other than simple restriction of access, require remediation objectives and standards to be set so that the short- and long-term "performance" of the remedial scheme can be audited. Site-specific remedial standards are developed by identifying the residual concentration of each contaminant which is acceptable relative to a specific level of risk.

The derivation of site-specific remedial standards (or contamination-related objectives) may be by reference to:

 (i) generic guideline values; or
 (ii) adaptation of generic criteria to take site-specific conditions into account; or
(iii) site-specific risk assessment involving back calculation following the source–pathway–target chain in reverse (see Chapter 10).

The remedial action objectives should provide the benchmark for verifying or monitoring the effectiveness of remedial action. In the UK, the latter has been one of the most neglected aspects of contaminated land risk management, to the extent that there is a relatively poor understanding of the effectiveness of different remedial schemes which have been implemented and the extent to which risks have been mitigated on some sites is not known.

2.5 UNCERTAINTY

As should be readily apparent from the above discussion, there remains considerable concern, debate and uncertainty over appropriate approaches to risk estimation. Ecological risk estimation, particularly for terrestrial ecosystems (as opposed to aquatic systems, for which there is an especially long history with experimental protocols), is still (at the time of writing) in its relative infancy. Guidelines are still in embryonic form and there is limited experience on which to draw. The USEPA Risk Assessment Forum report on ecological risk assessment (USEPA, 1991a) gives a useful summary of relevant uncertainties in relation to both the hazard assessment and risk estimation components. Uncertainties in risk assessment have three basic sources (Suter, 1993):

   (i) the inherent randomness of the world (stochasticity);
  (ii) incomplete or imperfect knowledge (ignorance); and
 (iii) mistakes in execution of assessment activities (error).

Much of the risk assessment literature has focused on uncertainty in the risk estimation and evaluation phases. However, in contaminated land risk assessment significant uncertainty arises from inadequate site investigations, in addition to the three basic sources identified by Suter. The failure to recognise the quality of site investigation as an important influence is the primary reason that this book was commenced and the reason that a large section of the book is devoted to discussing the optimisation of investigation.

Although there is widespread support for the concept of risk-based management, there are many who will view the practical problems of implementation as being so severe that quantitative risk assessment should not be used. In the context of the USEPA's Reducing Risk report (1990a) which attempted to set priorities for action in relation to different environmental risks, Garetz (1993) summarises widely held concerns about the use of risk assessment, including the incompleteness of risk information, uncertainty, non-comparability of risk information, and rigidity of risk assessment.

The US literature is replete with examples of overly cautious assessments, including examples where they have been reworked following more detailed site investigation work leading to a change in findings to indicate no significant risk, and where state criteria for acceptable soil concentrations have been found to be too high when a site-specific risk assessment has been conducted (e.g. the ASARCO smelter in Tarcoma Washington referred to in Anderson, Chrostowski & Vreeland, 1990; Paustenbach, Rinehart & Sheehan, 1991; Keenan, Finley & Price, 1994). Smith *et al.* (1996) discuss large expenditure on site remediation in the context of an assessment process which only addresses the incremental risk from sites with no consideration of background exposures from other sources, which may be larger than the potential exposure from an individual site. As discussed in Chapter 10, it is important to understand the relative contribution to risk of exposure to a contaminant from different sources. If background exposure is already large, then a decision will be required as to whether an additional exposure from a contaminated site is acceptable, or if the latter is small whether it should be accepted in the local circumstances. Any remediation criteria set for the site would take the background exposure into account.

Wildavsky (1995, p.184) summarised a discussion of the Superfund programme as follows:

In the administration of Superfund the contemporary inclination to "err on the side of safety" at points of uncertainty meets with a risk situation in which gaps in knowledge are legion: the result is hazard scenarios that multiply unlikelihoods and clean-up plans that sometimes cross the border from the prudential into the absurd. If we had limitless resources this would not be a cause for concern. But we do not of course and the billions of dollars we spend ensuring remote possibilities at abandoned waste sites must be seen as billions that might otherwise be invested, publicly and privately, in health, wilderness protection, education and other things that we value.

Such a potential loss of credibility of risk assessment in the regulatory process is damaging. While the desire for objectivity in decision-making has been part of the reason for the promotion of risk assessment in relation to contaminated sites, the outstanding problems in application and in uncertainty leave room for misuse unless the process is competent and transparent.

2.5.1 Worst-case assumptions

Worst-case assumptions are the primary means by which data and decision uncertainties can be handled. Thus, failures to identify key contaminants or critical pathways; inaccurate soil sampling and analysis; the complexity of the potential human factors influencing risks, and uncertainty about future uses of a site; model errors, and problems in defining dose-response relationships can all be provided for in the risk assessment process through the adoption of worst-case assumptions or scenarios.

The primary rationale justifying the use of such worst-case assumptions is to ensure that the risk to which targets will actually be exposed – the "true risk" – will be less than that which has been predicted by the risk assessment and which is regarded as acceptable during the evaluation. This argument is laudable, not least in terms of providing for long-term protection. In the political context of decision-making in relation to contaminated sites, the approach is highly attractive, providing for a degree of protection of a local authority or regulatory agency's decision.

Whilst overly conservative risk assessment has implications for the design and cost of remedial action, procedures can be followed to ensure that the risk estimates produced are both reasonable and realistic (Keenan, Finley & Price, 1994). The careful selection and use of models, and comparison between outputs of different models, forms a primary method. However, the use of statistical techniques (e.g. sensitivity analysis and Monte Carlo error analysis) can permit variations in the input values used to be taken into account in the estimation of risk (see Chapter 10).

At the time of writing there is a move away from risk assessments which result in a single, deterministic estimate of risk, often based on the MEI. Probabilistic, or stochastic, risk assessment (discussed in Chapter 10), using Monte Carlo simulations, provides a distribution of estimated exposure and risk across an

Table 2.3. Key stages in risk assessment and the consequences of poor investigation

Risk assessment stage/activity	Examples of consequences of poor investigation
(i) Hazard identification Identification of hazards	• Inadequate site history leads to failure to identify potential contaminants (e.g. anthrax spores, chlorinated solvents, flammable gas) putting the health and safety of the investigating team at risk and the responsible party at risk of prosecution.
Identification of potential migration pathways	Failure to recognise the presence of sand lenses in • predominantly clay soil (referred to in standard geological texts on the area) leading to failure to identify risks to a neighbouring population and failure to monitor for gas in surrounding areas. • Failure to recognise that the regional direction of groundwater flow may have been modified by local features such as a nearby extraction well, mounded waste tip or deep foundations leading to inappropriate location of sampling wells.
Identification of potential targets	Presence of protected species not identified resulting • in damage during investigation and subsequent prosecution. • Development plans for neighbouring greenfield site not identified so significance of possible migration of contaminants (gaseous or liquid) not identified.
(ii) Hazard assessment On-site investigation	• Trial pit or borehole penetrates low-permeability layer permitting contamination to reach underlying aquifer. Client sues consultant/contractor. Regulatory agency prosecutes land owner/consultant. • Trial pit encounters elemental phosphorus: risks to workers anticipated but fumes drift over neighbouring houses causing alarm etc. • Soil samples are left in hot sun promoting degradation of samples. • Soil samples placed in inappropriate containers leading to loss of components and/or reaction with container.
Screening of analytical results against guideline values	• Comparison not possible because analytical detection levels are not sufficiently low – money and time spent on investigation are wasted. • Quantity of data insufficient for valid comparison according to guidance attached to guideline values identified by regulatory agency at time of planning application, resulting in delay to consideration of planning application. Project delayed causing loss of money to developer. • Quantity of data insufficient according to guidance attached to guideline values for valid comparison. Identified by consultant acting for potential funder. Funding refused.

Table 2.3. (*cont.*)

Risk assessment stage/activity	Examples of consequences of poor investigation
Screening of analytical results against guideline values (*cont.*)	• Water samples not taken to laboratory within required timescale. Results for some parameters therefore of doubtful validity. Assessment rejected by regulatory agency. Work has to be repeated, resulting in additional costs and delays. • Laboratory employs inappropriate sample preparation method. Volatile or degradable components are lost. If not identified this will lead to serious underestimation of potential risks.
(*iii*) *Risk estimation*	• Inappropriate methodology used, resulting in underestimation of risks. Identified at time of planning application resulting in refusal of permission to develop. Project delayed, causing loss of money to developer. • Inappropriate assumptions used leading to underestimation of risks on school site. Errors discovered at later date, leading to concern among parents. • Leachable concentrations not determined so risks to water environment difficult to estimate.
(*iv*) *Risk evaluation*	• Uncertainties of the site investigation results and assumptions not sufficiently explained or stressed and neglect of societal aspects of consequences if risks become manifest lead to an unqualified narrative statement that the risks are low. Hostile response from those directly at risk and challenge to methodology from regulatory body.

exposed population. Probabilistic risk assessment is seen as making an important contribution to effective and consistent risk management, assisting in cost–benefit decisions (Moore & Elliott, 1996; Richardson, 1996). However, experience of its use in relation to contaminated land (particularly in the UK) is limited.

2.5.2 Uncertainty and site investigation

Table 2.3 lists the key stages in risk assessment and examples of the consequences of poor site investigations. These consequences concern the quality of the assessment of risks to humans and ecosystems, but also may present significant risks to the monetary and professional status of those responsible for the investigation. The latter are potential significant risks as identified in Chapter 1. Chapters 3–7 explore means by which the uncertainties inherent in investigation, and the liabilities and risks to the site investigator, can be minimised, while the validity of the overall assessment of risk is optimised.

2.6   RISK COMMUNICATION

From the preceding discussion it should be readily apparent that the communication of information is an inherent and critical component of the contaminated
land risk management system. Communication is often discussed in the risk
assessment literature as a part of the stage of risk evaluation, being interpreted by
risk assessment experts as merely the communication of the risk estimate by
them to other interested parties. However, such a focus on communication of
output has been to the detriment of understanding the perceptions of risk and
also that of attention to improving decision-making processes so that interested
parties can be involved in discussion of risk (Petts, 1996).

In relation to the identification and management of potential problem sites,
involvement of the public as active participants from an early stage is now
accepted as the key to effective communication and decision-making. The USA
and the Netherlands have public communication programmes specific to the
identification and management of contaminated sites. USEPA identified the
following main benefits of public involvement: (i) communities are often able to
provide valuable information on site history and site conditions, and (ii) the
identification of public concerns enables the EPA to identify remedial actions
which are more responsive to community needs (USEPA, 1988d). The US
Department of Defense has developed a demonstration programme for risk
communication which stresses the importance of risk communication training
for assessment teams, the involvement of representatives of the community and
interested parties in the risk decision, and proactive efforts to identify and
understand local perceptions and concerns about contamination risks (Klauenberg & Vermulen, 1994). The concept of two-way communication where public
interests and concerns are part of the decision process, rather than a one-way
process where the risk assessment results are merely communicated to the
community (or any interested party), is accepted as the primary requirement
(Covello, 1989; Van der Pligt & De Boer, 1991; Renn, 1992; Bradbury, 1994).

Chapter 11 discusses the components of effective communication. Communication needs awareness and planning should form the basis of all contaminated
site investigation and assessment, even where public or community involvement
is not apparent. Within the complex, multidisciplinary framework of contamination investigation and assessment, effective communication which considers the
information needs of different parties as a reflection of their interests (regulatory,
financial etc.) and knowledge is essential.

3

Site Investigation in Risk Assessment: An Overview

3.1 INTRODUCTION

Site investigation is the determination of the contamination status of a site and its environmental setting by the systematic collection of data through a variety of means, including: document searches, personnel interviews, remote sensing, ground investigation by intrusive means, physical inspection, sampling and testing (WDA, 1993). Site investigation is a component of the risk assessment of contaminated sites. It should provide for the identification, characterisation and measurement of the source–pathway–target framework (Chapter 1), i.e. it provides the primary hazard assessment information for the risk assessment. Site investigation has most frequently been discussed as something that happens before the risk assessment is conducted. Indeed, the most frequent discussion of hazards and risks in relation to site investigation has related to the health and safety risks to the investigator, rather than to the understanding of the risks presented to current and potential targets on the site or to sensitive off-site targets.

Viewing investigation as a separate component of the risk assessment of a site has the potential to result either in the collection of the wrong information relative to the source–pathway–target framework, or in collection of too much information. As introduced at the end of Chapter 2 (Table 2.3), the quality of the site investigation is a major determinant of the quality and robustness of the risk assessment.

Site investigation is conducted in five primary risk management contexts:

  (i) The purchase, transfer or divestment of land and property.
 (ii) Funding, insurance or valuation activities.
(iii) Establishment of potential legal liabilities by site owner or occupier.
(iv) Land redevelopment.
 (v) Implementation of environmental and public health legislation.

The scope of the risk assessment required to inform the management decisions in each of these cases is different and will affect the extent and nature of the site investigation. For example, the assessment of land pre-purchase is often limited to a desk study of available documentation about the site (see Chapter 4), whereas an assessment of a site suspected of presenting potential significant harm is likely to require a detailed and intrusive investigation.

The literature on site investigation for derelict land and contaminated sites is large (e.g. BSI, 1988; Leach & Goodger, 1991; ICE, 1993; Cairney, 1993; Harris, Herbert & Smith, 1995) and it is not the purpose of this book to replicate available guidance and texts. In the UK, the literature has grown from the civil engineering disciplines where site investigation is discussed in the relatively limited context of the determination of "the nature and distribution of con-taminants on and below the surface of a site" (BSI, 1988). Site investigation has often been taken to mean physical exploration on site by the excavation of trial pits or the sinking of boreholes with little attention to the objectives of, and strategy for, effective investigation (Hobson, 1993).

In practice site investigation is also non-intrusive, in that it has to inform understanding of the environmental setting of the site, including the nature and characteristics of any potentially sensitive targets, and it has to provide sufficient characterisation of the site and surrounding pathways to be able to determine the potential for a hazard to be realised. The concept of site investigation as a non-physical or non-intrusive exploration of a site may challenge the engineer-ing concept. It also has implications for the skills of the site investigators, as these do not all have to be based in an engineering discipline.

Site investigation also includes measurement and monitoring. An important trend in recent years, particularly in the USA, has been the development of rapid methods of on-site analysis (USEPA, 1995a; 1995b; 1995c) and an interest in real-time in-situ measurements (USEPA, 1995d; 1995e). The choice of such methods, rather than laboratory analysis, becomes an integral part of the investi-gation strategy.

Despite the different risk management objectives which site investigation and assessment have to inform, the primary objectives are the same:

(i) to determine the nature and extent of any contamination present, i.e. the hazard source;
(ii) to characterise the host media, i.e. the pathways;
(iii) to understand the nature of the potential targets and the relationship between the source and the effects (risks); and
(iv) to inform appropriate management and remediation decisions.

In practice, site investigation has suffered a number of problems which are addressed further in this chapter as a basis for more detailed discussion of site investigation in the risk assessment context in Chapters 4 to 7. These problems can be considered in terms of (i) data gathering, and (ii) management. The data-gathering problems have most frequently been based on an overconfidence in the ability of physical site investigation methods to identify all hazards, and an over-reliance on the identification of the contaminants rather than the potential pathways by which targets may be exposed to these hazards.

Furthermore, there has been a tendency to look for chemical contaminants for which generic guidelines exist to allow an assessment of risk to be made, instead of providing for an understanding of the site-specific characteristics. As identified in Chapter 1, the focus of site assessment, and hence of data and information collection through investigation, has been on human and physical targets, to the detriment of the consideration of other vulnerable environmental targets, including water and flora and fauna. Guidance on ecological assessment appropriate to the identification of potential ecological risks is primarily based on the US literature (e.g. USEPA, 1989b; 1989d; 1992b), although UK contaminated land guidance (e.g. Harris, Herbert & Smith, 1995) has begun to discuss ecological risk assessment, and texts relating to ecological assessment as part of environmental impact assessment may also be relevant (e.g. IEA, 1995). However, it is important to remember that the latter are generally focused on ecological assessment as a predictive process dealing with proposed actions, whereas ecological assessment for contaminated sites may be a form of retrospective assessment where both the source of the contamination and the contaminated environment are observed directly. This book focuses on human health rather than ecological risk assessment since, as yet, there is little experience (or indeed pressure) to undertake the latter. The data and methodological uncertainties involved in ecological risk assessment present significant problems.

Management problems have usually arisen from pressures of time and money often imposed by a client, and from insufficiency of knowledge and experience of site investigation in the risk management context. Site investigation is still largely an art, with abundant opportunities for personal interpretation and bias. Effective management is based on effective planning, and on a phased approach to investigation which optimises the use of resources and allows a targeted approach to the assessment of the most critical risks.

3.2   EXPECTATIONS OF SITE INVESTIGATION

Contaminants can be found in five phases or compartments of the subsurface: (i) solid grains, rock or waste; (ii) ions of molecules weakly bonded (sorbed) to solids that contaminate the solids and have the potential to be released to water or vapour; (iii) gas or vapour in the pore spaces around the solids (unsaturated zone); (iv) water in the pore spaces; and (v) non-aqueous phase liquid (NAPL), either as a mobile, continuous phase or as an immobile discontinuous residual. Contaminants in mobile phases (water, NAPL and vapour) can migrate. Contaminants from immobile phases (solids or residual NAPL) can contaminate subsurface water and vapour. No physical site investigation can explore more than a fraction of the volume of the materials which lie below a site's surface.

As discussed in Chapter 1, all industrial and urban sites are contaminated to some extent. Aerial deposition, the decay of galvanised materials and leaded paints, the deposition of ashes and other typical urban wastes may all be evident. Site investigation has to be able to identify and locate this type of contamination as well as that which can be directly linked to site activities.

The British Standards Institution (BSI, 1988) recommended that 25 investigation locations are necessary when a 1 ha contaminated site is investigated. These investigation points (if constructed as trial pits with surface areas each of about 3 square metres) would reveal the subsurface conditions of less than 1% of the whole site area. In such circumstances any claim of full understanding of the distribution and variation of the strata and conditions subsurface would be spurious. Only complete excavation, detailed examination of every horizon at close spacings and large numbers of laboratory tests on both representative and anomalous samples could provide such reassurance. Given the impossibility of fully appreciating all variations of ground conditions, it is not surprising that construction projects are affected routinely by site investigation deficiencies, and failures to predict anomalous conditions (ICE, 1991).

Undisturbed ("greenfield") sites are typified by a general regularity of the underlying strata. There will be some natural variations – for example, the gradual change of a uniform fine sand into a clay-rich gravel bed over a horizontal distance – which are consequences of original differences in geological depositional conditions. These, however, are generally of minor significance, and a regularity of soil horizons remains the norm. Yet even in these simpler conditions, site investigation failures can occur, because unexpected conditions have not been predicted: for example, the unknown existence of a large former pond, now floored with wet and weak silty beds, which may be too soft to accept any planned structural loadings.

In contrast to undisturbed land, "brownfield" sites (i.e. sites previously used for development or industrial activities) are marked by subsurface inconsistencies owing to the following:

- Infill with any available materials of quarries excavated for building materials. Thus abrupt and sudden variations occur on a scale usually not found on "greenfield" sites (e.g. Figure 3.1).
- Covering of disused tanks, pipes and vats (e.g. Figure 3.2). Often these older storage systems still hold liquids and residual liquors.
- Dumping of production wastes (e.g. Figure 3.3) when waste disposal practices were unregulated and waste collection services undeveloped.
- Site surfaces being obscured by a mantle of demolition rubble, generated when unwanted buildings were demolished and sites were "cleared" for re-use (Figure 3.4).

Such local variability poses greater investigation difficulties than is the case in undisturbed ground. A range of potential risks has to be assessed in relation to former industrial or urban land, that is:

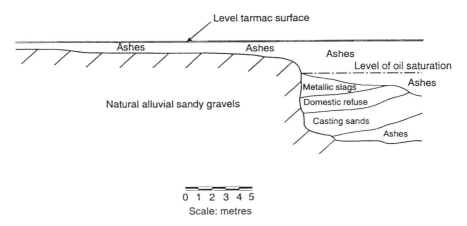

Figure 3.1. Conditions below a former road transport depot (West Midlands, UK)

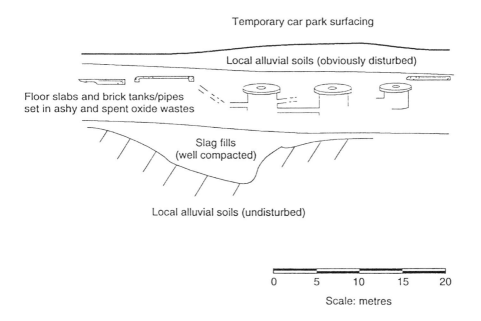

Figure 3.2. Conditions below a former gasworks (East Anglia, UK)

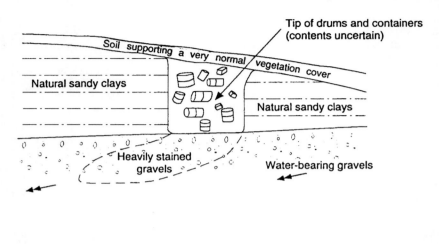

Figure 3.3. Tip of drums adjacent to a car component factory (West Midlands, UK)

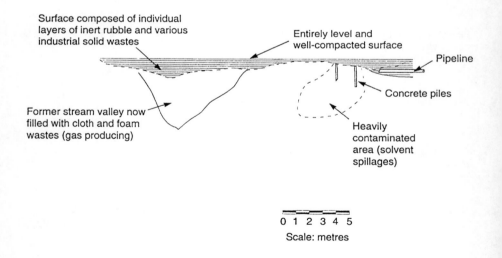

Figure 3.4. Conditions below a cleared clothing factory site

- risk of polluting surface waters;
- risk of polluting groundwaters;
- risk of creating area-wide air pollution;
- risks to sensitive ecosystems;
- risks of construction materials becoming degraded;
- risks of gases and vapours entering buildings;
- risks to human health.

Therefore, competent site investigation has to uncover the extent of the site variability, identify where the potentially most hazardous contaminants (surface and subsurface) occur, and establish risks to land users (current or future) or the wider environment. It is rare that an investigation is directed to proving the presence of contamination (this would require only one sample to be taken from a judiciously located point). The aim is usually to determine the nature and extent of contamination or to demonstrate with a given degree of confidence that contamination is not present to an unacceptable extent.

At this point it is relevant to note that some apparently "greenfield" sites may "hide" earlier industrial activities. For example, a site which had been orchards since the early part of the century had been classed as developable urban fringe land, and in the local development plan designated for housing. By chance, and when some of the new properties had already been built, chemical tests of soils revealed high lead levels relative to the current soil guideline value. Further examination of the maps of the area and of historical records revealed that the site may have been used for residue burial from brick-making activities on a site which had been over 0.5 km away at the beginning of the century. In practice the residue disposal had been taking advantage of a shallow depression in the land downhill and outwith the boundaries of the brick-making activity. Maps provided no direct clues that such activity had taken place on the site.

Despite the impossibility of revealing the full picture, it is possible to express well-based views on what constitutes good site investigation practice, and to recommend the approaches which are most likely to reveal the hazards of importance. The choice of words such as "most likely" is quite deliberate: subsurface investigation itself, like risk assessment, is an exercise in which uncertainty always exists to some degree. Site investigation is an exercise in probability evaluation. A competent investigator is one who (i) maximises baseline and preliminary information and thus has a greater chance of intersecting important contamination sources and pathways, and (ii) designs an effective investigation strategy, including appropriately located trial pits, inspection trenches and boreholes, and acquires representative samples without requiring an excessive number of "holes".

3.3 HAZARDS VERSUS PATHWAYS

The critical purpose of site investigation is to establish if a source–pathway–target chain is complete and unbroken. Until this is demonstrated, it is not justifiable to claim that deeper contaminants, from some prior use of an area of land, could pose predictable risks to current or future users of the land. Site investigators have to focus not only on the location of contaminated materials and their contaminant characteristics, but also on the pathways of potential migration.

In practice, site investigation has tended to be a mechanistic process with investigators focused on accurately locating their investigation points, logging the revealed strata variations and thicknesses in as much detail as possible, and collecting representative samples for later laboratory analysis and examination. The training of site investigators stresses these routine activities, and established guidance (BSI, 1981a) emphasises that they are essential. This is not to underestimate the importance of producing accurate geological logs and collecting adequately representative samples for analysis. The technical problems of obtaining representative water and gas/vapour samples and measurements are additional complexities (Chapters 6 and 7) which can easily obscure the importance of establishing critical contaminant migration pathways.

For example, consider an investigation and assessment for a potential developer. Merely informing him that metal-rich deposits lie at some depths below the local groundwater table is not enough, even if this statement is supported by analytical results from a comprehensive distribution of samples. The real question is whether these metallic materials pose predictable risks. To answer this requires an evaluation of possible migration pathways. Are the metals soluble? If so, they ought to be obviously present in the groundwater and might be carried up, by rising soil moisture in hotter seasons, to contaminate higher layers in the site's profile. Whether or not this is feasible will depend on the types of horizons which exist from the contaminated layer up to the site surface. If the metals are insoluble, this particular concern can be discounted. But if the metals exist as fine-grained deposits which are easily wind blown, and if landscape remodelling will allow these to occur at or near the final ground surface, then it could be possible to anticipate future hazards to human health if contaminated dusts could be inhaled or ingested.

Thus, contaminants *per se* are not the sole interest of contaminated land site investigations. What is equally important is whether these substances are mobile and, if so, whether pathways exist (or are predictable) via which targets could be brought into contact with contaminants, with resulting unacceptable harm.

3.4 INVESTIGATION, RISK AND GENERIC GUIDELINES

Linked to the insufficiency of recognising the relationship between contamination and pathways has been a tendency to look for contaminants for which generic guidelines or standards exist rather than investigating a site for the contaminants that may present a risk. There is little doubt that investigation and assessments have been considered to be easier and possibly more relevant where "official" figures exist indicating whether or not a site presents a risk. This has been most apparent where "soil" guidelines or standards have been directly available.

In the UK, this practice has been evident when requests have been made to laboratories to analyse for the contaminants on the ICRCL list (ICRCL, 1987), regardless of whether these contaminants might be expected to represent the primary and plausible hazard sources on a site. Most guidelines and standards in different countries represent incomplete listings of contaminants, even in relation to those soil contaminants which may present risks.

Guidelines have tended to provide insufficient guidance on the importance of metal speciation. Metals occur in a wide range of possible combinations in which their toxicities and bioavailabilities vary widely. Sampling and analysis are seldom adequate to allow this type of speciation detail to be assessed, with site investigators frequently asking only for the total content of a metal in the soil rather than whether the metal is in a hazardous form. Cadmium, for example, can have serious health effects if the contaminant can be concentrated in the kidneys or liver, but it can also occur in insoluble and non-bioavailable forms (in, for example, alkaline or hard water conditions), or as highly bioavailable complexes at very low or very high pH. Measuring total cadmium will not provide for a full understanding of the risks of a site (Cairney, 1995).

The ICRCL guidelines (ICRCL, 1987) presented two contaminants (copper, nickel) in terms only of phytotoxic hazards. However, some copper salts can be an irritant via inhalation or harmful via ingestion, and nickel can produce dermatitis and carcinogenic effects have been associated with occupational exposure (Barry, 1991).

Rigid and restricted use of guideline figures can lead to a failure to consider the effects of environmental factors which will affect whether or not contaminants are mobile and bioavailable (e.g. acidity, reducing or oxidising environments, the presence of clay minerals, organic soil content). Interactions between contaminants – i.e. synergistic effects (the increase in toxicity where more than one contaminant is present) and antagonistic effects (reduction of overall contaminant effects as a result of competition between contaminants) – are usually ignored.

In the UK, it is hoped that publication of new risk-based guideline values (Department of the Environment, 1997e), accompanied by detailed explanation of the basis of their derivation, will promote a more robust approach to the design of site investigation and assessment (discussed further in Chapter 8).

3.5 MULTI-STAGE SITE INVESTIGATION

Given the need for much greater predictive precision than is necessary for the investigation of "greenfield" sites, and the necessity to keep the source–pathway–target framework in mind, effective investigations of possibly contaminated sites are usually conducted as multi-stage exercises. Guidance stresses this point (e.g. IWM, 1990; NNI, 1991; WDA, 1993; Harris, Herbert & Smith, 1995; Cairney, 1995).

A range of differently biased and more or less comprehensive site investigations can be carried out to meet different risk management objectives (Table 3.1). Site investigation is best conducted as three distinct investigation phases:

(i) A preliminary, primarily non-intrusive phase.
(ii) An exploratory (limited intrusive) phase.

Table 3.1. Types of site investigations which may be required for contaminated land

Type	Necessity	Scope
Pre-purchase	Required by more prudent land purchasers, who wish to avoid acquiring excessive liabilities.	Limited. Often relies on desk-study data.
Preliminary	Necessary to establish the likely costs of reclaiming contaminated sites. Essential to main investigation design and to identification of potential source–pathway–target scenarios.	Usually based on desk studies, site reconnaisance and occasionally a small amount of exploratory investigation
Exploratory	To refine site zoning and confirm original hypotheses about site. To provide additional information to aid investigation design. Will not prove that site is not contaminated.	Trial pits, boreholes etc. at limited locations and located to minimise resource requirements.
Main	To prove whether risks exist and identify suitable remediation actions.	Scope is site specific, but is intensive and will also be extensive in terms of contaminants and media.
Supplementary	To resolve doubts left after main investigation and provide data to assist uncertainty estimation or to collect data to aid selection and design of remediation method.	Limited in overall scope, but often intensive in particular site areas or relating to particular problems; e.g. water or gas.
Method testing	To demonstrate effectiveness of an innovative remediation method	Often intensive and expensive. Usually involves a field trial of chosen method.
Performance testing	To demonstrate that remediation has achieved its objectives.	Usually limited to analyses of a few representative samples.
Environmental protection	Often required for land suspected of adversely affecting sensitive targets. Often required by major landowners wishing to concentrate limited remediation funds on priority sites.	Usually limited in scope. Focused on conditions at site boundary or in water.

(iii) The main or detailed phase dominated by intrusive ground investigation techniques.

The literature and guidance in the UK (particularly on environmental auditing) has used the terms Phases I, II and III with respect to these, similar to the US literature (ASTM, 1993a; 1993b). The International Standards Organisation refers to a four-phase approach to investigation, with Phase 1 limited to desk study, Phases 2 and 3 referring to increasingly intrusive site investigations and Phase 4 to more targeted investigation. In the UK, ecological assessments are referred to in terms of Phases I and II assessments (Nature Conservancy Council (now English Nature), 1990; IEA, 1995), with Phase I in essence a site reconnaissance and Phase II referring to more detailed investigation to identify community types, species richness and abundance, the structure of habitat and critical ecological processes operating in each habitat.

The three general phases suggested above involve several components of increasing intensity in terms of ground investigation, but not necessarily increasing relevance to the risk assessment:

(i) A *desk study* interpretation of historical, archival and current information to establish where particular past occupancies of the land were located, and where subareas of the site ("zones") exist in which distinct and different types of soil contamination can be expected to occur, and to understand the environmental setting of the site in terms of pathways and targets. This is the primary site investigation component essential to hazard identification.

(ii) A walk-over survey or *site reconnaissance* – this provides for the boundaries of postulated site zones to be located and modified as necessary and for information obtained from documentary records to be confirmed or updated. A range of biotic and abiotic indicators detectable by sight and smell provide for identification of contamination potential (Department of the Environment, 1994e).

(iii) An *exploratory* survey designed to test the truth of hypotheses and involving minimal subsurface exploration, designed on the basis of the site zoning, and chosen to confirm the accuracy of the zoning exercise. The exploratory investigation might be used to design the detailed investigation. However, in some cases in the risk management context, such as pre-purchase or pre-sale audit, an exploratory investigation may be sufficient to determine the risk of incurring liabilities but not necessarily the full scale of those liabilities. Such an investigation would not be sufficient to prove that a site presents no risk. Most guidance suggests that such a survey does not provide a valid basis for assessment of the site using dedicated generic guidelines or standards. Certainly such a survey would be insufficient for a site-specific estimation of risks (AGS, 1996).

(iv) A *main* (detailed or design) investigation to reveal additional detail, and to enable the hazard assessment and risk estimation stages discussed in Chapters 8–10 to be completed. In certain circumstances preferred remediation strategies may already be apparent and the main investigation has to be able to contribute to discussion of the acceptability of these decisions.

(v) A *supplementary* phase when further targeted investigation is undertaken to assist with uncertainties in the risk estimation, to confirm the extent (temporal and spatial) of contaminant effects, or to gather information relevant to selection and design of the remediation strategy developed following the risk assessment.

(vi) A *post-remediation*, targeted re-investigation designed to confirm that remediation work has attained the standards and quality required.

The desk study and site reconnaissance are treated in detail, as a single subject, in Chapter 4. They are qualitative, rather than quantitative, assessments of a site's conditions and its environmental setting, and are fundamental to the design of all intrusive sampling activity.

The main investigation phase (discussed in Chapter 5 in relation to soil investigation) must be planned on the basis of information revealed in the earlier investigation stages. The exploratory investigation forms the link between the desk study and reconnaissance and the main investigation, forming the first part of the latter. Unlike the earlier stages, this later work is more quantitative in emphasis.

Multi-phase site investigations are sometimes viewed as unnecessarily complicated and expensive. There has been a tendency for investigations to proceed direct from a desk study (sometimes only minimal in extent) to a main investigation. However, in such a practice it is often the case that money and time can be wasted. Adopting a multi-phase approach optimises the valuable benefit of time. The multi-phase approach allows the investigator to consider what data have been gained, and to develop a conceptual model of the site's environmental setting and subsurface variations. This is a critically useful tool, since it prevents the significance of individual elements of information being overlooked, and highlights anomalies to which additional attention ought to be given.

3.6   INVESTIGATION PLANNING AND MANAGEMENT

3.6.1   Objectives and integration

Site investigation must be planned on the basis of a clear set of objectives (Hobson, 1993). Practice, certainly in the past, often has been too focused on getting on to a site as fast as possible so as to collect data. In the UK, it is undoubtedly the case that a focus on contaminated land investigation as part of the development process, rather than as part of the identification and assessment of problem sites, has been the primary reason for this experience.

The setting of objectives is likely to include identification of a main objective and of subsidiary objectives related to specific components of the work, i.e. relating to the source (contamination), hydrology, geotechnics and geology, and target and pathway characterisation. The main objective may simply be to provide sufficient information on the source, pathways and targets to allow an assessment of risk to be made so that any required remedial actions (immediate or longer term) can be identified. Example secondary objectives are listed in Table 3.2.

The integration of investigation activities which address soil, water and gas contamination with geotechnical aspects brings both management, technical and risk assessment advantages, but again has in the past not been common practice, largely because of the different skill bases. Project management and health and safety management requirements can be simplified. Geotechnical boreholes may be able to be used for subsequent installation of gas or water monitoring equipment. Understanding of contaminant behaviour may be enhanced by data relating to geotechnical and hydrological factors.

However, integration is not always acceptable. Harris, Herbert & Smith (1995) note that it would be unacceptable to move a borehole which is being used to intercept a contaminated groundwater plume to provide geotechnical data on another part of the same site where construction is proposed. Furthermore, it is not acceptable to use large disturbed samples collected for geotechnical testing for subsequent detailed organic analysis.

3.6.2 Investigation management and quality control

There are various management issues which arise during any investigation, particularly:

- compliance with legislative requirements concerning physical planning, rights of way, environmental protection, protection of archaeological remains, protection of worker and public safety, and waste disposal;
- obtaining samples which are representative of the soil or other medium being sampled;
- obtaining appropriate and reliable real-time measurements;
- ensuring proper documentation of activities and recording of site and sample information;
- appropriate handling, transport and storage of samples to ensure preservation of chemical, physical or biological characteristics and identification through the "chain of custody".

Good quality assurance/control (QA/QC) procedures are important to all stages of an investigation, from the desk study data collection, through to sampling strategy design, to obtaining samples, through to reporting. Quality control failures can result in poor-quality data which do not adequately

Table 3.2. Example investigation objectives (after Harris, Herbert & Smith, 1995). Reproduced by permission of CIRIA

Factor	Objectives
Source, pathways and targets (potential)	To determine as part of the initial conceptual model: • Potential contaminants on site. • Existing or proposed use of the land. • Potential human targets on and in the vicinity of the site. • Proximity to, and sensitivity of, water bodies. • Proximity to, and sensitivity of, flora and fauna. • Nature of building materials and services. • Nature of building structures and confined spaces (relevant to gas hazards). • Potential physical and manmade pathways.
Contamination	To determine as appropriate: • The nature, extent, form (solid, gas, liquid etc.) and distribution of contaminants in a range of media – soil/fill; water; air; biota; containers (drums, tanks etc.) – both on- and off-site (including contaminants migrating on to the site). • Ground temperatures and level of microbial activity. • The health of existing on-site or off-site ecosystems.
Water environment	To determine as appropriate: • Geological strata, structure and composition. • Groundwater levels, direction of flow, flow rates, pressures • Abstraction and recharge effects on the site. • Chemical and mineralogical quality of ground and surface water – background and related to the site. • Permeability. • Propensity of site to flood. • Tidal fluctuations.
Geotechnics	To determine as appropriate: • Physical characteristics of the ground (services, buried structures etc.). • Topography of the site and surroundings. • Physical characteristics of the contaminated matrices (mineralogy; moisture content; permeability; particle size distribution etc.). • Presence of old mine workings.
Source, pathways and targets (actual)	To determine in relation to above information: • Plausible sensitive targets (human, water, flora and fauna, buildings, economic resources). • Plausible exposure pathways.

inform the required risk assessment. In the worst-case scenarios this might either result in a failure to assess potentially significant risks or to an overcostly and inappropriate remediation strategy. With heightened liability awareness there are increasing pressures on site investigation and assessment teams to ensure quality.

The elements of a QA programme for contaminated site sampling relate to the following (Hobson, 1993; Harris, Herbert & Smith, 1995):

- *Sampling* – records of sampling; sampling numbers; the use of techniques compatible with the contaminants of concern and analytical methods to be used.
- *Sample handling* – e.g. use of appropriate containers, preservation etc.
- *Sample preparation* – e.g. use of methods that will not affect the results, recording of procedures etc.
- *Instruments and techniques* – e.g calibration testing of reproducibility on split samples, running standard samples.

Canadian guidance (CCME, 1993a; 1993b) provides checklists of subjects to be covered in planning sampling programmes and also minimum requirements for documenting environmental sampling. However, the guidance warns that QA will never ensure that samples are taken which are representative of the source (or of the source at the time the sample was taken).

The greatest source of variation in soil analysis is soil sampling procedures (as opposed to analytical procedures). Moving the sampling point by only a few centimetres may produce a profoundly different result (e.g. see Smith & Ellis, 1986; West *et al.*, 1995). Yet despite this fact the focus of error discussion continues to be on laboratory and data-handling sources, perhaps because these are the easiest to control. An important consideration in planning for sampling and analysis of contaminated sites is the type and number of quality control samples to take. Software is available to help in the selection of the proper types of quality control samples to complement sampling and analysis for the production of known quality data (Keith, 1991a).

Within the laboratory, quality control protocols are employed to demonstrate the reliability of sample analysis data and compliance with documented sample handling, storage, preparation and test methods (general discussions of quality in respect of environmental data can be found in Davies, 1980 and Keith, 1991b). After analyses are completed the third-phase data handling and reporting begin, with a focus on the production of data that are accurate, precise, complete and representative (Gaskin, 1988). However, there are many unmeasurable factors which can severely bias data (Maney & Dallas, 1991): e.g. sampling the wrong area, or the wrong matrix; mislabelling sample containers; incorrectly preserving the sample; using the wrong method for analysis.

There is considerable emphasis on using "accredited" laboratories, i.e. those whose quality control procedures for specified analytical or test methods have been independently inspected and certified. The main sources of guidance on the accreditation of laboratories are ISO standards EN 45001 and 45002 (ISO, 1989a; 1989b). However, simply using an accredited laboratory is not sufficient since the schemes are more concerned with consistency than with accuracy. Those using laboratories need also to evaluate their quality management arrangements and their performance in proficiency testing programmes. In the UK, the main inter-laboratory proficiency programme is the CONTEST scheme (Rix, 1994). For a general guide on quality in the analytical laboratory see Pritchard (1995).

Finally, it is important to stress the need for appropriate skills among site investigation and assessment contractors. Attempts have been made to suggest minimum skill requirements for those taking various levels of responsibility for contaminated land work (ACE, 1993). Requirements for investigation personnel are given in UK and Dutch standards for site investigation (BSI, 1988; NNI, 1991). With the more specific focus on risk assessment of contaminated sites, there is no doubt that even among site investigators with considerable on-site experience there is a need to update knowledge and appreciate the information demands of the risk assessment process.

3.7　MANAGING SITE INVESTIGATION RISKS

3.7.1　Health and safety of on-site personnel

While site investigation is undertaken primarily to assess the risks which a site could pose to current or future site users or to ecological, water or building targets, site investigation personnel themselves could be exposed to immediate risks. Those authorising, designing and supervising site investigations have a joint responsibility for safety which extends beyond the site fence and workforce to the general public.

On a contaminated site, hazards are most frequently associated with:

● the contamination – e.g. chemicals, biologically active agents such as anthrax spores or *Leptospirosis*;
● the physical condition of the land – e.g. the presence of unstable ground;
● moving vehicles and equipment.

However, other "hazards" may also need to be considered, such as adverse weather conditions or even the presence of macro-fauna such as grazing cattle or dogs.

Achievement of a safe site will require the adoption of a formal safety policy and operating frameworks which will require and permit (BSI/ISO, 1995b):

● the identification of hazards and evaluation of risks;
● avoidance of risks wherever possible;
● failing this, control of the risk through the adoption of appropriate operating procedures;
● failing this, or in addition, the protection of individuals against unavoidable risks (for example through wearing protective clothing).

It will be necessary to provide training, to keep records of procedures adopted and of any incidents (often required of national legislation). It may be necessary to establish health screening and surveillance programmes. Table 3.3 provides an overview of relevant UK legislation relevant to risk management in site investigations. The recent Construction (Design and Management) Regulations 1994 (HSE, 1995a) have added further statute force to prediction of foreseeable risks, the devising of a safety plan and the development of this plan by the principal contractor before any site work is commenced.

Health and safety should be a fundamental consideration in the design and selection of investigation methods. Initially investigation techniques (e.g. driven probes) which minimise exposure of the investigation team to contaminants can be employed. Where hazardous, reactive or volatile contaminants are expected, consideration should be given as to whether trial pits and other open excavations are an appropriate exploration method. Phased investigations can make as important a contribution to safety as they can to the risk assessment of the site. For example, where either substantial gas or leachate migration is expected, initial investigation around the boundary of the site is preferable as direct interference with the source could result in an uncontrolled release of toxic, flammable or strongly smelling gases or vapours.

A few examples of what can go wrong may help to make the point. The first concerns investigation of a brick pit over 20 metres deep filled with a mixture of domestic and industrial wastes. Information gathered during the desk study stage indicated the presence of solvents, paint residues and other substances that might give off vapours. The site was known to be actively gassing, i.e. there were uncontrolled emissions of methane and carbon dioxide. During the drilling of the first shell and auger hole (light percussion drilling) through the overlying clay cap, the drilling crew were overcome by vapours of toluene and other volatile organic compounds – one member was hospitalised overnight.

In another case, during the drilling of a shell and augur hole through filled land, an underground pool of highly phenolic water (containing about 10% total phenols) was encountered. There was nothing in the site history to suggest such a hazard. The drilling crew spotted the hazard and stopped work. Work was only resumed once appropriate protective clothing was available. Contact with the water could have caused burns and possible systemic effects if, for example, clothing became soaked.

Failure to plan ahead can also put the public at risk. A site under investigation was an old gasworks currently in use for car parking – the surface had received minimal preparation for such a use and gasworks clinker etc. was exposed on the surface. Senior staff from the investigation company arrived in late afternoon to find a number of open trial pits filled almost to the top with water with a floating oily/tarry layer. The on-site investigator did not know how to proceed, since backfilling of the pits would merely have spread the water over the surface. The area of the site close to the trial pits was occupied by itinerant travellers' caravans and children were playing in the area. Emergency measures had to be instigated,

Table 3.3. Overview of primary-site investigation safety risk management legislation

Legislation	Relevant points
Health and Safety at Work Act 1974	The HSW Act applies to all persons engaged in site investigation. It places all employers under a duty to ensure so far as is reasonably practicable the health, safety and welfare of all their employees. The place of work itself, the means of access and egress and the working environment all have to be kept safe and without risk to health.
	The employer is required to prepare a safety policy where five or more persons are employed. This document should describe the arrangements for ensuring the health and safety of all work undertaken.
	The Act also imposes duties on employees, and self-employed people, to conduct their undertakings in such a way that persons other than their own employees are not put at risk.
	Persons controlling premises (e.g. an investigation site) used by others who are not their employees have duties to those who, for example, work at, visit or trespass on the site.
	Employees must take reasonable care of themselves and of others who might be affected by what they do or fail to do at work.
Control of Substances Hazardous to Health Regulations 1995	The COSHH Regulations require the employer to adhere to the following principles:
	● Avoid exposure to substances hazardous to health, and only when this is not reasonably practicable should methods of controlling exposure be employed.
	● If exposure has to be controlled, personal protective equipment should only be used as a last resort, i.e. other forms of control such as ventilation should always be considered first.
	If contamination is suspected a COSHH assessment must be carried out.
Management of Health and Safety at Work Regulations 1992	Requires all employers to undertake a risk assessment of all activities which may present a risk to employees and to the public and environment. The assessment requires an analysis of work activities to identify hazardous substances and situations and an estimate of the risks which may arise.
	The assessment should lead to a review of all existing and possible control options and action priorities.
	Arrangements must be made to provide information to all employees, including subcontractors, and where necessary to provide health surveillance.

Table 3.3. (*cont.*)

Legislation	Relevant points
Construction (Design and Management) Regulations 1994	The CDM Regulations apply (with some exceptions) to a wide range of construction activities, including site investigation, involving five or more people on-site at any one time, and to all design work and demolition work. They place duties on the client, the designer, the planning supervisor, the principal contractor, contractors and the self-employed.
	The client must be satisfied that only competent people are appointed as planning supervisor, main contractor, designer and contractors; and as far as possible that sufficient resources, including time, are allocated to enable the project to proceed safely.
	The planning supervisor has overall responsibility for co-ordinating the health and safety aspects of the design and planning phase and for the early stages of the health and safety plan and the health and safety file.
	The principal contractor should take account of health and safety issues when preparing and presenting tenders and has also to co-ordinate the activities of all contractors to ensure that they comply with health and safety legislation. They have a duty to check on the provision of information and training for employees and for consulting contractors and the self-employed on safety.

including warning parents to keep their children inside, fencing for the trial pits and a tanker brought on-site the following day to pump out the trial pits.

Safety responsibilities also extend to consideration of laboratory staff. An engineer who phoned a specialist consultant to discuss sampling on an old tannery site assured the consultant that he knew about the risks of anthrax spores being present on the site. Shortly afterwards, however, samples from the site were received by a laboratory through the post in insecure containers and with no indication of their origin or potential biological hazard.

In the UK, the Health and Safety Executive advises that where possible details of contamination and outline precautions should be included within tender documents for site investigation contracts (HSE, 1991). Site investigation guidance and technical literature provide further detail on health and safety provision (BDA, 1981, 1991; USDHHS, 1985; USEPA, 1988c; HSE, 1987, 1989; BSI, 1988; AGS, 1992a, 1992b; Smith, 1992; O'Brien, Steeds & Law, 1992; ICE, 1993; BSI/ISO, 1995b; Steeds, Shepherd & Barry, 1996).

3.7.2 Environmental protection

Since the intrusive investigation of contaminated sites involves physical disturbance, there is the potential for uncontrolled release of potentially hazardous materials to the environment. For example, contaminated water or soil brought to the surface may cause ground contamination or be translocated to a neighbouring water course. An injudiciously placed and designed borehole might create a path for contaminants to enter an underlying aquifer due to penetration of a low permeability layer that has hitherto protected it.

3.8   REPORTING

Reporting of site investigations has often been viewed as an end result of investigation activity designed to report only the findings of the main investigation. Yet it can be argued that if effective multi-phase investigations are to be conducted (i.e. those that are only taken as far as is appropriate to the risk assessment relevant to the particular source–pathway–target characteristics) then reporting should be a component of each stage, allowing an opportunity for reflection on findings before proceeding (if necessary) to the next stage.

Contaminated land reports are no different from any other report: they need to be written with the requirements and interests of the reader in mind. This requires appreciation of who the reader could be and what the interests of different readers might be. In practice a wide and technically variable readership will usually be identifiable. For example, consider a report on a site which is to be developed. The direct client for the report may be the developer. However, the potential readers could include the planning authority, agencies with statutory responsibilities who are consultees during the planning process (for example, in England, the Environment Agency, English Nature etc.), funding institutions who will be interested in potential long-term liabilities, and members of the public, either in the vicinity of the site who will become aware of the proposals to develop the site or those who may become residents on the site.

Reports must:

- distinguish between factual information and the author's interpretation and conclusions;
- cite precisely data and information sources used;
- discuss where predictable variability (for example, future groundwater conditions or other obviously non-static parameters) could occur, and the consequent effect on the investigator's conclusions.

With the reports of preliminary site investigations, where a good deal of interpretation of historical or archival data is inevitable, care has to be exercised to avoid assessments which cannot be justified by the available information. For example, it would be unwise if a report stated that an underlying glacial clay horizon would be an effective seal (capable of preventing any downward migration of large volumes of oily liquids found in a site's surface fill layers) unless the investigator had actually determined that this indeed was the case. Proving such a contention would call for permeability measurements at several points over the area covered by the glacial clay; for geological logging which showed that the clays were physically consistent enough for the permeability measurements to be representative; and for chemical analyses which confirm that no downward oil movements had taken place in any of the explored locations. However, no investigation sufficient to provide this proof would take place in a preliminary site investigation. Thus, in a preliminary investigation report, it would be much safer for the investigator to state merely that the glacial clay horizon exists, and that this (subject to additional more detailed investigation findings) could be an adequate safeguard restricting deeper movement of oily contamination.

When main site investigation reports are being written, a much greater body of detailed and factual information will invariably exist, but there will necessarily be report sections which summarise an investigator's personal interpretations. These could well be disputable by other specialists. Separate reports which (i) produce a factual account detailing the investigation carried out and the information acquired, and (ii) offer an interpretive statement of the investigator's judgements, on the significance of the factual information, are appropriate.

Site investigation as part of risk assessment has a potentially long-term effect, not least in terms of legal action. For example, years after a site investigation has been carried out, a client could decide that the investigator's reports had misguided him. If this were to occur, and if the factual and interpretive sections of the reports were intermingled, it could be possible for a court to find that "the duty of proper professional care and diligence" had not been observed. However, if the factual and interpretive report elements were clearly separated, and if the factual investigation work conformed to good professional practice, it would be unlikely that a court could find against the investigator.

4

Preliminary Investigations for Hazard Identification

4.1 INTRODUCTION

Preliminary investigation, comprising a desk study, site reconnaissance and some limited investigation, provides the information for the hazard identification stage in the risk assessment process. It is also essential to the other components of risk assessment and particularly to the design of site investigation. The objectives of the hazard identification stage are to: (i) produce a qualitative understanding of the potential for a site to present a risk, (ii) highlight those sources of risk which will require detailed assessment (including investigation), and (iii) enable the assessor to discount those sources not requiring further assessment when the source–pathway–target chain may not be complete. The preliminary investigation allows the assessor to develop a *conceptual model* of the source–pathway–target chain.

The hazard identification stage also has a fundamentally important management role in the whole risk assessment, in that it should ensure that any required investigations are adequately designed so that they are focused on the hazards of concern (i.e. first the plausible hazards and second the critical hazards); that zones of sites which may present the greatest hazard are identified for appropriate investigation; that investigations are conducted with due regard to any potential risks to personnel and to the environment from the investigation operations; that the potential for acute risks (for example, arising from loose or friable asbestos, or from explosive or highly toxic vapours or gases) are identified; and that any immediate or interim remedial works (such as fencing the site to prevent access) can be identified.

As discussed in Chapter 3, there has been a tendency to view site investigation as a primarily intrusive activity, with considerable pressures exerted to get out on to a site as quickly as is possible to construct grids of trial pits and boreholes to determine the characteristics of a site. Experience suggests four underlying reasons for this view.

First, the investigator will usually not have been aware of a site's existence prior to being asked to undertake an assessment of the risks it presents. The reaction is often that the site is an unknown quantity, so that only taking samples to determine the contaminants present will reveal the risks. Secondly, visits to sites often provide few obvious clues to the subsurface conditions. On former industrial sites, surfaces are often obscured by buildings and hardstanding, by the remnants of foundation slabs and road sub-bases, or by large cappings of levelled and compacted demolition debris.

Thirdly, where generic guidelines indicate contaminant concentrations which suggest that either a site is uncontaminated or it is not, the pressure is to determine the relationship of the measurements from a particular site to these guidelines, so that apparently formal and "official" judgements about risks can be made as quickly as is possible.

Fourthly, document review and desk studies are sometimes thought to be time consuming. The fact that they can rarely provide a complete under-standing of the potential hazards of a site has led to a questioning of the value of more than a cursory examination of maps. Hazard identification has tended to focus on understanding the potential contaminants which may be present as a result of previous or current site uses (i.e. source information), to the detriment of collecting information which may indicate potential path-ways (permeable strata, water features, service runs, mineshafts, soakaways, old stream beds etc.) by which different targets may be at risk or potentially sensitive targets, not only those which may come on to the site as a result of development but those in the vicinity. This has reflected a dominance of interest in sites as redevelopment sites, with a failure to understand potential off-site liabilities.

A focus on understanding what industrial and other potentially contamina-ting activities may have existed on a site has also been to the detriment of understanding the background levels of contaminants in an area. Yet for the purposes of site-specific estimates of risk (discussed in Chapter 10), it is import-ant to have this information. First, it provides a means of understanding whether contaminant levels measured on a site (which may be above generic guidelines) are actually in line with the ambient soil conditions in the area. Secondly, it provides the means of understanding the contribution of the site to contaminant exposure compared to the local background.

As regulatory and liability assessment activity has moved to the proactive identification of sites which may present a risk, hazard identification has come to form an important basis of the ranking of sites (discussed in Chapter 9). Here, readily available information about the characteristics of a site and its environ-mental setting are used in place of directly measured and observed information to score the hazard potential of a site or group of sites.

Liability assessment relative to individual sites also recognises the potential of the hazard identification stage. In the USA, most commercial transactions must now be accompanied by a Phase 1 assessment (ASTM, 1993a; 1993b). This is required to identify and avoid any potential environmental liability on any property or transaction and provide for worker and public safety. It has to identify any hazardous materials that might be released, and provide information about the environmental conditions on a site. This is based on a document review and inspection of the site.

UK guidance stresses that "the importance and value of the preliminary investigation should not be underestimated" (Harris, Herbert & Smith, 1995). This chapter identifies information requirements and sources. It discusses the role of the preliminary investigation in allowing for the "zoning" of a site so that different contaminant areas can be understood and their relative impact on the risk from the site determined. This zoning provides the important link from the hazard identification to the site investigation (discussed in Chapters 5–7).

4.2 INFORMATION REQUIREMENTS AND SOURCES

4.2.1 Information requirements

Hazard identification includes the collation and review of any available information which allows the characterisation of potentially relevant contaminant sources, identification of the targets which may be harmed if they come into contact with contaminants and the pathways by which such contact could be made. UK guidance (BSI, 1981a; 1988) discusses the nature of information required at the preliminary stage and lists potential sources.

Evidence which will characterise the source in terms of potential impact will include:

- The current or previous owners and uses of a site, particularly where potentially contaminating or hazardous materials may have been used, stored, processed, disposed or produced.
- Records of the nature of the processes used, including raw materials and reagents.
- Records of residues, consented discharges and waste disposal.
- Information on the size and layout of a site at each stage of its history, including boundary locations; locations of storage and disposal; materials handling; underground structures and services; storage tanks, disposal lagoons, pits etc.
- Information on the current topography of the site, areas of flooding etc.
- History and evidence of previous pollution incidents, fires, accidents or spillages.
- Records or verbal reports of the deposit of biodegradable wastes.
- Visual evidence of vegetation die-back, uneven ground, pits or lagoons, odours, discoloration of soils, process plant structures etc.
- Reported adverse health effects, water pollution or nuisance incidents (odours, dust).

The history of site use must be tracked back as far as its original undeveloped status. For many industrial sites in the UK, this could be in excess of 100 or even 200 years. Lead and iron mining and smelting have been practised since pre-

Roman times. Not only have uses of sites changed over such periods, but there has also been extensive land reclamation and drainage. In relation to smelting and mining, much of the very early activity was intensive and small scale. In most cases sites will have been reworked intensively in the last 300 years.

Primary documentary sources (including map sources) are often only available from the mid-1800s. Historical records (industrial archaeology) relevant to site uses before this time often exist, particularly in relation to significant industries and areas (e.g. the iron and steel industries of south Wales). However, greater effort to acquire relevant information may be necessary.

Particular industries often made use of a limited range of raw materials and produced wastes containing characteristic contaminants. Perhaps the best researched and most written about (in many countries) are coal carbonisation sites (Anon, 1995; Department of the Environment, 1987a; USEPA, 1988b). Various attempts have been made to collate such information (Barry, 1985; Bridges, 1987; Department of the Environment, 1995/96; LPC 1992; Steeds, Shepherd & Barry, 1996). However, care is required when using such compilations lest they inhibit a critical examination of available site information. It is also important to bear in mind that certain compounds are fairly ubiquitous (e.g. lead, zinc, mineral oils, chlorinated solvents) and are likely to be present at least in restricted locations on most sites with an industrial history or in an urban setting.

Within the risk assessment the principal question which the hazard identification will need to answer is (Department of the Environment, 1995a):

Which of the identified properties of the substance, organism, operation, process or undertaking *could* lead to adverse effects on the environment?

This will require consideration of:

- toxicity, immunotoxicity, pathogenicity, mutagenicity, teratogenicity and carcinogenicity;
- potential for long-lived presence in the environment, including the potential to bioaccumulate and bioconcentrate;
- potential for effects on environmental processes such as photosynthesis, the nitrogen and carbon cycles;
- potential for effects on air, water and soil;
- potential for affecting ecosystem function, such as changes in population numbers of the species in the ecosystem;
- potential for causing offence to people or adverse effects on them; and
- potential for accidents.

Given that no site can present a contamination risk unless there is a source of contamination, this information is paramount. However, source information in itself is not sufficient (although it has tended to dominate preliminary investigation data gathering). It is also important to gather information relating to targets (including those who might come on to the site and those in the vicinity) and pathways. The hazard identification stage could result in decisions not to take any further "action", i.e. exploratory or detailed site investigation or any emergency action to deal with potential acute risks. Therefore, the information gathered at this stage has to be sufficient to ensure not only that any required action is taken and adequately designed (see Chapters 5–7), but also that a decision not to take action is justifiable. Information must be gathered on the sensitivity of potential targets (e.g. children versus adults, potable water supplies versus a recreational source, protected ecosystems, building services etc.); the nature and severity of possible adverse effects; the scale of possible effects (numbers of individuals, area of land, significance of a water body); and social, political and legal factors (e.g. stress among residents at identification of a hazard, potential liabilities arising from a failure to identify a hazard).

Undoubtedly there has been a tendency to focus on human targets and more recently in the UK on groundwater resources, to the detriment of flora and fauna except in highly sensitive and protected areas. Building structures and services which may be at risk from corrosive or explosive substances have often only been considered in terms of those on a site with insufficient attention paid to surrounding areas. For example, fuel tanks within the vicinity of a site could be at risk from the migration of explosive substances.

Some guidance suggests the identification of targets within defined distances of sites. UK guidance on the control of development in the vicinity of closed landfills suggested a distance of 250 m as a guide to the consideration of risks to buildings and services from landfill gas (Town and Country Planning General Development Order 1988). While useful as an indicator of potential zones of influence, if such guide distances are followed rigidly potential targets at risk may not be identified. For example, in relation to landfills the scale of gas production at a site and the nature (i.e. permeability) of the surrounding strata could lead to gas migration beyond 250 m. However, where waste deposition was completed many decades previously, a gas hazard may not even exist at the landfill site boundary.

Many of the ranking schemes discussed in Chapter 9 use defined distances to provide for the identification of vulnerable targets – residential areas; groundwater; surface water and drinking water supply; sensitive or protected ecological areas. However, few provide an explanation as to why the specific distances have been selected.

A UK report on the prioritisation of sites (Department of the Environment, 1995b) suggests that if land within 50 metres of a contaminant source is used for residential, open space or amenity purposes, this provides evidence to support the existence of a risk to targets. No explanation for the distance is provided. It is ot meant to imply that residential areas at 75 m from a contaminated site are "safe" (indeed residential and similar development out to 250 m is placed in a secondary risk group). Any such guidance must be used intelligently in the light of an understanding of the contaminant source and the potential pathways.

The relative sensitivity of current and potential (e.g. proposed residents on a development site) targets to contaminant risks is a relevant issue in relation to the assessment. Guidance often suggests the need to identify different land uses, either existing or which might be planned on and in the vicinity of sites, for example hospitals, old people's homes, schools, housing, commercial land uses, open space and allotments. However, within the risk assessment process, and particularly in the consideration of the site-specific risk estimation, it is the nature and activities of the targets (i.e. young children playing in gardens; people playing sport; the growing of vegetables for home food consumption etc.) which are the key to sensitivity rather than the land uses *per se*. The latter may only provide indicator information relevant to the sensitivity.

Information on potential pathways will relate to the characteristics of the ground on the site itself: for example, areas of degraded ground and exposed soil could provide a direct or air-borne pathway to human exposure. However, it is also important to look at potential natural and manmade pathways in the surrounding environment: for example, the presence of watercourses flowing around or away from a site which could carry contaminants to human or animal targets, or permeable strata or road sub-bases or sewers which could provide for migration of leachates or gases. The pathways of exposure are also linked to the activities of the targets.

4.2.2   Information sources

Information sources for hazard identification are of two types:

- documentary;
- non-documentary.

In the UK, with a failure proactively to identify contaminated sites and a reliance on identification related to redevelopment activities, there are few sources of collated information on potentially contaminated sites. This is a different situation to many countries (as discussed in Chapter 1) which have proactive policies to identify sites and collate national and regional registers. A survey by the Welsh Office (1988) represents one of the few collated "lists" in the UK, although it only related to vacant sites (to avoid the potential problems of blight that might arise from the identification of already occupied sites) over 0.5 ha in size and therefore cannot be relied on in seeking to understand past uses. Since the late 1980s a number of local authorities have collated information on potentially contaminated sites in their areas, but on an *ad hoc* basis.

The Environment agencies (previously through the local waste regulation authorities) collate information on potentially gassing landfills (active and those closed within the last 30 years). However, it is accepted that not all sites may have been recorded and some, where supposedly "inert" wastes were deposited and were not thought to have the potential to be gassing, have proved to be problems. Between 1991 and 1993 a large number of landfills were "closed" to avoid having to meet new controls over operating sites (Department of the Environment, 1995c).

UK guidance describes in some detail access to, and the use of, different documentary sources (BSI, 1981a; 1988; Department of the Environment, 1994c; 1994e; 1995/6; Harris, Herbert & Smith, 1995) and it is not the intention to repeat this guidance. Information on the past uses of a particular area of land, and on its likely subsurface conditions, is not difficult to obtain from published sources and various public authority records. Table 4.1 lists some of the sources of documentary information in the UK, noting their value and limitations. The most important point is that no one source can provide all relevant information. While maps always act as the primary source, they usually need to be backed by supplementary information.

Maps, aerial photographs and remote sensing imagery all require skilled interpretation. For example, from maps the recognition of place names, site names etc. which may indicate particular industries; the recognition of symbols denoting quarries or tips which are often not accompanied by descriptive names; the disappearance of cut features between map editions which indicate potential infilling; the transference of boundaries of sites from one edition and/or series to another. Some sites are difficult to identify from maps, such as older inactive, tips and slag heaps which may have "merged" into the topography; small-scale trades and businesses; illegal tipping sites; overhead and underground features.

Specialised map sources are important for the understanding of hydrology, soil and geology and relevant particularly to hazard identification of impacts on the water environment (discussed further in Chapter 6). The combination of hydrological map data and soil map information can be used to establish the potential for runoff. The HOST (the hydrology of soil types) scheme in the UK (discussed in Department of the Environment, 1994c) provides for classification of soil runoff.

The use of aerial photography and thermal scanner data for the identification of old landfills has been documented in remote sensing texts (e.g. Erb *et al.*, 1985). However, the collation of all relevant aerial photographs to cover a time period can be difficult and they may have been taken for different reasons and from different heights etc. False colour infra-red photography may indicate an area of stressed vegetation caused, for example, by landfill gas or phytotoxic compounds in the ground, although it should be noted that stress may be due to natural causes such as drought, flooding or disease. The technique requires expert interpretation but is comparatively cheap, especially when cameras are flown in model aircraft.

Obtaining such data is both relatively easy and inexpensive. Local and public libraries, regulatory and local authority offices, including county record offices, museums, industrial archives, business archives and record offices usually all provide for both telephone access as well as visits and photocopying services.

The caveat is that historical research can be continued almost indefinitely, with more and more obscure records being sought and consulted. This usually is neither necessary nor productive, and it is generally sufficient to consult more readily accessed records and attempt from these to establish the main features of a site's history. Should this analysis reveal a potentially important gap in the history, it would then be justifiable to spend more time locating other information sources.

Documentary sources which may relate to the current, or most recent, activities on a site and which may increasingly become relevant are records of environmental audits. These will not only be valuable in confirming the nature and location of site activities, but also identify pollution events and incidents.

The final type of documentary information is that available from previous site investigations. These should be interpreted with care and critically as they may have been undertaken for different purposes: they may have focused on specific source–pathway–target scenarios, or may simply have been poorly conducted.

Non-documentary sources can be invaluable in clarifying or even identifying site activities which may have led to contamination but which are not apparent from maps etc. People who worked on sites in the past, residents who have lived in an area for a long time and officials in regulatory authorities can all be invaluable sources of information. Waste disposal practices, the presence of processes which only had a short timespan, storage of materials in particular areas and the locations of buried services may all be revealed. Sometimes, local knowledge can be wrong but influential. An example is provided by the person who phoned a site investigation manager with the information that mustard gas might have been stored on a military site during the Second World War. All site activities had to be stopped until the information was checked owing to the potentially serious nature of any unexpected discovery. No records or problems were revealed.

4.3   UTILISING AVAILABLE HISTORICAL AND DOCUMENTARY INFORMATION

4.3.1   The value of multiple historical records

A 6.6 hectare former dock area (Figure 4.1a) was of interest to a developer who hoped to construct a large retail centre. The presence of good road communications, adjacent densely populated urban areas and a large and cheap site area suggested that the retail park re-use would be successful. All that was uncertain was whether the site's subsurface conditions would be suitable.

The site was a levelled tarmac area as a result of the previous docks being infilled in the mid-1960s during a period of urban renewal. A site visit revealed no indications of what might lie below the pervasive tarmac paving. However, in only two days a significant amount of historical data was made available by the public library, the local waste regulation authority and the current site owner. The information included:

Table 4.1. Value and limitations of different documentary sources

Source	Value	Limitations
Maps – Ordnance Survey	Wide coverage Historical time-frame Reasonable level of accuracy	Detail may be omitted or generalised Requires skilled interpretation For most recent site uses may be out of date
Maps – Geological Survey for planning purposes, soil survey and land research centre	Identify unstable ground, mining and infill areas	Restricted area coverage Soil maps too small scale
Maps – enclosure maps; property maps; tithe maps; estate maps; town plans	Specific to areas and can provide site-specific details	No continuity over time
Directories – e.g. Kelly, post office, local city	Identify individual premises Historical uses – back to about 1830	Urban locations only Interpretation of specific trades may be difficult Street names may have changed
Waste Disposal Registers (Environment Agency)	Location of sites and types of waste deposited	Only since 1976 Licence may not correspond with actual disposal
Scrap Metal Registers (local authorities)	Register of premises used for scrap metal storage	Since 1964 – only includes active sites
Consent Registers – Discharge to Water	Records of consents for discharges to surface water and sewer	Only since 1976
Disposal of diseased animals (local authorities; MAFF and water companies	Record sites where animals buried	
Disposal of sludge (Environment Agency)	Records of sites for disposal of sludge	Only since 1989
Authorisations (Environment Agency)	Authorisation of prescribed processes Authorisations of processes controlled for air pollution	Only since 1991
Explosives storage (local authorities)	Register of premises on which explosives waste kept	
Radioactive Substances Registers (Environment Agency)	Premises regulated under Radioactive Substances Acts	Since 1960
Hazardous Substances (local authorities)	Record of consents issued to sites storing hazardous substances under Planning (Hazardous Substances) Act 1990	Since 1982 original registrations with Health and Safety Executive under Notification Regulations; refers to current active consents

Table 4.1. (*cont.*)

Source	Value	Limitations
Aerial and satellite photographs	Coverage of most of Britain since 1946	Need skilled interpretation
	Good for determining boundaries of quarries, landfill etc.	Satellite imagery may not identify individual areas
Local archives, museums, societies	Good for information on local industries, processes etc.	Secondary source only to maps and directories

- various Ordnance Survey maps (1850, 1870, 1889, 1910 and 1976 editions);
- a technical paper describing the dock construction in the 1870s;
- various engineering detail drawings (included in the technical paper) which gave the constructed levels of the various dock units and the original dock platform;
- waste disposal and local authority correspondence describing how parts of the docks were infilled with wastes;
- the current landowner's records of the final infilling and tarmac capping of the site, to convert it to a large car park.

From these, it was possible to establish the main historical events (Figure 4.1b):

(i) A natural, cliff bounded shore-line existed until the 1870s. Geological Survey maps identified the rocks as massively banded sandstones with subordinate shale horizons.

(ii) The construction of a tidal dock basin, together with two graving docks (for the repair of ship hulls), took place in the 1870s. This work was known (from the technical paper) to have required the blasting and cutting out of large volumes of the cliff face to produce a level dock platform, the adjacent dock basin and the graving docks. It can be assumed that the rock waste was used to raise dock platforms and form dock walls, in the areas away from the original cliff line.

(iii) The docks continued in use until the 1960s, and while various alterations were made, none of these (apart from a local fuel oil store) entailed large-scale storage of contaminating liquids or solids or the deposit of wastes. Thus, no unusual ground contamination was likely to have arisen from the dock usage period.

(iv) With the decline of sea transport in the 1960s, the dock fell into disuse and finally became derelict. The potential dangers concerned the local authority to the extent that it was decided to infill the graving docks, and part of the dock basin, with various wastes, generated mainly by adjacent urban renewal works. From the waste regulation authority's records, it was possible

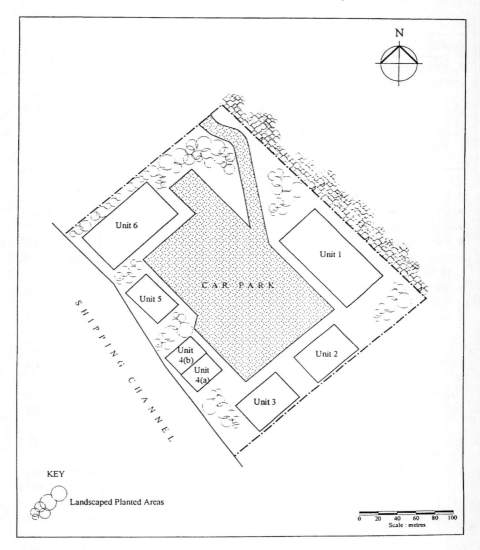

Figure 4.1a. Proposed dock site retail development

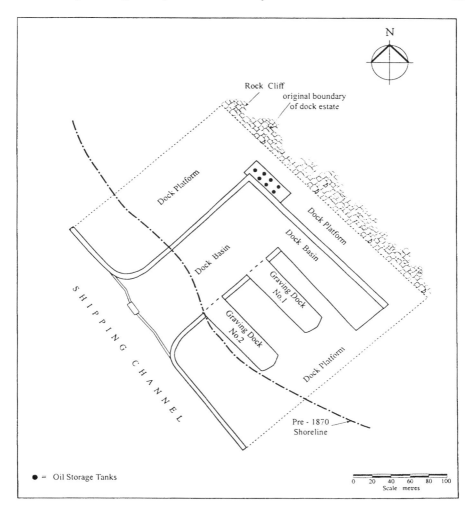

Figure 4.1b. Original dock layout (1870s to 1960s)

to identify where the two main types of solid wastes ("inert" demolition rubble and domestic refuse) were tipped.

(v) Finally, the remainder of the dock was filled (with sands dredged from the adjacent shipping channel) by the current owner, and the site was capped and converted into a car park.

The dock site illustrates the importance of a combination of several different strands of evidence. In this case, previous map editions, a published technical account and waste disposal records together gave the necessary information. No one data source is ever likely to be complete, and "snapshots in time", such as Ordnance Survey maps which are published only infrequently, can omit some very important short-lived events, such as waste dumping.

4.3.2   The problem of over-reliance on a single data source

An executive housing development was planned on a large site adjacent to a long-established village. The site was visited and proved to be supporting a healthy grass cover established in a clay soil. Conditions appeared to be entirely natural. Ordnance Survey maps (1857, 1897, 1912, 1957 and 1981) all showed the site as a series of fields. No indications of industrial or of non-farming activities were apparent. On this basis, the site was acquired as "greenfield" land.

The maps, however, had failed to record that quarrying of sands and gravels (to 13 m depths) had occurred in 1961 and that the quarry void had then been infilled with "inert" wastes, mainly from road building works, until 1969. The wastes, which were loosely compacted and so offered poor foundation conditions, were also pervasively contaminated by oils and tars. While the Ordnance Survey maps did not identify either the quarrying or the later tipping, the local waste regulation authority had records of the tipping. Additionally the local Electricity Board's plans showed the extent and depth of the quarrying, since this was adjacent to one of that Board's transmission power lines.

4.4   SITE ZONING

An important reason for the preliminary investigation is to allow the investigator to "zone" the site. Zoning is no more than dividing the land into discrete subareas, each of which has distinct and different prior histories, from which distinguishably different soil contamination is likely to have arisen, and/or different pathways and targets may be important. Zoning allows the conceptual or theoretical (Hobson, 1993) model of a site to be developed.

Details of the zones into which the dock site might be divided are listed in Table 4.2 and zone locations are shown in Figure 4.2.

While the dock site example was uncomplicated, different investigators could easily have devised zone distributions and numbers which differ from those shown in Figure 4.2. This could have been because it seemed reasonable to

Table 4.2. Former dock site zoning

Zone	Characteristics
1(a)	Original dock platform area
	Underlain by either solid bedrock, or by tipped rock debris, derived from the original construction of the docks. No alien (i.e. imported) material is likely to exist.
1(b)	Original dock platform (fuel oil store)
	Underlain by either solid bedrock or by tipped rock debris, derived from the original construction of the docks.
	The fuel oil store, which existed from before 1910 until the 1960s, could have leaked and given rise to oil/tar contamination of the underlying rock and rock debris.
2	Graving dock No. 1/dock basin main area
	Infilled with loosely tipped "inert" building debris (1962–64). Depth of tipping is known to be between 3.2 m and 6.4 m. Types of tipped materials are not known in detail.
	Below the tipped demolition rubble, dock silts are known to have existed in the main dock basin. Thicknesses are believed to be of the order of 1.5 m, at most. These local dock silts are likely to be contaminated both with metals and with oil/tar residues.
3	Graving dock No. 2
	Infilled (1965) with domestic refuse to thicknesses of some 3.2 m.
	Infilling carried out by the local authority to allow refuse disposal at a time when no other landfill facilities were available.
	Landfill gas and metallic contamination are possible.
4	Seaward area of the main dock basin
	Filled (late 1970s) with up to 6 m of dredged marine sands, prior to the entire area being capped and converted to a car park.
	No high levels of contamination are suspected, although metals, sulphates and chlorides may exist at enhanced concentrations in the dredged sands.

subdivide the dock platform (zone 1(a)) into a number of smaller zones. Obviously this platform once housed various buildings and storage sheds, and it would be entirely defensible to highlight some of these, which might be thought to have been locations of contaminating activities (such as, for example, the store where toxic paints to prevent the fouling of ships hulls were kept), as separate investigation zones.

The basic point is not that the selected pattern of site zones is correct (only actual and later site investigation will prove or disprove this), but that a logically based conceptual model of the site has been devised. As subsurface data are gathered, the chosen zones can be modified without difficulty. Until this proves to be necessary, a framework exists into which the discrete items of subsurface information can be inserted. The significance of particular items of exploration information is unlikely to be overlooked. For example, had the dock site not been zoned in the manner explained, it could be quite possible for the importance of oily rock rubble, in a single trial pit sited at the northern corner of the former

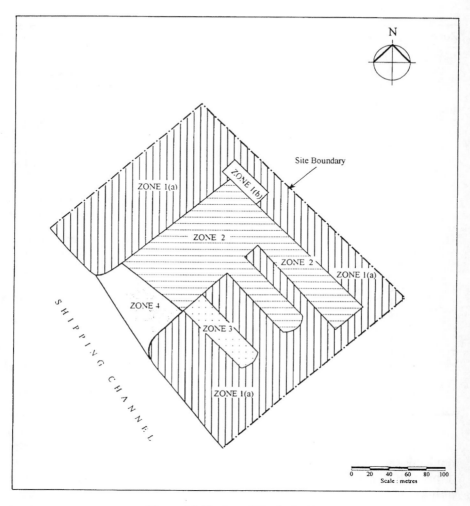

Figure 4.2. Zoning of the dock site

dock basin, to be dismissed. However, knowing that this trial pit existed on the edge of zone 1(b) gives the information much greater significance, and should be enough to persuade the investigator to sink additional, and adjacent, trial pits, to establish if a wider migration of spilled oils (from the former oil tanks) has taken place. Any widescale oil contamination could generate unacceptable levels of volatile organic compounds, and pose risks to a large retail store.

While the former dock site, and similar long-established manufacturing units, are likely to feature on older maps and in various archives, a good many areas of land, including areas used for waste disposal, would be less readily identifiable.

For these, much less comprehensive histories might be definable, and in some cases historical research would reveal little or nothing of any value to aid site zoning.

An example is an area of land (10 hectares) adjacent to a large sewage-treatment works. On all Ordnance Survey maps since the 1950s this had been shown as arable land or as an unoccupied area of undulating topography. Local authority and library archives revealed little useful information (other than that it had been owned by the same sewage authority since the 1920s). The sole helpful source was the local water company, which was able to advise that the land had been used for "sewage sludge disposal and for the tipping of inert wastes". Discussions with former managers of the sewage-treatment works then disclosed that much of the tipping had been of industrial wastes, rather than sewage sludge, in the period immediately after the Second World War, and that these tipped materials could have been contaminated.

After several days spent trying to establish the site's history, all that was factually known was that up to 500 000 cubic metres of variable fills had been tipped over the site's surface, in a period of 40 or so years. Zoning such a site was impossible, and the investigator had no other option but to construct boreholes and trial pits to establish subsurface conditions. Given the unknown conditions, there had to be a multi-stage site investigation (introduced in Chapter 3), and was more expensive than if the site had been able to be fully characterised and zoned.

Site zoning, if achievable, allows a targeted and cheaper contaminated land investigation. Where it is not possible to define earlier site uses to a reasonably complete level, then site investigation is slower, more expensive and much less likely to identify the more significant subsurface conditions.

4.5 SITE RECONNAISSANCE

Much information will be gathered by desk-based study. However, inspection and walk-over of a site is important not only to provide visual and other clues and confirmation of the documentary evidence, but also to confirm that site conditions and environmental setting are as suggested by maps and records, and to confirm the zoning. It is important to understand that, while maps are not generally inaccurate, they are rarely up to date given the pace of development in urban areas.

The first phase of the site reconnaissance is usually to walk over the site, establish landmarks, and identify precisely the boundaries of the postulated zones.

The US Phase 1 site reconnaissance checklist (ASTM, 1993b) identifies the following features:

- Evidence of abandoned storage tanks.
- Stained soil, distressed vegetation, groundwater seeps and standing pools, which may be evidence of leaking underground tanks or hazardous wastes.

- Pits, ponds or lagoons which may have been used for waste disposal.
- Hazardous materials or oil storage drums.
- Unidentifiable containers.
- Drains and sumps.
- Significant odours.
- Abandoned electrical transformers and hydraulic equipment indicating possible polychlorinated biphenyls (PCBs).
- Artificial fill and altered topography.
- Friable asbestos, including wall, ceiling and pipe insulation.
- Service ducts and trenches.
- Septic tanks and drain fields.

A site visit is essential in the evaluation of potential or existing contamination to water resources (Department of the Environment, 1994c). Water flows, from pipe discharges, in culverts or of water courses, may not always be visible and may vary with process operations, weather conditions or seasons. Water bodies and small water courses present on maps need to be verified in the field as land drainage may have altered flow paths.

A range of abiotic and biotic indicators are relevant to site reconnaissance (Department of the Environment, 1994e). Abiotic indicators include debris and structures on the site, anomalies in topography and soil between the site and adjacent areas, and the presence of characteristic colours and odours. For example, lead deposits may be confirmed by white, yellow or black coloration; green coloration may confirm use of copper-based timber treatments; arsenic deposits are mostly white. Severe cases of landfill gas contamination will be evidenced by orange/brown staining at the soil surface. Few abiotic indicators are as reliable as "blue billy" (spent oxides coloured by complex iron cyanides), indicative of gasworks waste. Soil which has been brought on to a site as cover material may be different in colour and structure from that in surrounding areas: it may or may not hide underlying contaminated material. Signs of heating or combustion will assist in identification of disposal areas.

Changes in drainage on a site, perhaps with flooded areas, may indicate that disturbance or infilling has taken place. Discoloration of water and sediments, odours associated with water or the presence of sewage fungus or foaming will all provide clues to pollution.

Biotic indicators include the flora and fauna species present; the relative diversity of plants compared with surrounding (possibly uncontaminated areas); symptoms of effects of contamination such as vegetation staining, discoloration or die-back; and the condition of the soil. Plants can be useful biotic indicators as they root directly into the soil medium and discontinuities and damage (such as leaf yellowing), evidence of growth inhibition, thickening of primary and secondary leaves are often directly visible. However, plants are seasonal, and their distribution and abundance may reflect other ecological and environmental factors. Extremes of soil pH can be indicated by the degree of diversity of plant species. Guidance (Department of the Environment, 1994e) suggests that 30–40 species may be supported by a high soil pH, whereas no more than 5 may be supported by a low pH.

Some species are tolerant of particular contaminants and a community occurring on a contaminated site may be dominated by tolerant species. For example, brown and common bentgrass (*Agrostis canina* and *Agrostis gigantea*) are metals tolerant. Different species tolerate different ranges of contaminants, and some have a wide range of tolerance, not only of contamination, but also of physical and chemical conditions. A capped colliery tip, for example, will be colonised by shallow rooting species adapted to conditions of adverse strata. The presence of indicator species, even one of the universal indicator (i.e. most tolerant) species – alpine pennycress (*Thlaspi caerulescens*) – needs to be evaluated in conjunction with all other visual indicators. For high biochemical oxygen demand (BOD) discharges and seepages into water, aquatic invertebrates are useful indicators.

It is the site reconnaissance which is also important in obtaining information for planning access to a site for investigation purposes, confirming the location of footpaths etc. across a site which may have to be diverted, the condition of buildings and structures which may have to be demolished or fenced to prevent access. Making a photographic as well as graphical and textual record of the site during reconnaissance, including general site conditions and layout of important features, is important to investigation design.

A limited amount of sampling may be undertaken from surface deposits, or existing boreholes, or surface water, or of gas by spiking (see Chapter 7). However, this should not be confused with exploratory investigation which can only be undertaken once the preliminary investigation evidence has been evaluated and conclusions drawn about the nature of the source–pathway–target chain.

4.6   HAZARD IDENTIFICATION INPUT TO INVESTIGATION DESIGN

The output from the hazard identification should be a report that collates and summarises all of the available information; describes the methodology used and any remaining uncertainties about the site arising from lack of information; provides clear conclusions about the contaminants potentially present on the site or in particular areas of the site, the hydrological regime and the geological setting; and identifies the potential hazards and targets at risk (immediate and long term). The role of the report is then to make the link to any required investigation, including exploratory work required prior to design and implementation of a detailed investigation.

The information (including the zoning) allows hypotheses to be developed as to the possible distribution and nature of contaminants. It might provide for the hypothesis that part of the site is uncontaminated. Dutch guidance (NNI, 1991) stresses the need to have a hypothesis regarding the nature and distribution of contamination, and other important factors such as the heterogeneity of strata and the hydrogeological regime. Hypotheses about the spatial distribution of contaminants should take into account their properties (e.g. solubility), how they entered the soil (point or diffuse source), the hydrological regime, the possible characteristics of the soil (pH, redox potential, organic matter content, cation exchange capacity), depth to groundwater, age of wastes, and soil stratification. The guidance suggests four possible hypotheses: (i) the site (or part of it) is uncontaminated, (ii) that contamination is heterogeneously distributed, (iii) contamination is homogeneously distributed, (iv) contamination is heterogeneously distributed without known point sources.

Thus, the hazard identification will have allowed a preliminary judgement about risk to be made. If the judgement is that the site is uncontaminated or that even if contaminants are present the pathway–target chain cannot be complete, then the risk assessment can stop at this point. However, if a source–pathway–target chain appears to exist then the hazard identification will provide for important decisions about:

- the need and scope for further work;
- the precautions required before intrusive investigation can be undertaken (e.g. health and safety access, decommissioning);
- the need for any immediate or interim remedial work (e.g. fencing of potentially unsafe areas; the removal of drums or containers; covering of areas of loose/friable asbestos prior to appropriate removal);
- the focusing of investigation on the hazards, pathways and targets of concern;
- any additional hazards, pollution or nuisance that might be generated by investigation.

The hazard identification is often followed (relatively quickly) by an exploratory investigation to test the truth of hypotheses/confirm the conceptual model and to amend them if appropriate. In the exploratory phase locations for boreholes and sampling points etc. are designed to minimise resources used, but to optimise the ability to prove the model rather than build a full picture of the site condition. The exploratory investigation allows more detailed investigations to be designed in a cost-effective manner. For example, where it has been possible to zone the site, and the exploratory investigation can confirm which zone(s) present a potentially "high risk", detailed investigation can be targeted. Chapters 5–7 discuss sampling strategies, methods and sample analysis relevant to the exploratory and detailed investigations. Chapter 5 includes discussion of the exploratory investigation conducted for the docks site introduced in this chapter.

In conclusion, the hazard identification phase involving the desk study and then site reconnaissance provides for a model/hypotheses about a site to be derived and, if appropriate, tested by exploratory investigation. The hazard identification provides essential information in the risk assessment. A site zoning (if possible), while biased by individual preferences, will provide a logical framework for a phased investigation which will be more cost-effective (and safer) than merely rushing on to a site to sink trial pits and boreholes.

5

Site Investigation in Risk Assessment: Soil Contamination

5.1 INTRODUCTION

5.1.1 Site investigation and risk assessment

Site investigations are conducted to provide sufficient information in the risk assessment to characterise the contamination present within a site and off-site from migration (actual and potential), and the relationship between the contaminants and the media in which they are found. This relationship is the key to determining the behaviour of contaminants and hence the potential exposure of targets. The investigation has to provide adequate information to inform the assessment of hazards and if necessary the estimation of risk.

Chapter 3 provided an introduction to site investigation as part of risk assessment. Chapter 4 stressed the role of the hazard identification stage in providing the conceptual model of a site which may first be tested by an exploratory investigation. This chapter is concerned primarily with investigation (exploratory and detailed) of the soil medium and with methods for obtaining samples of soil and similar solid materials for subsequent chemical analysis or physical characterisation. Chapter 6 addresses the water medium. Chapter 7 addresses investigations of gas. The generic investigation issues discussed in Chapter 3 relating to planning, health and safety and quality control underpin all of the discussion in these chapters.

It is not the intention to provide a detailed manual on how to carry out investigations. Extensive guidance and texts are available (e.g. BSI, 1988; ASTM, 1993a; 1993b; CCME, 1993a; 1993b; 1994; MVROM, 1993a; 1993b; Boulding, 1994; BSI/ISO, 1995a; 1995b; 1995c; 1995d; Harris, Herbert & Smith, 1995; ISO, 1996a). However, much of this guidance has not been written in the risk assessment context. This and the following two chapters highlight key aspects that may affect the robustness and representativeness of investigations and of the data collected, and, hence, the validity of the risk assessment.

The value of guidance is often lost when inexperience leads to over-rigid adoption of suggested strategies to the detriment of understanding of the investigation which is appropriate to the source–pathway–target scenarios relevant to the site. In the UK this has sometimes been the case in use of the sampling guide in DD175 (BSI, 1988).

As discussed in Chapter 3, it is frequently advantageous to carry out site investigation in a number of phases. In particular, as introduced in Chapter 4, first carry out an exploratory investigation (Phase 2 investigation in some terminologies, Phase 1 being the preliminary investigation for hazard identification), before embarking on a detailed (main) investigation (Phase 3). The main investigation will itself benefit from being carried out in two or more stages. No particular investigation is the precise model on which every design should be based: every investigation should be site specific. Two case studies conclude the chapter to illustrate the importance of the phased approach and means to optimise the information required in the risk assessment.

5.1.2   Soil properties

It is important to understand the transport and behaviour of contaminants in soils to design an effective investigation for risk assessment. The following introduction presents some key considerations related to soil processes as a basis for the discussion of collecting representative data in the investigation. Soil and environmental pollution texts (e.g. Ross, 1989; 1994; Alloway, 1990; Wild, 1993; Alloway & Ayres, 1993) provide detailed discussion. It is important to recognise, however, that contaminated land risk assessment is often related to artificial fill materials (particularly arising from waste disposal) as much as to the consideration of contaminant behaviour in natural soils.

Soils consist of minerals, organic matter, water, gases and biota. Soil characteristics are determined by soil-forming processes and are dependent on the nature of the primary geological source, organisms living in and on the soil, erosion, groundwater levels, flooding, wind, rain, solar radiation etc. With time soil-forming processes modify the original material, contributing to the horizons visible within soil profiles, and producing a wide variety of soil types both across and between areas. Within one soil type large variations can occur within a short distance. Soil layering has implications for subsurface migration and the fate of contaminants (Petts & Eduljee, 1994).

When contaminants reach the soil surface they are either absorbed or washed down through the surface layer into the soil profile. Because contaminants tend to accumulate, soil acts as a sink. Adsorption processes which bind inorganic and organic contaminants with varying strengths to the surfaces of soil colloids. This adsorption inhibits leaching downwards, reduces bioavailability to plants, and in the case of organic contaminants affects their rate of decomposition. The extent to which adsorption reactions occur is determined by the composition of the soil (particularly the amounts and types of clay minerals, hydrous oxides and organic matter); the soil pH; the initial chemical forms of the contaminants; redox status (a measure of oxidizing-reducing potential of a chemical system which has important consequences for solubilities at a given pH); and the nature of the contaminants. Ionic pollutants such as metals, inorganic anions and certain organic molecules are adsorbed onto soil colloids. Non-ionic organic molecules (i.e. hydrocarbons, most organic micropollutants and pesticides) are adsorbed on soil humic material. Some organics such as solvents tend to be easily leached. Decomposition is strongly affected by the redox conditions and the predominant types of micro-organisms (Alloway & Ayres, 1993).

The complexity of soil contaminant reactions and transformations is the reason that it is difficult to predict bioavailabilities, mobilities and retentions. The fate of metallic contaminants in soils has been studied to a much greater extent than that of organic contaminants, and workers such as Ross (1994) have shown that soil metal retention processes are generally more important than leaching effects. It is because of this that significant metal pollution of ground-water is an unusual discovery even in relation to land with high concentrations of metallic contaminants. Ross claims that "the risk of metal leaching to ground-water is generally small".

The cation exchange capacity (CEC) of soil is an important factor affecting adsorption reactions. The CEC is a measure under defined conditions of cations that can be displaced from a soil. Sandy soils with low contents of both organic matter and clay tend to have low adsorptive capacities and pose a threat for contaminants infiltrating down to the water table. Non-ionic and non-polar organic pollutants are normally absorbed on soil humic material. Since most of the soil humic material is found in the surface horizons, there is a tendency for organics to be concentrated in top soil, although deeper peaty beds invariably display enhanced metallic concentrations because of this effect. Migration of organics downwards only occurs to any marked extent in highly permeable sandy or gravelly soils with low organic matter contents.

Organic contaminants are decomposed by soil micro-organisms, but the rate at which this happens depends on the compound (i.e. its toxicity to micro-organisms), the pH, the nutrient status of the soil and its adsorptive properties. Chlorinated solvents are relatively mobile in soils and some tend to be degraded quite rapidly, to the extent that they may not reach groundwater (Alloway & Ayres, 1993).

5.2 SAMPLING STRATEGIES

5.2.1 Introduction

Prior planning is essential if an investigation is to be effective. Sampling should only be undertaken once a preliminary investigation has been completed (Chapter 4) from which hypotheses have been deduced about the nature and distribution of contamination, and about the underlying factors governing their potential impact such as human activity, geology and hydrogeology. It may, for example, have led to the conclusion that the site is "uniformly contaminated", or that serious contamination is likely to be largely restricted to a few locations such as the location of below- or above-ground storage tanks, or that the site can be divided into zones which are likely to differ in the nature and distribution of contaminants or their physical attributes. If a site is divided into two or more zones, the investigator is dealing with a number of contiguous, possibly inter-acting, "sites". The principles for effective sampling discussed below should be applied to each zone in the way that they would to a single site.

A sampling strategy for soils has to address a number of questions, including:

- where to take samples in terms of both location within the site or zone (the sampling pattern) and the depth from which samples should be taken;
- the type and size of samples required;
- the analytical strategy in terms of what to analyse for and where to carry out the analysis.

Decisions on the sampling pattern must take into account what is known and postulated about the site and contamination, the objectives of the investigation, and the confidence with which it is desirable to be able to say that the objectives have been met. Issues of cost will inevitably arise: the client may not want to spend the money or the money may not be available to achieve the required standard of investigation. To ensure clarity as to what can be achieved, proper assignment of responsibilities and loss prevention on the part of the investigator, and the limitations resulting from any arbitrary restrictions on expenditure should be made explicit prior to work commencing and in any subsequent report (AGS, 1996).

Sampling and testing are usually two separate activities: (i) collection of a sample in the field, and (ii) testing (including chemical/physical/biological ana-lyses) in an off- or on-site laboratory. However, field testing (e.g. on-site measure-ments of soil gases, water or air quality) is a combination of the two activities and both sampling and analytical aspects must be addressed at the same time.

5.2.2   Approaches to sampling

There are two principal approaches to sampling:

(i) *targeted* or judgemental sampling focused on a known, or suspected, con-tamination point source, with the extent of migration of more mobile con-taminants being established by locating exploration activities (e.g. trial pits or boreholes) along the more probable migration routes;
(ii) *non-targeted* sampling aimed at characterising a defined area or volume (site, zone etc.) and possibly identifying unsuspected point sources and areas of contaminant concentration ("hot-spots").

In addition, composite sampling may be useful, either to examine the condition of a whole site or zone, or as a component of a targeted or non-targeted investigation.

Exploratory investigations are usually based on targeted sampling. Most detailed investigations use a combination of the two approaches. Where a site has been zoned, the balance between the two approaches may differ for different zones.

5.2.3 Targeted sampling

Targeted (judgemental) sampling involves concentrating sampling around known or suspected point sources, although during an exploratory investigation it might also comprise locating one or two exploratory holes (e.g. trial pits, boreholes, probeholes) in those zones where contamination is considered most likely to be present. In a large-scale investigation (e.g. of an industrialised area) the point source might be considered to be a particular site (e.g. old landfill).

Within a site or zone point sources might include: a leaking above- or below-ground storage tank; a leaking supply or drainage pipe; a liquid-handling area subject to spillages; an atmospheric discharge such as a chimney; or the end of a pipe where contaminated effluent was discharged to the land surface (found, for example, on old sewage farms).

Whatever the size of the point source, the general principle is to place exploratory holes around it, taking into account the anticipated direction(s) of movement of more mobile contaminants and any features that are likely to channel such movement, such as drainage runs and ditches. In the absence of evidence leading to a different approach, holes should be constructed at regular distances along lines radiating from each likely contamination point source. The spacings are best chosen as the anticipated distances over which mobile contamination could have spread. A staged approach, in which an initially wide trial pit or borehole spacing is reduced wherever this seems reasonable, is usually preferable.

5.2.4 Non-targeted sampling

As indicated above, the aim of non-targeted sampling is usually to characterise a defined area or volume, to identify unsuspected point sources or "hot spots", and to demonstrate, as far as practicable, that those parts of the site believed to be uncontaminated are in fact free of contamination.

Experience and theoretical studies indicate that a sampling strategy that is effective in identifying a "hot spot" will also be effective in characterising the whole area. Characterisation refers to the types of contamination present, their physical distribution and statistical characteristics such as mean and maximum concentrations.

Three fundamental questions must be addressed:

(i) What is a "hot spot"?
(ii) What size of hot spot is of interest?
(iii) What confidence is needed that the site characterisation is reliable and that a "hot spot" has not been missed?

The size of the hot spot is not a fixed parameter and is to some extent in the control of the investigator. The size of the target area may vary, therefore, depending on the definition of a hot spot. A hot spot may be regarded as either an area of contamination within an otherwise uncontaminated site, or as greater contamination within a site that is generally contaminated. It might also be defined as an area with contamination above a guideline value.

Department of the Environment research (1994b) addressed the number of sampling points needed to detect a hot spot of contamination (defined as a percentage of the total area of the site) with a specified degree of confidence (e.g. 95%), such that if a hot spot exists it will not be missed and that if contamination is not found, a hot spot of at least the specified size does not exist. According to the research the size of hot spot to be targeted should be based on consideration of the largest hot spot "that could be dealt with economically (and without unacceptable health risks) were it to be missed during the site assessment".

The area of contamination considered will be related to the pathway–target chain of concern. For example, in relation to exposure of humans in a residential area, the primary risk is often considered to be to small children ingesting soil while playing in their own gardens. Therefore it is appropriate to devise a sampling strategy that will detect an area of contamination equivalent to a small garden (which may be < 50 m^2 or just 0.5% of a 1 ha site on a modern housing development). The Dutch soil quality criteria require investigators to obtain enough well-located samples to typify an area of 7 m $\times$ 7 m $\times$ 0.5 m (i.e. 24.5 m^3) (MHSPE, 1994). The emphasis on sampling to 0.5 m depth makes clear the near-surface health risks of concern.

There are theoretical grounds for suggesting that an efficient sampling pattern should satisfy four conditions (Ferguson & Abbachi, 1993; Department of the Environment, 1994b):

(i) it should be stratified (i.e. the area to be sampled should be partitioned into regular subareas);
(ii) each stratum (subarea) should carry only one sampling point;
(iii) it should be systematic;
(iv) sampling points should not be aligned.

The *simple random pattern* (Figure 5.1) satisfies only one (iv) of these requirements and is not recommended because it can result in very uneven sampling. The *square grid* (Figure 5.2) and *stratified random patterns* (Figure 5.3), as

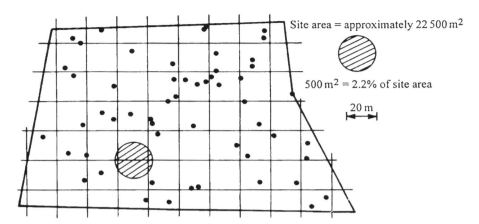

Site area = approximately 22 500 m^2

500 m^2 = 2.2% of site area

20 m

Figure 5.1. Simple random sampling pattern (points located randomly on 0.5 m grid across whole site)

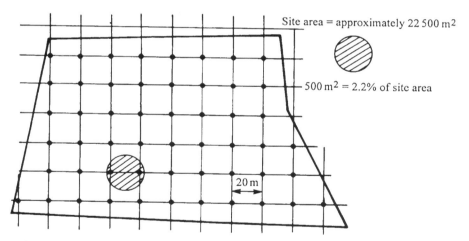

Site area = approximately 22 500 m2

500 m^2 = 2.2% of site area

20 m

Figure 5.2. Square grid sampling pattern (points located at intersections of 20 m grid)

recommended by the BSI (1988), both satisfy three out of the four conditions. The disadvantage of the former is its reduced ability to detect elongated targets that happen to be parallel to the grid directions. The latter tends to produce uneven sampling.

The *herringbone sampling* pattern (Ferguson, 1992) (Figure 5.4) was devised to overcome these disadvantages and yet to retain the practical advantages of being fairly easy to set out on site: it can be formed from four overlapping square grids

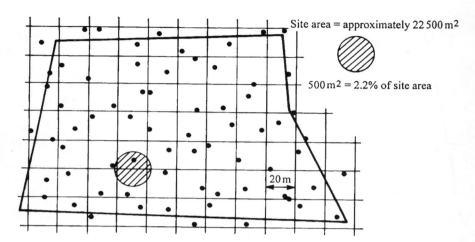

Figure 5.3. Stratified random sampling pattern – 1 point per stratum (points located randomly on 0.5 m grid within each stratum)

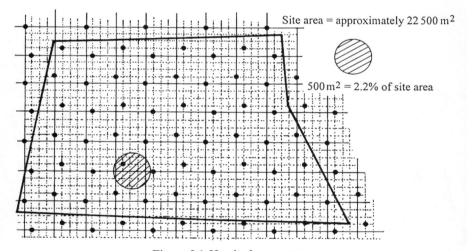

Figure 5.4. Herringbone pattern

of the same size and orientation. It has been the subject of some experimental investigation and limited application. In practice the square grid has been favoured.

Figure 5.5 shows the ability of different sampling designs to detect a circular target occupying 5% of the total site area. Figure 5.6 shows the comparative performance of square grids and herringbone patterns for different target shapes: the regular grid might be judged satisfactory (provided that sufficient samples are taken) unless there is an elongated target aligned with the grid. In practice, the

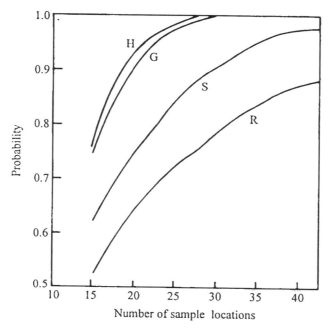

Figure 5.5. Performance of five sampling designs for detecting circular target occupying 5% of total site area (Department of the Environment, 1994b). Key: R = simple random; S = stratified random; U = stratified systematic unaligned; G = regular (square) grid; H = herringbone. Reproduced with the permission of Her Majesty's Stationery Office

probability of detecting a hot spot of defined size can be improved by use of multi-stage sampling.

The guidance on hot spots covers a defined percentage of the area, whereas in practice the concern is usually for a dimensioned area. To locate (with 95% confidence) a circular target area of contamination of 100 m² employing a herringbone would require about 110 sampling locations. This contrasts with the 25 sample locations suggested in the British Standards draft guidance (BSI, 1988); 25 sample locations on either a square grid or herringbone pattern might be sufficient to locate a circular target of 500 m² with 95% confidence for a specified site area and investigation hole spacing (say 20 m grid).

If the target area were elliptical (e.g. aspect ratio 1:4) the probability of locating a 500 m² target using a regular square grid could fall to less than 70%. For a 100 m² elliptical target and 25 sampling points, the probability of locating the target would be less than 30% for both sampling patterns.

The British Standards Institution (BSI, 1988) indicates that investigation points should increase in number as the area to be investigated becomes larger (up to a maximum of 5 ha). Essentially this guidance calls for the spacing between investigation holes to be some 18 m (for areas of 0.5 ha or less) to some 24 m (for

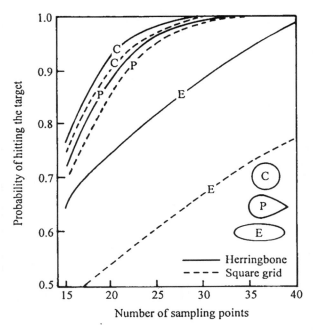

Figure 5.6. Performance of square grid and herringbone sampling designs for detecting targets of various shapes (relative size of each target is 5% of total site area) (Department of the Environment, 1994b). Reproduced with the permission of Her Majesty's Stationery Office

sites up to 5 ha in size). Similar guidance exists in relation to the redevelopment of gasworks (ICRCL, 1979) which allows for 10–25 m spacings between investigation holes on smaller sites, and for 25–50 m spacings on larger land areas.

The reality of site investigation is always an issue when considering sampling strategies. Figures 5.7 and 5.8 show that not all sites are easily accessible. In the case of the site shown in Figure 5.8, only *ad hoc* sampling where machinery could reach would be possible until the site is cleared.

5.2.5   Sampling depths

The sampling strategies described above apply only to a single contaminant in a single plane. The distributions of different contaminants on a site may vary because they have different origins and, if from the same source, because they behave differently in the ground. Therefore, appropriate strategies for sampling at depth must also be developed.

There is little authoritative guidance on this subject. The number of samples and depths from which they are taken depend on the objectives of the investigation. However, the basic premise is that it is cheap to take samples, relatively

Figure 5.7. Difficult site – former steelworks (photo M. Smith)

expensive to analyse them, and very expensive to have to repeat a sampling exercise.

For a typical filled site, a sampling programme would need to take samples: (i) from the immediate surface layer (10 to 100 mm) being indicative of child exposure or surface runoff to an adjacent river, (ii) from within the fill taken at fixed, often 0.5 m, intervals, (iii) from within the fill to reflect changes in appearance, and (iv) in natural ground beneath the fill – the first from as close to the boundary within the fill as possible and then at fixed intervals. Taking samples of natural strata is always beneficial: if uncontaminated, these will indicate the local background chemical conditions and are essential to the evaluation of risk and decisions on remediation values to adopt.

Sample depths should reflect what is known about intentions for the site (i.e. the targets which may be at risk) and the probable pathways by which contaminants can enter the environment. For example, on most housing developments excavation to at least 1.5 m is likely to be required to install services and strip foundations. Deeper excavations may be required for the installation of main sewers. Thus, the construction workforce may encounter materials to this depth and materials from this depth may be brought to the surface and either become spread about or have to be taken off-site for disposal.

Figure 5.8. Difficult site – scrap yard (photo M. Smith)

If samples are taken at regular depth intervals, a cross-section through the site will reveal a "flattened" (say 0.5 × 15 m) grid of sampling points. By analogy with the sampling patterns investigated by Ferguson (1992) and others for two-dimensional sampling, it should be possible to say something about the probability of missing a lozenge of contamination that would fit within the three-dimensional grid. However, since contamination might be concentrated in a layer only a few centimetres thick, there is no substitute for on-site observations or careful examination of an extracted core.

The need for adequate data is illustrated by the following case. The assessment concerned risks arising from surface and near-surface soils. In the first investigation of the approximately 10 ha site, 45 samples were taken of which only 10 could be regarded as "near surface". Of the 45 samples, only 7 were analysed for PAHs. In a second survey, an approximate 30 m grid was applied yielding 103 "near surface" samples for use in the assessment. Use of the guidance (Department of the Environment, 1994b) suggested that, even with this increased sampling frequency, there was less than a 50% probability of finding a circular hot spot 500 m² in diameter, i.e. there was a high probability that a more intensive sampling would have found higher concentrations than those recorded in Table 5.1, which records the maximum observed concentrations for a number of analyses. It should have been obvious that the first limited survey would give a poor picture of the potential hazard.

Inspection of the data from the second survey showed an approximate log-normal distribution. Because it was thought that the data were now sufficient in

Table 5.1. Comparison of data from limited exploratory investigation and fuller, more extensive investigation

Analyte	First survey	Second survey	
	Maximum observed concentration in all samples irrespective of depth of sampling (mg/kg)	Maximum observed concentration in near-surface samples (mg/kg)	95% UCL of the mean*
Antimony	30	1 100	27
Arsenic	58	300	44
Copper	990	150 000	974
Lead	2900	23 000	3874
Mercury	16	90	1.6
Nickel	340	340	46
Zinc	1100	8 800	1388
Total PAHs	185	375	61

*Calculated for log-normally distributed data.

quantity, it was concluded that it was safe to make the primary estimates using the 95% upper confidence limit (95% UCL) of the mean, with parallel calculations using the maximum observed values and the 90 percentile values providing comparative estimates to illustrate the uncertainty of the calculations.

5.2.6 Composite sampling

Composite sampling is common to soil investigations for other purposes such as soil quality surveys. In such environmental sampling a composite sample is prepared from either (i) numerous, approximately equal, subsamples (typically 50 or so) taken from a defined area of land from roughly equally separated points on a predetermined pattern, or (ii) from a relatively small number (perhaps five to nine) of approximately equal subsamples taken over a small area (possibly 1 m^2) around a sampling location on a targeted or non-targeted sampling pattern.

A composite sample of the first type is commonly used to evaluate soil quality for agricultural purposes where a measure of the average quality is required in a field or similar area. It could be prepared by mixing together samples taken to one of the patterns described earlier, but custom has been to use a number of generalised patterns that are easy to employ in the field – for example, the *W* pattern illustrated in Figure 5.9.

Composite samples of the second type are employed typically where large-scale surveys are being carried out to determine local background concentrations in an urban area. They are regarded as likely to be more representative than a single spot sample.

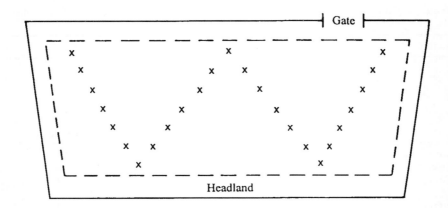

Figure 5.9. *W* field sampling pattern: *X* marks points where subsamples taken. (Note: soil should be taken from at least 25 spots in the area to be sampled. This applies irrespective of the size of the area)

In investigations of contaminated land the issues are somewhat different and spot samples are usually regarded as preferable. Compositing may have a place in well-structured and staged site investigation, but it is essential to recognise the impact that compositing would have in risk assessment against compliance criteria. A composited sample may indicate conditions closer to the average contamination across an area (for example a garden), but "low" and "high" concentrations are effectively lost.

Blacker and Goodman (1994) report the investigation of a site where surface soil had been contaminated by dioxins from waste oil which had been used as a dust suppressant on dirt roads. An investigation using standard USEPA methods would result in the site being split into units or zones of less than 5000 ft^2 each, mainly long, narrow units paralleling the suspected hot spots (i.e. the dirt road) or along contours near streams and gullies. Each area would be subdivided into a regular square grid of 50 points and at each point three samples taken and placed into a separate composite sample. The three analytical results would be averaged and the 95% confidence limits calculated. If the UCL was less than 1 $\mu g/kg$ (microgram per kilogram) the soil would be passed as clean. If it exceeded 1 $\mu g/kg$, the surface soil would be excavated from the whole 5000 ft^2 (464 m^2) area to a depth of 4 in (100 mm). The composite sampling would provide no indication of the patchy nature of the contamination and the remediation strategy may be to deal with "clean" areas, with the inevitable cost implications.

A pilot study was carried out to develop a statistical model of the variation in contamination across the site. This was used to develop a sampling and remedi-ation strategy which enabled just 14 ft × 14 ft (18 m^2) cells to be remediated to leave a residual average contamination in a 50 ft × 100 ft zone of less than 1

μg/kg. If the 95% UCL of three composite samples from one of the new 5000 ft² zones was less than 1 μg/kg, no further investigation was carried out. If the limit was exceeded, the zone was divided into 14 ft × 14 ft cells and a composite comprising nine subsamples taken from each cell. Starting from the most contaminated cell, sufficient soil was then removed to yield a 95% UCL of <1 μg/kg for the whole zone. The survey carried out during implementation of the new design confirmed the assumptions and extrapolations that went into the design. Total costs, including those for the pilot sampling exercise and study team work, were less than 50% of the original estimates.

Some studies have been carried out to compare the effectiveness of different sampling patterns, including area-wide composited samples for determining the average composition of soils within an area. For example, Argyraki, Ramsey & Thompson (1997) report a study to determine measurement uncertainties in site investigation. There are two types of measurement errors:

(i) *random errors* related to the precision of repeated measurements which expresses the closeness of agreement between independent test results obtained under prescribed conditions (random errors when normally distributed can be related to the standard deviation); and

(ii) *systematic errors* related to the closeness of the agreement between the result of the measuring procedure and the accepted reference value (systematic errors are quantified with bias which is the difference between the expectation of the test results and an accepted reference value).

The study was based upon a 1.76 ha metal contaminated site which was used for lead smelting from 1300–1550 and which had not been cultivated since 1948. The study involved:

(i) the researchers testing different sampling strategies;

(ii) a number of organisations applying the same strategy, and

(iii) a number of organisations being able to choose their own strategies and employ their own analytical methods.

The study was not concerned with identifying hot spots, but with obtaining an overall characterisation of the site by, for example, estimation of the mean concentrations of contaminants. The effect of sampling methodologies on the estimation of metal concentrations in the soil was assessed, focusing on the estimation of the systematic error. The measurement of sampling bias was initially considered by estimating the bias arising from the application of different sampling protocols. These findings were then used to design a study to estimate the sampling bias between different organisations. In the final stage, nine organisations estimated the metal concentrations on the study site using sampling protocols and analytical methods of their own choice.

In the first-stage study, four sampling protocols were applied to sample the topsoil (0–150 mm):

(i) Simple regular grid of 20 m grid design.
(ii) Similar grid but with five augur subsamples being taken for compositing in an X pattern over 1 m^2 at each location.
(iii) Herringbone.
(iv) Stratified random design.

In all the designs the sampling precision was estimated by taking sampling duplicates 2 m distance from the initial sampling point in a random direction, in order to make a realistic estimate of the sampling repeatability variance. The samples were analysed by ICP-AES, with duplicate samples to enable estimates to be made of sampling and analytical precision.

The statistical interpretation of the results from the first part of the study focused on two elements with contrasting properties: lead and copper. The results are summarised in Table 5.2. The frequency distribution of lead approached log-normal distribution with a geometric mean of 6000 mg/kg. The distribution of copper was closer to normal with a mean of about 27 mg/kg. Lead showed a high geochemical variability, while copper showed little variation. The natural heterogenicity (geochemical variability) of the elemental concentration over the site is a factor limiting the size of bias that can be detected. The minimum detectable biases for the study site for lead and copper were 726.5 and 0.8 mg/kg respectively.

It is not possible to say in this situation that any one of the four protocols performed better than the others, since the true mean is not known. In addition, a

Table 5.2. Lead and copper concentrations estimated from the four different sampling protocols and % bias from the grand mean of 6229 mg/kg for lead and 27.9 mg/kg for copper (n is the number of samples for each design)

Sampling design	n	Mean (mg/kg)	Median (mg/kg)	Std.dev. (mg/kg)	% bias from grand mean
			Lead		
Regular grid (comp samples)	40	6322	4320	5633	+ 1.4
Regular grid (single samples)	40	6046	3915	5321	− 3.0
Herringbone	39	5680	4470	4420	− 8.9
Stratified random	40	6868	4860	6837	+ 10.2
			Copper		
Regular grid (comp samples)	40	27.4	26.1	4.9	1.8
Regular grid (single samples)	40	26.9	25.7	5.3	− 3.6
Herringbone	39	28.2	27.3	5.6	+ 1.1
Stratified random	40	29.4	28.3	7.3	+ 5.4

repeat exercise could give different results of relative bias from the grand mean. The results, however, suggest that all four methods give a reasonable estimate of the mean when analytical variations are minimised.

In the second stage, a substudy area of 60 × 150 m was selected. Nine organisations sent samplers to the site to take two independent samples according to a *W* sampling design. The 18 composite samples were analysed in the same laboratory. The results were used to assess errors between samplers etc.

In the final sampling proficiency test each organisation was asked to design and implement a sampling protocol for the estimation of lead and copper in the topsoil. Organisations took samples and analysed them using their own choice of method. The statistical analysis of the results is complex, but the results presented in Table 5.3 serve to illustrate the wide variation that can occur from different combinations of sampling pattern, sampling method, sample preparation and analytical methods.

In recent years there has been work to develop software tools to assist in the design of sampling strategies. Groh & Pahl (1993) report a tool to provide local authorities with assistance in designing investigation programmes. In the UK, work on "Site Assess", a tool to aid investigation design, has been proceeding at Nottingham Trent University. The USEPA's Environmental Sampling Expert System (ESES) (Cameron, 1991) has been developing in sophistication to integrate a geographic information system and site description system (including data quality objectives, quality assurance and quality control and site description) with a knowledge frame manager (for analysis, interpretation of data and report preparation). However, at the time of writing there appears to be little experience (at least in the UK) of the use and potential value of such tools.

5.2.7 Deterministic, classical statistical or geostatistical estimation

Assessment of the extent of soil contamination involves interpolation between sampling locations. This process of designating an "expected value" to any

Table 5.3. Mean concentrations reported for subarea by different organisations using their own sampling and analytical protocols

Organisation	Reported Pb (mg/kg)	Reported Cu (mg/kg)
1	4 595	29
2	4 650	< 10
3	6 532	31
4	5 185	27
5	5 936	27
6	11 571	29
7	5 968	—
8	4 203	31
9	6 900	31
Grand mean	6 171	27

unsampled location can be carried out in a deterministic fashion, i.e. on the basis of the modelling of process parameters using experimental distributions. This requires the modelling of a series of experimentally determined functions that are accepted as adequately describing the variable behaviour of the contamination being studied (BSI/ISO, 1995a). Such an approach might, for example, be appropriate where the contamination takes the form of an expanding plume from a leaking underground tank, or from a linear source (e.g. by aerial deposition), where there is some form of correlation between concentration and the distance from the source. Predictions from the constructed model can be tested by a second round of sampling and testing.

However, polluting processes and resulting soil contamination are often of such a complex nature that it is difficult to establish a number of deterministic functions that can be accepted as sufficiently detailed to explain the observed (or potential) distribution of soil contamination. It is therefore better to use a probabilistic approach. This involves the application of statistical or geostatistical methods. Such methods, however, can only be used if sufficient data have been gathered adequately to describe the statistical distributions or geostatistical (spatial) structures (BSI/ISO, 1995a).

During the preliminary investigations no analytical data are available and it is only possible to guess the type of distribution to be expected: for example, a homogenous or heterogeneous distribution (e.g. presence of a point source). An exploratory investigation may provide sufficient analytical data which, together with other information, may make it possible to estimate the parameters of a statistical distribution. For example, it may be possible to use classical statistics and sampling theory to estimate averages and the chances that percentiles may be exceeded between specific sample locations. However, exploratory investigations, as described in this book, are intended only to provide limited data and to test hypotheses drawn from the preliminary investigation. It is more likely that the first stage of a detailed investigation will yield the necessary data for application of geostatistical techniques which can then be employed to enhance the technical value and economy of the second stage (Department of the Environment, 1994b).

During the detailed site investigation the amount of sampling carried out to establish the extent of the contamination may provide sufficient information for the definition of spatial "structures". In such cases it is possible to use geostatistical methods to verify the validity of the sampling strategy, and to calculate estimation errors and probability intervals related to the extent of the contamination. In addition, spatial structures that can be defined from the analytical data often highlight correlations in the soil which were not understood before and which lead to a better understanding of the polluting processes (BSI/ISO, 1995a).

It is important for risk assessment that the available data are studied in such a way that statistical distributions and possible spatial correlations are established and used to calculate uncertainty intervals. Sampling theory should be used to establish the chances that threshold values will be exceeded.

The benefits of multi-stage sampling and the application of geostatistical techniques in this context have been discussed by Ferguson and Abbachi (1993) and in a UK Department of the Environment report (1994b). It is considered that sampling in two or more stages almost always results in much better spatial definition of contaminants for a given total number of samples. Two-stage sampling allows use of a formal geostatistical procedure for demonstrating how the sampling density and configuration affect the precision with which contaminant concentrations at various points can be estimated. Careful analysis of first-stage results is particularly important when multiple small hot spots are anticipated. Explanations are provided by these authors of how the degree of spatial dependence between measured concentrations in different samples can be determined using a *semivariogram*. Having constructed a semivariogram, the form of a regionalised variable is estimated by the method known as *kriging*. At its simplest, kriging is no more than estimating the weighted average of the contaminant concentration within an area or volume, the weighting being chosen to ensure that the estimates are unbiased and have minimum variance. The weighting depends only on the configuration of the sampling points and on the semivariogram, not on the observed values. So if the semivariogram is known from first-stage sampling, it is possible to estimate how the precision in a second stage of sampling will vary with sampling density. First-stage information therefore allows the design of a second-stage sampling scheme to satisfy a given data quality objective. The theory is mathematically complex and those wanting to pursue the topic should consult standard texts (e.g. Journel & Huijbregts, 1978) and the numerous conference papers that deal with particular applications of geostatistics and modelling techniques in relation to contaminated land.

It is claimed (Department of the Environment, 1994b) that the use of semivariograms and kriging can lead to substantial savings in analytical costs, especially on large sites, and that this is particularly true when there are several key contaminants that are themselves correlated. The example quoted is that of gasworks sites where key contaminants include phenols and polycyclic aromatic hydrocarbons (PAHs), the soil concentrations of which are strongly correlated. However, phenols are relatively inexpensive to analyse for, compared with PAHs. A two-stage sampling plan can be devised in which phenols are analysed for all first-stage locations, while PAHs are only analysed at a subset. The phenols semivariogram can then be used in conjunction with the phenols–PAHs correlation to derive a PAHs semivariogram that can be used in turn to guide second-stage sampling density. This technique is known as *co-kriging* (Journel & Huijbregts, 1978).

5.3   ON-SITE ACTIVITIES

5.3.1   Sample collection methods

Improper use and selection of sampling tools may result in data that are not representative of the soil environment being sampled. Choice of appropriate soil sampling devices should take into account the depth of the sample to be taken, the soil characteristics, and the nature of the analyte of interest (e.g. organic or inorganic; relatively immobile, highly mobile water-soluble substances, volatile). During sampling the equipment used should not contaminate the sample (e.g. oil and paint from machinery, adhesives used in borehole construction); absorb contaminants or allow contaminants to escape (e.g. sample containers); and be kept clean to avoid cross-contamination between samples (Harris, Herbert & Smith, 1995).

Methods for obtaining soil and groundwater samples are listed in Table 5.4, together with advantages and disadvantages. The most commonly used in the UK are trial pits and light percussion boreholes. For a full description of each technique reference should be made to standard texts (e.g. CCME, 1993a; BSI/ISO, 1995a).

Surface soil samples are taken using hand-held tools such as stainless steel trowels. Sometimes when using scoops and trowels it may be easier to use a separate device for each sample. A soil punch or other thin-walled steel tube device is more suited for obtaining reproducible samples at the soil surface or shallow depths. These devices are pushed into the soil to a desired depth and retain a sample. Some thin-walled tube samplers are designed as a combination sampling and shipping device, since the ends of the sampler can be sealed for shipment after the outside of the device is decontaminated.

Augers come in a variety of forms intended for use in different soil types, and may be hand or machine operated. In suitable soils it should be possible to penetrate to over a metre using a hand auger: in unsuitable conditions, to no more than about 100 mm. Power augers mix soils and preclude collecting a sample from a particular horizon in a reasonably undisturbed state.

If augering is not carried out with care, cross-contamination is possible caused by loose material from shallow depths falling to the bottom of the hole. An initial survey of an old sewage-sludge disposal area using hand augers suggested contamination with metals such as cadmium to depths of 1.5 m. A subsequent survey using trial pits failed to detect contamination in the dense clay underlying the friable topsoil (up to about 300 mm thick), except in a few isolated fissures. The contamination was dealt with simply by stripping the topsoil rather than excavating to 1.5 m or more.

Trial pits are cheaper than percussion boreholes (about one-twentieth of the borehole costs – up to 20 can be constructed in a day at a cost of about £300), and allow an investigator to see the relationships between different strata and soil layers. Depending on the size of the machine used, depths of up to 5 m can be reached: the common wheeled excavator is typically used to achieve depths of 3–4 m. Trial pits which are extended in length become a trench, which can be the best form of exploratory investigation since it permits the lateral extent of a visually distinct layer to be determined.

Trial pits require careful locating and forming. Figure 5.10 shows a trial pit dug to expose the buried clay bund constructed around a landfill site to increase the volume of tipping space (further soil fill was subsequently tipped against the outside face). The water table in the tip was very close to the surface, but in the adjacent filled ground was several metres down. Inadvertent breach of the buried bund could have led to large leakage of contaminated groundwater from the tip area and creation of a ready path for gas migration.

Care should also be taken when excavating in spoil that the stability of the ground is not adversely affected. In a pit with a depth of more than 1.2 m, shuttering should be in place if the pit is to be entered. Cross-contamination can be a significant problem, arising where samples of the original ground are taken after samples of fill or overlying contaminated soil (e.g. from the base of the pit) or if loose material from the upper part of the pit falls to the base. The walls of the trial pit may become coated with materials from above and material from the excavator bucket may contain material from several different depths. Experience suggests that site assessors should be wary of samples taken from buckets.

Boreholes are required when investigations have to go to depths in excess of 5 m, when disturbance of the ground must be minimised, and when deeper groundwaters have to be monitored and sampled (see Chapter 6). Gas surveys also usually employ borehole installations (see Chapter 7). In the UK, boreholes are usually formed using cable-percussion methods. However, there are a number of different methods of borehole formation which provide contrasting opportunities to obtain relatively undisturbed samples and for the potential release of contaminants into the environment (BSI/ISO, 1995a).

Driven probes and related techniques are gaining in popularity. Among their advantages are speed of operation, minimal disturbance of the ground and, in some cases, the ability to extract an "undisturbed" sealed core for sampling in the laboratory under controlled conditions. This can be valuable to minimise pos–sible loss of volatiles.

Geophysical techniques can be employed to detect the boundary between strata having different physical properties. They measure differences in physical properties such as electrical conductivity (resistivity), bulk density (gravimetric), velocity of shock waves (seismic) or magnetic susceptibility. They can only detect a boundary where a distinct change in physical properties exist. Anomalies, such as near-surface disturbances, may limit their usefulness.

Geophysical techniques may be of value in contamination investigation (Reynolds & McCann 1992; McCann, 1994; Wajzer & Glover, 1995) in order to:

(i) detect the boundaries between ground with different levels of contamination (where this affects bulk properties of the soil – which here includes any groundwater present);

(ii) detect the boundaries between materials with different physical properties that incidentally have different levels of contamination (e.g. the boundary between fill and natural ground or two types of fill); and

(iii) detect buried drums, tanks or other artifacts.

Table 5.4. Methods of exploration (after CCME, 1993b; Harris, Herbert & Smith, 1995; BSI/ISO, 1995b)

Methods	Application to sampling design	Advantages	Disadvantages
Surface sampling using scoop or trowel	Soft surface soil and debris	Easy to use and decontaminate. Allows assessment of immediate hazards.	
Surface sampling using bulb planter	Soft surface soil	Easy to use and decontaminate, in suitable soils. Preserves soil structure and may be useful for VOCs.	Limited to depth of about 150 mm, not useful for hard soils.
Thin-walled soil punch or similar device	Soft soil in surface and at shallow depths	Easy to use. Pushed into soil (e.g. side of trial pit) to capture and retain sample. Some designed as combined sampling and shipping device when sealing caps fitted to end.	
Hand augering	Relatively soft soils to 1 m plus	Allows examination of soil profile and collection of samples at pre-set depths.	Limited depths achievable. Ease of use very dependent on soil type – difficult to use in stony, dry or sandy soil. Can lead to cross-contamination if not done with care.
Hand-operated power auger	Soil 0.15 to 5 m	Good depth range.	Destroys soil core. Requires two operators. Difficult to decontaminate. Requires petrol-powered engine (potential for cross-contamination).
Trial pits and trenches	Varied strata to about 5 m	Allow detailed examination of ground conditions. Ease of access for discrete sampling purposes. Rapid and inexpensive.	Limitation on depth of exploration. Greater exposure of media to air and greater risk of changes to contamination. Greater potential health and safety impacts.

Technique	Advantages	Limitations
Boreholes	Permit greater sampling depth. Provide access for permanent sampling/monitoring points. Less potential for adverse effects on health and safety, or above-ground environment (but note potential risks to groundwater). Smaller volumes of waste to dispose of. May permit integrated sampling for contamination, geotechnical and gas/water sampling.	More potential disruption/damage to site. May generate wastes for disposal. More potential for escape of contaminants to air/water. May need to import clean material to site for backfilling (to ensure clean surface). Reinstatement may be difficult. More costly and time consuming than trial pits. Less amenable to visual inspection. Limited access for discrete sampling purposes. Depending on the technique may be disturbance to samples and therefore loss of contaminants. Potential for contamination to an underlying aquifer. Potential for groundwater flow between strata within aquifer.
Driven probes	Minimal disturbance of site – no need to remove material from the hole. Some soil properties can be determined during penetration. Undisturbed samples can be recovered. Variety of measuring devices can be installed once hole is formed. Fewer health and safety and above-ground environmental implications.	No opportunity to inspect strata. High mobilisation costs for most powerful equipment.
Flow-through sampler (window sampler)	In ideal conditions can provide samples to 8 m.	Best suited to dry, cohesive soils.

Figure 5.10. Trial pit trench across bund (photo M. Smith)

The use of geophysical methods is attractive in areas of high contamination such as landfills, since they are non-intrusive and will not penetrate a capping layer giving rise to release of gases or ingress or water. However, the exact position of a boundary cannot always be identified because properties may change within a transition zone, rather than at a given point, or only over large distances where they are associated with migrating contaminants. The best results are obtained when ground conditions are uniform and simple, with large differences between the physical properties of the various formations. In groundwater studies, electrical resistivity has been used widely because the flow of electrical current in the ground is largely a function of the electrical properties of the fluids present. Electromagnetic methods, such as ground conductivity measurements, are also applicable since the conductivity of leachate is usually significantly different to that of uncontaminated groundwater.

Geophysical investigations must be carried out as an integral part of the overall investigation process and all available information (i.e. the results of preliminary investigations and intrusive investigations to date) should be made available to the geophysicist. This is essential to ensure that appropriate geophysical methods are selected and the investigation is properly designed.

5.3.2 Sampling procedures

Sampling procedures depend on the exploration technique(s). However, there are common requirements to ensure representative data:

(i) Sample size should be a minimum of 1 kg – analysis is usually done on only a few grammes but collection of the larger amount reduces sampling errors and enables preparation of a homogeneous analytical sample in the laboratory. Larger samples may be required for special tests, for example, 25 kg for detection and characterisation of metallurgical slags and 5 kg for combustion susceptibility tests (Cairney, Clucas & Hobson, 1990).

(ii) Disturbed samples (for chemical and similar analyses) should be collected separately from those required for other purposes (e.g. geotechnical testing).

(iii) Sample containers should be wide-necked, sealable and made of materials that will not react with or absorb contaminants, such as glass or stainless steel; polythene bags are seldom suitable as they may be permeable to some contaminants and can become damaged or be difficult to seal.

(iv) All containers should be properly cleaned before use – in general containers should not be reused, but even newly manufactured containers may contain detritus and dust etc. that may distort the analytical results.

(v) Sample containers should be filled to capacity to minimise the volume of air.

(vi) Where the presence of volatile substances is anticipated, particular care is required in the collection and handling of samples – undisturbed samples should be taken wherever possible (e.g. using a sealable coring device that is only opened under laboratory conditions).

Obtaining reliable analytical results for volatile organic compounds is difficult because of the losses that may occur during sampling and because of the marked heterogeneity over time and space (variations in concentrations can be 10 to 100 times over distances of less than 1 m, West *et al.*, 1995). Even under ideal laboratory conditions no more than about 50% of volatile substances may be retained, and under worst-case conditions (e.g. disturbed samples placed in polythene bags) virtually all the organics may be lost (see Table 5.5) (Siegrist & Jenssen, 1990; Liikala *et al.*, 1996). *In situ* measurements may be more effective.

An example of where the choice of the wrong sampling method and containers led to false results is provided by a petrol service station where a number of leaks had occurred from feed pipes to and from the storage tanks. The site was immediately adjacent to housing. The contamination was first detected by smell in the garden nearest to the service station; however, measurable quantities also entered surface drainage and were subsequently detected at a school about 1 km away. Odours began to appear inside the houses which had suspended wooden floors. Samples were taken by a contractor appointed by the service station owner to a depth of about 0.6 m from approximately 300 mm diameter holes drilled using an auger designed for insertion of fence posts. They were then

Table 5.5. Results of experimental studies on sampling soils for volatile organics
(Siegrist & Jenssen, 1990)

Contaminant/sampling method/sample treatment						
	A	B	C	D	E	Original*
Disturbance	yes	no	yes	no	no	
Headspace	low	high	low	low	low	
Container	bag	glass	glass	glass	glass	
Methanol	no	no	no	no	yes	
Measured level of contaminant (all values mg/kg)						
Methylene chloride	<0.4	1.75	6.10	4.90	7.2	24.5
1,2 dichloroethane	<0.1	5.15	5.15	6.70	18.72	22.6
1,1,1-trichloroethane	<0.1	0.20	0.28	0.36	1.87	6.6
Trichloroethylene	0.1	0.32	0.42	0.55	2.27	4.7
Toluene	0.06	0.37	0.39	0.49	0.70	1.7
Chlorobenzene	<0.01	0.56	0.58	0.69	0.76	1.5

Key:
A = Disturbed sample (taken in 7–10 aliquots with stainless steel spoon) in empty laboratory-grade plastic bag with zip closure and headspace of about 40% of volume.
B = Undisturbed sample (core) in Teflon-sealed glass jar with headspace equal to 85% of total volume of jar.
C = Disturbed sample (taken in 7–10 aliquots with stainless steel spoon) in Teflon-sealed glass jar with headspace volume equal to 40% of jar volume.
D = Undisturbed sample in Teflon-sealed jar with headspace volume equal to 40% of jar volume.
E = Undisturbed sample immersed in methanol (about 40 total volume) in Teflon-sealed jar.
* = Best estimate of concentrations in laboratory-prepared material that was sampled.

placed in polythene bags. No petroleum hydrocarbons were detected in the samples by the off-site laboratory.

The local authority engaged its own investigation company. Samples were taken to about 1.2 m using a 50 mm hand auger and compacted into small tins of the type used for paints etc. The laboratory was able to detect significant quantities of benzene, toluene etc. The site owner was then persuaded to employ a competent investigation firm. This took soil cores for analysis, soil-vapour measurements and groundwater samples (the groundwater was about 1.5 m from the surface) from the gardens and the service station site. The consultant was able to produce concentration contour plans allowing appropriate positioning of horizontal soil-vapour extraction wells and groundwater extraction wells to remediate the site.

Appropriate sample collection must be accompanied by appropriate marking and recording, if contamination is to be located and mapped. For example, knowing that a sample contains elevated contents of arsenic, and then being unable to decide where the sample originated, does not help in understanding a site's conditions. Every sample should have a unique number which relates to the site investigation location and depth. Additionally, it is good practice to add a brief material description (e.g. "black ashy gravel"), since there will often be a relationship between material type and the analysed contents of contaminants.

An investigator's notes of the soils found at each trial pit must clearly indicate where a sample was collected, its physical description, and the number it was given.

5.3.3 Sample preservation, storage and transport

The most frequent changes in soil, sediment and water samples are loss of volatiles, biodegradation and oxidation. Low temperatures reduce biodegradation and, sometimes, volatile loss, but freezing water-containing soil samples can cause degassing, fracture the sample, or cause a slightly immiscible phase to separate. Insulated containers specifically designed for transport of soil and water samples should be used – recreational coolers do not give sufficient temperature control (Lovell, 1993). Anaerobic samples must not be exposed to air (BSI/ISO, 1993). Air drying is generally appropriate for metals and other non-volatile analytes. Volatile organics would be lost or reduced in concentration if they were present in soils subjected to air drying.

Naphthalene is an example of a semi-volatile compound that may be lost during air drying or mishandling of the sample on-site or during transit. The importance of this is illustrated on a site where there were repeat surveys over about three months. The samples were all analysed for sixteen individual polycyclic aromatic hydrocarbons (PAHs). Naphthalene was scarcely present in the first survey, whereas in the second survey it accounted for about 20% of the total PAHs present. The primary concern in the site-specific estimation of risks was the carcinogenic properties of some PAHs. Comparison with a guideline value for total PAHs, for example the ICRCL threshold trigger value (ICRCL, 1987) of 50 mg/kg, gave a false conclusion of risk acceptability. The measured value of PAH was given as 48 mg/kg, but the true value allowing for loss of naphthalene could have been 60 mg/kg, which would have been above the ICRCL value and which would have suggested the need for further investigation or remediation. It was not possible in this particular case to be certain where the loss of naphthalene had occurred, but the most likely reason was air-drying: the samples from the second survey were analysed on an "as received" basis.

Unlike water samples, the addition of chemical preservatives or stabilising agents to soils is not common because a single sample is usually used for a large number of determinations and has to undergo pretreatment (milling, drying etc.), during which the preservatives may undergo unwanted and unquantifiable reactions.

5.3.4 Recording site information

The fuller the information collected in site investigations, the more soundly based will be the risk assessment and risk management decisions. The investigator has to ensure that

ENVIRONMENTAL CONSULTANTS

CLIENT:	PROBE HOLE RECORD
Date:	Site:
Probe hole No: 2	Sample Nos: SB2773 - SB2782

Water depth m	Casing depths m	Depth m	No	T	Legend	from m	to m	Ground level above ordnance datum / Description of strata
		SB2773	1	D		G.L.		**MADE GROUND: brown sandy gravelly soil**
		0.0 - 0.15					0.15	
						0.15		
		SB2774	1	D				**MADE GROUND: brown sand and gravel,** brick fragments, concrete fragments
		0.15 - 0.45					0.45	
						0.45		
		SB2775	1	D				**MADE GROUND: brown black ash, sand,** tarry, pockets of green foul lime
		0.45 - 0.8					0.8	
		SB2776	1	D		0.8		**MADE GROUND: green orange foul lime**
		0.8 - 1.0m					1.0	
		SB2777	1	D		1.0	1.1	**MADE GROUND: black tarry ash**
		1.0 - 1.1m				1.1		
		SB2778	1	D				**MADE GROUND: brown black ash, peaty,** and foul lime, very oily
		1.1 - 1.7m					1.7	
		SB2779	1	D		1.7		**MADE GROUND: bright green foul lime, oily**
		1.7 - 1.9m					1.9	
						1.9		
		SB2780	1	D				**MADE GROUND: black tarry peaty ash**
		1.9 - 2.6m					2.6	
						2.6		
		SB2781	1	D				**Brown CLAY with black** mottle and pockets of black sand
		2.6 - 3.1m				....	3.1	
		SB2782	1	D				**Black SAND, oily, becoming** orange mottled
		3.1 - 3.7m				....	3.7	

Remarks:

Gases: CO_2 = 0.5 % O_2 = 20.0 % CH_4 = <1.0 %

Type of boring WINDOW SAMPLER
D - disturbed sample

Figure 5.11. Trial pit log

(i) All exploration locations are well enough identified that they can be re-established at a later date if this is needed.

(ii) Surface ground levels are known.

(iii) Clear descriptions of all the materials encountered are recorded (colour, texture, smell, appearance and vertical and lateral variations are important and the presence of "anomalous" material calls for specific mention) – the descriptions and material thicknesses have to be detailed enough to allow an accurate geological log to be drawn (Figure 5.11).

(iv) Depth, flow rate and appearance of any groundwaters are recorded (for example, smells arising from waters entering trial pits can be usefully informative).

(v) All sample locations, material types and sample numbers are included in the investigator's notes.

(vi) Any *in situ* testing results (e.g. those from gas spike tests driven into the bottom of trial pits) are recorded.

(vii) When in-ground instrumentation has been installed the type and location of the installation is noted, together with the details of any backfilling (e.g. bentonite plugs) employed.

Multiple soil profile descriptions can provide a large amount of information that may be useful in evaluating the variability of soil properties, and the directions and potential for transport of contaminants in the subsurface. USEPA's field guide on sampling of soils (1991c) expressed concern that soil profile descriptions were not being commonly used. Horizons, texture, colour, roots, surface features and sedimentary features (in zones of increased permeability), moisture conditions and water table are the suggested parameters for field description in the USEPA guidance.

Since site investigation is primarily conducted to identify where risks might arise, investigators also have to note those potential pathways by which buried contaminants could interact with human or other targets. As pathways will only be visible when trials pits and trenches are open, a major responsibility for ensuring that the later components of the risk assessment are properly based rests with the individual responsible for supervising and logging a site investigation.

5.4 ANALYTICAL AND TESTING STRATEGY

5.4.1 Introduction

The analytical and testing strategy comprises: (i) the range of analyses to be carried out, the methods to be employed and detection limits required, (ii) the selection of samples to be analysed, (iii) decisions as to whether testing is to be performed *in situ*, on-site or off-site, and (iv) the quality management measures to be applied.

The factors that have to be considered when developing analytical and testing strategies are listed in Table 5.6. Selecting analytical methods is an integral part of the sampling planning process and can strongly influence the sampling protocol. For example, the sensitivity of the analytical method directly influences the volume of sample required to measure analytes at specified minimum detection (or quantification) levels. The analytical method may also affect the selection of storage containers and preservation techniques.

Hazard identification (Chapter 4) provides the primary information when designing the analytical and testing strategy, whilst not becoming so focused that there is no opportunity to find less probable forms of contamination. The analytical strategy takes into account the objectives of the investigation, previous investigation findings, visual and olfactory observations on samples, and other site observations.

In Chapter 3 the dangers to the risk assessment of analysis which is restricted to those contaminants covered by generic guidelines and standards were introduced. A comprehensive approach is likely to involve testing for ubiquitous contaminants, such as lead, zinc, mineral oils and solvents, and testing for substances that are specific to the past and present use of the site as indicated by the preliminary investigation and the plans for the site (if known). However, over-reliance on literature which lists "common contaminants" related to different types of use (e.g. Barry, 1985; Bridges, 1987; LPC, 1992; Department of the Environment, 1995/96) can lead to contaminants being missed (as the listings are not always exhaustive), and a failure to consider contaminants which have

Table 5.6. Analytical strategy design parameters

Parameter	Variable
Scope of the analytical programme	Range of tests
	Numbers of samples
	Types of samples
Use of screening techniques	Field based
	Laboratory based
Sample preparation needs	As received, air dried, other
	Size reduction
	Size fraction
	Extraction
Detection method	Sensitivity
	Reproducibility
	Turnround time
	Reliability
	Cost
Quality control procedures	Sample logging
	Blank samples
	Spiked samples
	Recovery/accuracy performance
	Storage of samples
	Disposal of samples

migrated from a neighbouring site. If the future use of the site is not known then assumptions about potential targets will need to be made.

In the case of elemental analysis (e.g. for cadmium), there may be little advantage in restricting the analytical suite whatever the history of the site or the planned use, since modern automated analytical methods are capable of providing determinations on a large number of elements with minimal additional time required and at little additional cost.

An example of the results of failing to plan an analytical strategy is provided by a timber treatment yard that had used copper-chrome-arsenic treatment agents. It had been subjected to a reasonably full investigation during which copper, chrome, arsenic and hydrocarbons had been detected. There were, however, some gaps in the investigation in terms of sampling locations and depths, and an area likely to be retained for open space (the site was to be developed for housing) had received little attention. A second company was called in to do additional work. Forty trial pits were dug, but only 35 samples sent for chemical analysis. These were analysed either for arsenic or hydrocarbons, but not both. Only three surface samples were analysed from the intended open space area. The local authority insisted that a third, better designed investigation was carried out. A misguided attempt to save the client money resulted in additional costs and in project delays.

Other factors influencing the detailed scope of the analytical or testing strategy include (Harris, Herbert & Smith, 1995):

- the use of specific extraction regimes to provide measures of "total", biologically "available" or water-soluble contaminant concentrations (see below and WRC/NRA, 1993);
- sample preparation needs specific to different types of contaminants;
- the accuracy and detection limits of the analytical methods used.

If generic guidelines are to be used in the assessment, the analytical methods specified in the guidelines must be employed. In the absence of specific guidance, preference should be given to standard methods for soils (or other media) specifically designated for the purpose. This is essential since, for example, methods that work well to determine trace quantities of elements such as cadmium in "clean" agricultural soils may not be reliable when applied to industrial fills. In addition to standards developed or issued by national bodies, a range of analytical methods for use to assess the quality of soils, including those from old industrial sites, is being developed by the International Organisation for Standardisation (ISO).

In principle, and at least initially, it is often better to undertake a smaller range of key tests on a large number of samples than elaborate and expensive testing on only a few samples. A staged approach can also be effective and economic. For example, "total" concentrations may be used as initial indicators of the presence of particular contaminants (i.e. as a screening mechanism), with analysis for specific, and perhaps more reactive or mobile, chemical forms triggered only when specified total concentrations are exceeded. Similarly, the detection of a contaminant by "total" analysis could prompt determination of that proportion available to the water environment, to plants, or in the human gut. These subsequent analyses are sometimes called "dependent analyses" or "dependent analytical options".

In the case of elemental analysis the values reported are rarely the true total concentrations. Usually they are a result of extraction of the sample with a concentrated strong acid such as nitric acid or aqua regia (a fuming mixture of concentrated hydrochloric and nitric acids) (e.g. BSI, 1995). Such methods are called empirical because they measure only that portion which is soluble in the chosen extractant under the test conditions. The results obtained may also depend on the method used for the determination of the concentration in the extractant (e.g. atomic absorption spectrometry). Measurement of "total" aluminium by empirical methods may only recover 40% and potassium only 20%; cadmium, chromium, copper, iron, and lead typically achieve recovery of 70–90%. To obtain true total concentrations requires specialist techniques such as fusion and hydrofluoric acid and/or X-ray fluorescence analysis. For general discussion of analysis see, for example, Davies (1980) and Keith (1991b).

"Total" applied to organic chemicals, e.g. phenols, usually means that an analytical technique has been employed that cannot distinguish between similar compounds: in the case of phenols say between monohydric, dihydric or trihydric compounds (i.e. compounds with one, two or three hydroxyl groups attached to a benzene ring). Note that "phenol" is applied to both the compound C_6H_5OH and to the class of compounds in which one or more hydroxyl groups is attached to a benzene ring. Again, the term total may be misleading: not all phenols may be detected by the method employed (for example, there may be limitations in terms of molecular weight or the number and size of other groups present on the benzene ring, and different methods may give different results).

"Total" may also be used misleadingly to distinguish between, for example, the amount of an anion such as sulphate dissolved in an acid of prescribed strength, and the amount soluble in water: often inaccurately termed "soluble". In practice, since some compounds (e.g. gypsum – $CaSO_4.2H_2O$) have limited solubility in water, the ratio of solid to water should also be given: for example, water-soluble concentration based on extraction with water in a ratio $1:5$, "soil: water".

If the basis of a contamination risk assessment is to establish the potential risk via ingestion and the foodchain pathway, "total" levels may not be the primary concern, as many heavy metals are held in complexes and mineral structures that render them relatively immobile with respect to plant and animal uptake. A significant pathway of exposure is direct ingestion. Risks are usually expressed by comparing a daily dose calculated using "total" soil concentrations, of say lead, with tolerable daily intake (TDI) or other reference dose (see Chapter 10). It is often suggested that this overestimates the risk, since not all the toxic element may be soluble in the human gut. The actual availability will depend on the mineralogical form in which the contaminant is present and the precise conditions in the stomach at the time of ingestion (when food enters the stomach the pH usually rises for a time). Various proposals have been made for extractants that will simulate stomach acids and absorption in the gut. Gasser *et al.* (1996) suggest that extraction with saline (i.e. NaCl), 0.1M HCl for one hour at 298K (25°C), may be a sufficient indicator of *bioaccessibility*. However, Ruby *et al.* (1996) have proposed a means of estimating the *bioavailability* of arsenic and lead using a "physiologically based extraction test".

There is often confusion between methods described as producing "extractable" metal levels and those methods giving "plant-available" metal levels. Weak acids and salts, such as ammonium acetate, calcium chloride, ammonium nitrate, EDTA (ethylenediaminetrata acetate) and acetic acid, are frequently used as extractants for estimating the metals available for plant uptake. These methods have generally been developed for use with soils and numerous studies have shown them to be good predictors of plant uptake, although none is universally applicable. They may be of doubtful value (ICRCL, 1990a) when applied to some soils and to other materials such as metalliferous mining wastes because of the variations in pH, mineral composition, weathering rates and metal speciation; hence care must be taken in any comparative interpretation of results.

Water-soluble sulphate, free cyanide and monohydric phenols are all examples of determinands commonly tested as "dependent options". However, decisions to use such "triggers" (and the concentrations at which they apply) should be made for each site in the light of what is known about the probable nature of the contamination. It is inappropriate to include such triggers in standard specifications.

Whereas the inorganic analyst is primarily concerned with the analysis of a defined number of elements and anions, the organic analyst is often interested in looking for any chemical which might be present. The detection of "adventitious" or unexpected substances, particularly when complex mixtures of organic chemical species are present, requires the use of analytical screening methods. Techniques such as gas chromatography/mass spectrometry (GC/MS) and inductively coupled plasma spectrometry (ICPS) can provide data on a wide range of compounds and elements and be a useful means of identifying more specific analytical requirements (Taylor, 1993).

Analytical laboratories sometimes offer analytical packages (see Table 5.7) covering a range of common contaminants: these should be checked carefully to confirm that they are consistent with site-specific requirements as determined by the hazard identification.

UK practice owes much to pioneering work on the investigation and assessment of contaminated land by the Greater London Council in the 1970s (GLC, 1976). It has been adversely influenced by the restricted list of contaminants covered by the ICRCL trigger values (see Chapter 8) and the failure of investigators to think for themselves and properly relate the analytical strategy to the

Table 5.7. Typical analytical package for soils

"Elements"	Anions	Other determinands
Cadmium	Chloride	pH
Lead	Sulphate	Phenols
Arsenic	Sulphide	Toluene extractables
Chromium		Cyclohexane extractables
Zinc		Coal tars
Copper		Mineral oils
Nickel		Sulphur
Boron		Polycyclic aromatic hydrocarbons (PAHs)
Mercury		Electrical conductivity
		BTEX (benzene, toluene, ethylbenzene, xylene)
		Total hydrocarbons

source–pathway–target scenarios of concern. A prime example is the use of the toluene extractable matter (TEM) and cyclohexane extractable matter methods. These were first developed by the Greater London Council to enable mineral oils, tars, animal fats etc., to be distinguished. For example, the TEM would be redissolved in toluene and then used in a thin layer chromatographic (TLC) method to give a semi-quantitative determination of "mineral oils" and/or "coal tars". Or it might be used as a trigger for more specific analyses (e.g. PAHs). However, the method is not specific: toluene will also extract elemental sulphur, humic matter in soils and peat, and organic substances from coal etc. In practice, little is known about what toluene will extract and under what conditions.

In addition, whereas PAHs at concentrations of a few tens of mg/kg or less may be of concern, the "detection level" for TEM may be several hundred mg/kg depending on the conditions employed, i.e. the TEM is not a satisfactory means of triggering more detailed analyses. Experience also shows that there is little standardisation of the method and the results obtained by different laboratories may differ by a factor of two or more (Smith & Ellis, 1986).

The wrong choice of analytical methods can give misleading results about the risks arising from the site, the applicability of a particular remediation method or the success of that method if it is applied. For example, dynamic headspace analysis, or "purge and trap", is widely used in the USA for volatile organic compounds (VOCs) in soils because several USEPA methods specify it. A report by the US National Research Council (Travis & MacInnis, 1992) showed that conventional purge and trap methods for determination of the concentrations of volatile organic compounds may not give a reliable result because the purge gas removes only those contaminants present in air within the pores in the soil. It cannot remove those trapped in the soil matrix and not readily available. This possible underestimation is of concern both in the initial estimate of the quantities of contaminants present and in the assessment of performance of, for example, soil-vapour extraction systems. It has been shown (Voice & Kolb, 1993) that direct (static) headspace analysis of soil samples can provide superior results for several common VOCs and soil types.

5.4.2 On-site laboratories and on-site testing

Considerable effort is being devoted to the development of reliable rapid methods of analysis that can be employed, if not actually at the sampling location, then in an on-site laboratory. In addition, real-time methods of analysis are being developed for some lower volatility compounds (e.g petroleum hydrocarbons) to considerable depths.

On-site laboratories are most likely to be used when an investigation involves only a limited number of analytical determinands selected on the basis of site history and earlier investigation, or requires rapid analysis of indicator species as a means of controlling on-site works with subsequent confirmatory testing in an off-site laboratory.

Field testing and on-site laboratories can reduce turn-round times and provide data that can be used to direct site works (e.g. extend the depth of a borehole or length of a trench). However, analytical reliability may be reduced, and there may be a need to employ surrogate methods that have been shown to provide an adequate indication of the concentrations of key contaminants. A balance has to be struck between rapid, on-site methods of analysis offering less precision but a greater number of samples for the same cost and time, and fewer samples subjected to more precise laboratory analyses.

The relative accuracy of the different methods is of concern (Stock, 1995; Ramsay, 1995). Rapid on-site methods of chemical analysis were reviewed by Montgomery, Remeta & Gruenfeld, (1985). On-site analysis of hazardous wastes is addressed by Simmons (1990) and specialist conference proceedings (ACS, 1992). Recently developed on-site methods include stripping analysis for soil and sediment (Olsen *et al.*, 1994), immunoassay methods (Van Emon & Gerlach, 1995) and gas chromatography methods designed for field use. Studies under the Innovative Technology Programme (USEPA, 1995a; 1995b; 1995c) show the need for independent evaluation of such methods in carefully controlled evaluation studies (e.g. ASL, 1995; Marvan & Herbert, 1996).

Soil-vapour measurements provide an indirect and relative measure of the concentration of volatile contaminants in the soil, either immediately surrounding the sampling point or at depth. When sampling is carried out on a regular grid, isopleths (lines of equal concentration) can be developed showing the magnitude and direction of movement of a pollution plume. These plots can be used to help locate suitable positions for other types of sampling installations (e.g. deep probes or boreholes).

Although the measurements obtained may be only relative to the total concentrations of the substances in the soil or groundwater, they are likely to be of comparable reproducibility and reliability to measurements made on laboratory samples, given the difficulty of obtaining representative samples and the potential for loss of volatiles during sampling and subsequent handling.

Volatile organics (VOCs) may occur as traces in landfill gas (either as an inherent component or as an indicator of the disposal of organic substances), or from contamination of the ground with the substance in question or a complex mixture of substances (common materials such as petrol and fuel oils contain a variety of substances varying greatly in volatility). The presence of VOCs in the latter case may be indicative of either a near-surface spill or of pollution at considerable depth being transported on the surface of the groundwater table.

Measurements are typically made using near-surface sampling points rather than specially constructed and (deep) boreholes, although narrow diameter probes can be forced into the ground to considerable depths.

Sampling and measurement techniques may permit immediate identification and/or measurement of the contaminants present, or integrate emissions over a period (often several weeks). A range of different sampling point types may be used in the first (most common) application, but shallow probes and various instrumental techniques are most often employed. Sampling devices can be placed at depths of 30 m or more using driven probes. Samples for analysis may be withdrawn under suction or by flushing the gas from a sample chamber into which it has entered through a gas-permeable membrane of high diffusion impedance (Robitaille, 1992).

Measurements are frequently made on-site, although samples for off-site analysis can also be collected using either gas containers or absorption tubes. Substance-specific detection tubes may also be employed down exploratory holes. An indication of the presence and quantity of bulk gases can be obtained using portable instruments. These are easy to use and cost-effective for safety control purposes and for initial identification and location of the hazard, although their selection, use and interpretation require special care (Crowhurst & Manchester, 1993) – see Chapter 7.

5.5   THE DOCKS SITE EXPLORATORY INVESTIGATION

In Chapter 4 the docks site was introduced and a zoning of the site produced on the basis of the hazard identification (see Chapter 4, Figure 4.2). It was apparent that some of the postulated site zones (i.e. zone 1(a), zone 1(b), parts of zone 2, and zone 3) could be investigated by trial pits: the latter being appropriate for excavation depths up to 3 m (see Section 5.3.1). Boreholes of greater depth, however, would be needed to confirm whether deeper oil migration had occurred in any area of zone 1(b); the types of fills in the deeper areas of zone 2 (i.e. the former dock basin area); the nature of the silts beneath these fills; and the strata sequences and contamination conditions in zone 4. The docks site provides an example of successful targeted sampling.

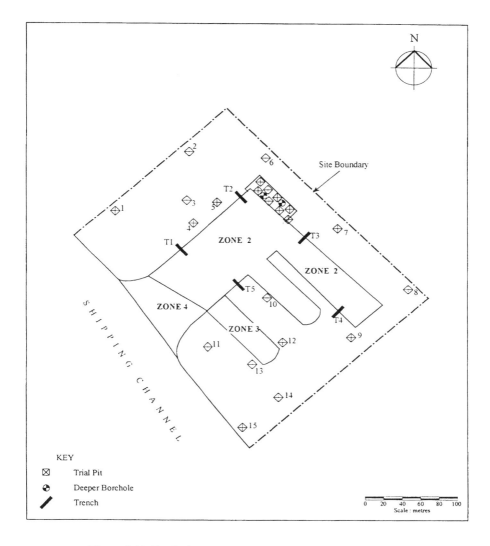

Figure 5.12. Dock site zones 1(a) and 1(b) investigation strategy

Figure 5.12 shows the exploratory site investigation which was thought necessary to prove the conditions in zones 1(a) and 1(b). Five trenches successfully confirmed the edges of the original dock platform, and showed that the distinctions between zones 1(a) and 2 were soundly based. A relatively coarse trial pit grid (at investigation centres of about 100 m) was enough to prove that either solid bedrock or local rock debris occurred throughout zone 1(a). This trial pitting density was increased on the boundaries of other zones (i.e. zone 1(b) and zone 3) from which mobile contaminants (i.e oil spillages and landfill gases) could be migrating to affect zone 1(a), and also around buildings where anti-fouling paints had been stored.

In zone 1(b), where localised oil spillages might have existed, the initial trial pits were sited quite close together (15 m centres) to give a greater probability of locating any oil-contaminated materials. Two deeper boreholes (to depths of 6.3 to 7.4 m) proved later to be necessary, to establish the depths to which oil contamination had migrated, and to permit volatile organic vapour measurements in more oily horizons.

In total fifteen trial pits and five trenches over zone 1(a), eight trial pits in zone 1(b), and two boreholes in zone 1(c) were used. This work proved adequate to confirm the zonings. Zone 1(a) was found to be underlain largely by fractured sandstone bedrock, without obvious signs of any ground contamination. Chemical analyses of a few (eight) selected rock samples confirmed that these were quite uncontaminated. The sole abnormality occurred at trial pit number 13 (adjacent to a storage unit which is known to have been used to hold anti-fouling paints), where oddly discoloured rock debris was found saturated with an oily pale liquid. Later chemical analyses of samples from this locality revealed concentrations of soluble metals (including organotin), which suggested that leakages of anti-fouling liquids had created significant soil contamination. No migration of oils (from zone 1(b)) or of landfill gases (from zone 3) appeared to have adversely affected the dock platform area.

In zone 1(b) most trial pits revealed conditions identical to those in zone 1(a). Oils were generally absent, and chemical analyses revealed no measurable hydrocarbon contents. However, in two trial pits visible pools of oily residues were seen. These had collected over siltier bands in the local bedrock, but had also migrated to greater depths via natural fractures in the rock. When this was appreciated, the two trial pits were pumped out (to a waste liquid tanker) and two deeper boreholes were carefully drilled. Periodically, during the drilling, volatile organic compound (VOC) concentrations in the boreholes were measured *in situ* using a photon ionisation detector, recognising problems of "losses" during sampling discussed earlier.

In zones 2, 3 and 4 (Figure 5.13) a similar targeted site investigation was implemented. Seven trial pits, supplemented by three boreholes, were adequate to establish that the fills tipped in zone 2 were predominantly inert demolition wastes, without any obviously biodegradable contents. However, blue asbestos lagging was found to exist in layers at the base of the fills. The underlying dock silts (only intersected by the three boreholes) were proved and found to be both metal rich and heavily contaminated with tarry residues.

A closer distribution (25 m centres) of 15 trial pits was thought to be necessary in zone 3, since older domestic wastes (once biodegradation has taken place) can be quite variable in their final nature. To supplement these trial pits, a landfill gas spike test survey (at 10 m centres) was carried out to establish the levels of methane, carbon dioxide and oxygen gases in the waste layers. This gas survey

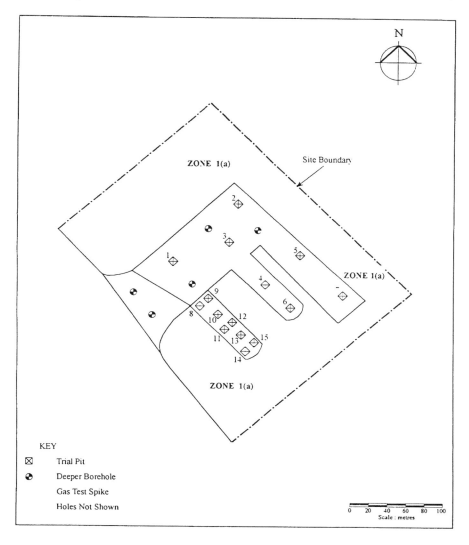

Figure 5.13. Dock site zones 2–4 – investigation strategy

revealed that explosive levels of methane gas did occur in various areas of zone 3, and that carbon dioxide concentrations were usually greater than 5% by volume (see Chapter 7 for survey discussion).

Finally, two boreholes were sunk through the dredged sand infill in zone 4, to establish if any meaningful high levels of contamination existed. In the event, these sands proved to be essentially uncontaminated.

5.6   TIMBER TREATMENT SITE

5.6.1   The case site

The detailed investigation of a former timber storage and treatment yard serves to illustrate and emphasise the essential principles discussed in this chapter: i.e. that each site investigation has to be focused in the context of the specific risk assessment objectives and all the possible design choices (e.g. number of investigation holes, investigation depths, sample selections, choice of analytical strategies, sample preservation and investigation specification) have to be made to ensure that this is achieved.

The site was on the edge of an urban area of flat topography. A desk study showed that until 1895 the land was in agricultural use. From 1895 until 1985, a timber storage and treatment yard existed on the site. Until the 1930s this was quite small and occupied only the southern corner of the site – later, the entire area was used for various timber trade activities. Timber treatment took place using "Protim" and "Cellcure" processes (see below).

A three-part zoning of the site was thought to be appropriate (see Figure 5.14).

A limited exploratory survey confirmed the general adequacy of this zoning, and revealed the essential features of the identified zones, as set out in Table 5.8. No evidence was found that activities likely to have generated the combustion ashes found in zone I ever took place on the site. It was therefore assumed that these were deliberately imported to provide a dry surface when the timber yard was established and extended.

The details of the Protim and Cellcure timber treatment processes were easily found in standard timber technology texts (e.g. Wilkinson, 1979). The Protim process employed a white spirit solvent which, at various times in the evolution of the treatment method, carried the chemicals pentachlorophenol (PCP), lindane or tributyltin oxide, into the wood being treated. The Cellcure process used a mixture of water-soluble chemicals – copper sulphate, sodium dichromate and arsenic pentoxide – which when injected into wood become essentially non-leachable (such treatment methods are known generically as copper-chrome-arsenic (CCA) methods).

Employee accounts suggested that timber treatment only took place after wood had been cut into the required lengths, and so all sawdust was expected to be free of treatment agents. However, it should be noted that on many timber treatment yards treated timber was left to drain, with surplus agent producing extensive surface contamination.

5.6.2   Hazard identification

The intention was that the entire site should be used for domestic housing purposes and the developer required a detailed site investigation to confirm the site zoning, identify possible environmental and human hazards, and provide enough data to permit cost-effective reclamation choices to be made.

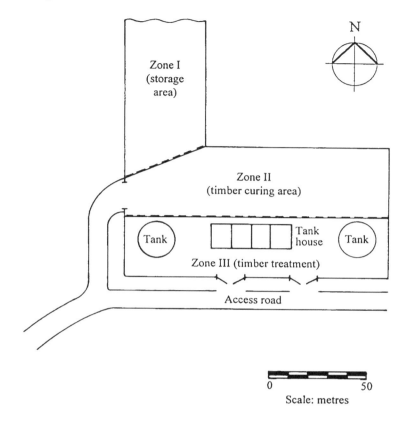

Figure 5.14. Timber treatment site zoning

In general four discrete risks relating to housing development are plausible as a result of chemical contaminants, i.e. risks to:

- human health;
- human health if gases or vapours enter dwellings;
- plant populations;
- buildings and building materials.

Resulting from the development process itself an additional four risks are plausible:

- pollution of surface waters;
- pollution of groundwaters;
- risks of widescale air pollution;
- risks of contaminants migrating off-site.

Table 5.8. Zoning of timber treatment site

Zone I	Hummocky ground surface due to the presence of low waste heaps of timber scraps, sawdust and demolition rubble.
	Below this is a pervasive (0.5 m) capping of combustion ashes overlying alluvial clays with subordinate water-bearing gravels.
	No buildings or foundations occur, and past records suggest that Zone I was employed only as a timber storage area (1940s to 1980s).
Zone II	Flat area, obscured by abundant concrete foundation slabs and roadways.
	Excavation revealed a thicker (1.5 m) capping of combustion ashes above alluvial clays and gravels.
	Past records reveal that timber drying and preservation treatment took place here, using the "Protim" and "Cellcure" processes.
Zone III	Flat area, entirely covered with concrete surfacing, in which large vats and tanks (for timber preservation) could be seen.
	Limited trial pitting revealed a much thicker (3 m or more) ash surfacing, the presence of oily liquors and obvious chemical odours.
	Information from past employees revealed that Zone III had been the main timber-preservation area (1930s to 1980s), and was an area where spillage occurred frequently.

These last four would result in breaches of UK statute law (the Environment Act 1990, the Water Resources Act 1991 and the Environment Act 1995). The others would give rise to unsafe conditions for future residents and could be breaches of statute law in worst cases, or could expose the developer to common law litigation.

PCP, tributyltin oxide and lindane are toxic to most life forms, but are essentially insoluble except in oils. In contrast, the copper-chrome-arsenic salts are water soluble and can be expected to add high concentrations of water-soluble metals to those areas where they are spilled. If treated timber is burnt the ash may contain percentage concentrations of copper-chrome-arsenic, a large proportion of which may be soluble. Arsenic is toxic and a proven human carcinogen. Hexavalent chromium (the form present in timber preservatives) is toxic and attacks the skin. Copper is toxic to plants at low concentrations and to humans at high concentrations. Soluble sulphate is also toxic to plants. As the treatment agents are designed to prevent microbial decay, adverse effects on micro and macro soil fauna are also to be expected.

The information available on the site indicated that the most likely risks were:

- to groundwater quality;
- off-site migration of contaminants through groundwater movement;
- to human health from ingestion and contact, and possibly from entry of vapours into buildings;
- to plant populations;
- to soil ecosystems;
- subsurface combustion of clinker or sawdust.

These risks seemed feasible mainly because of the known past usage and spillage of wood preservatives and, to a lesser extent, because ashes tend to have elevated metal concentrations which could harm human health or plant cover.

As emphasised in Chapter 4, commencing a site investigation with such a preliminary risk assessment is useful. Attention is focused on the subsurface conditions which might make a risk more probable – such as conditions along the site's boundaries where off-site migration of toxic fluids can be looked for. If and when unexpected materials are found, consideration of their significance in the pre-identified risk assessment context will make it easier to decide whether further investigation and analyses are required.

5.6.3   Questions to be addressed

From the hazard identification, exploratory investigation and preliminary risk assessment, a number of questions to be addressed were apparent. In zone I the questions included:

- Is the thin ashy surface contaminated?
- Are the sawdusts free of Protim and/or Cellcure contamination, as has been suggested?
- Are the underlying alluvial strata (and groundwater contained in these) clean, or have leachates from ashes and other contaminants migrated downwards?
- Are any other contaminants of concern hidden below the pervasive ashy capping?
- How best can this area be investigated (trial pits, boreholes, other methods such as geophysical surveying)?
- What contaminant hazards could adversely affect the planned reuse of the site?

Essentially the same questions had to be answered for zone II, the main difference being whether particular health and safety concerns should be addressed (because of past frequent spillages of wood preservative fluids).

Similarly, the same questions had to be addressed for zone III. However, the apparently much greater thickness of the ashy cover and the presence of oily liquors (possibly the residues from Protim preservative spillages) required, in addition, consideration of how sampling was to be done and what health hazards could affect investigation personnel.

5.6.4   Investigation strategy

It was decided that in zones I and II trial pits would be adequate, since contamination (on the evidence available) was thought unlikely to extend below depths of a few metres. Trial pits were also considered suitable for use in zone III, although a concrete breaker unit would have to be available on the excavator to penetrate the pervasive concrete surfacing. It was decided, however, that boreholes would probably be necessary in a later phase of the investigation, if preliminary results from trial pits indicated that wood preservative fluids had penetrated to substantial depths into the ground.

It was decided that the more important questions listed above could be answered by taking at each sampling point:

- a sample for chemical analysis of the surface sawdust;
- a representative sample of the ash for chemical analysis, with additional samples being taken to reflect differences in appearance of the ash;
- a sample of the alluvial soils for chemical analysis taken from just below the ash layer, to provide an indication of whether or not vertical leaching or migration from the overlying ash had taken place;
- a sample of groundwater from within the alluvial strata to show if the site's industrial history had resulted in contamination;
- additional samples for analysis if unusual soil coloration or smells were encountered, or if pools of oily liquors were found.

In addition, because it was thought that the surface ashes could pose subsurface smouldering and burning risks if heated to moderate temperatures, samples were taken for combustion testing.

It was expected that between four and six samples would be collected from each sampling location.

A sample storage protocol for ashes suspected of containing oil-based wood preservatives was prepared (see Table 5.9). The use of glass containers was considered essential to avoid the interactions that were likely with plastic containers.

The site provides an example of the critical importance of accurate on-site recording of site observations. If in the trial pitting of clays underlying the ashy fills in zone II, the investigator merely noted that the clays were "soft and very moist", and had collected samples which proved to be both chemically clean and very impermeable, a colleague, summarising the investigation findings for the risk assessment, could conclude that no risks of deeper migration of soluble or mobile contaminants were likely. This would prove to be entirely incorrect if

Table 5.9. Sample storage protocol for ashes suspected of containing volatile oil-based wood preservatives

Sample size	1 kg
Sample container	Pre-cleaned screw-top glass jar supplied by the test laboratory
Filling required	Total volume of the jar
Storage arrangements	In an insulated cool box and out of direct sunlight
On-site storage period	8 hours maximum
Receipt by analytical house	On the same day as sampled

unrecorded lenses of permeable sands were present within the clay and providing preferred leakage pathways for chemicals to enter deeper groundwaters.

The most obvious contaminant hazard to workers was the likelihood that different pesticides and wood preservatives would be found within the ashy fills or as pools on the surface of the underlying alluvial clays. Since these chemicals are designed to be biotoxic, the obvious concern was that site investigation personnel could be at risk. Appropriate protective measures were taken. It was considered particularly important to avoid contact with chemicals which might directly affect the skin (e.g. hexavalent chromium) or penetrate the skin (e.g. pentachlorophenol).

5.6.5 Analytical strategy

The ICRCL trigger values (ICRCL, 1987) relevant at the time provided no guidance on PCPs, lindane or tributyltin likely be present on the site and, for the metals copper, chrome and arsenic, gave guidance only in terms of the "total" metal concentrations. On the wood storage yard, where water soluble metallic salts were known to have been spilled, "total" metal concentrations were thought to be of little use on their own. Water-soluble contaminants were judged more relevant to assessment of risks to groundwater quality within and off the site and to the possibility that recontamination of the site's surface could occur if dissolved contaminants were to migrate upwards (through the action of soil suction) when the site surface was dried out in a summer drought. Thus, the strategy adopted was to analyse for:

- the contaminants expected to occur in ashes (arsenic, boron, cadmium, copper, lead, mercury, nickel, zinc, toluene extractable fractions, phenols, sulphates and pH);
- those anticipated from wood preservative fluids (pesticide residues, water soluble arsenic, copper, chrome, zinc and light oils);
- those more likely to harm a built development and gardens (the pH, sulphate, water-soluble metals already listed *plus* the chloride content and the plant-available concentrations of copper, nickel and zinc);
- other contaminants (e.g. asbestos fibres) which might blight the planned housing development.

If budgets had been inadequate to undertake this testing on all samples, then a viable reduced programme, still sufficiently diagnostic of site conditions, might have been to expose every second or third sample to the full analytical programme, and restrict the testing on the remainder of the samples. What would not have been defensible would have been to tailor the analytical programme to a budget limit imposed by a client, and ignore contamination (such as the possible presence of water-soluble metallic salts) which could prove to be especially harmful to the development project.

5.6.6   Modifying the site investigation design

Although the investigation was carefully planned, the basic design rested entirely on the desk study and site reconnaissance findings, and on the contamination zones which were delineated. It was necessary to allow in the specification and the instructions to the on-site investigator for the possibility that these did not adequately represent the actual subsurface conditions (i.e. the hazards).

For example, if instead of clays underlying the ashy surfacing, permeable sands were found in one area where obvious spillages of an oily liquid (possibly Protim wood preservative fluids) were identified, the person in charge of the site investigation would have to designate a further contamination zone and investigate this. Likewise, if asbestos wastes were found to have been concealed in a pit, excavated into alluvial clays, then investigation would have to be carried out to define the plan area and depth of the encapsulated wastes, and specific additional health and safety precautions would be necessary since air dispersal of asbestos fibres presents a risk to human health. The need to react appropriately to such events places demands on the knowledge and experience of both on-site personnel and those supervising the work.

Even a thorough main site investigation is likely to leave some questions less well resolved than is desirable. This arises largely because chemical analytical results take time to produce and are usually only available after drilling and trial pitting have been completed. Usually it is groundwater pollution concerns which call for a later site investigation, since it is difficult properly to design and locate groundwater monitoring wells until strata sequences and contaminant distributions in a site are known (see Chapter 6).

On the timber site, a later groundwater investigation seemed likely to be required, particularly if water samples collected in the main site investigation proved to be contaminated with metals or pesticide residues. Additionally, if pools of the Protim preservative were found in hollows in the alluvial clay beneath the ash surface of the site, the risk that volatile vapours could rise up through the ground in warmer weather and pose hazards to homeowners would become of greater importance. Thus a volatile organic compound survey, using a portable gas chromatography, might be necessary.

5.7   CONCLUSIONS

Understandably, many site investigators would prefer to have access to prescriptive protocols which could ensure that contaminated land investigations were conducted to the same consistently robust standards. Unfortunately, the complexities of soil environments invariably preclude such prescriptive guidance. The varying fates of different contaminants in various soil types alone ensure that no simple cause (contaminant condition) and effect (predictable risk to a sensitive target) relationship will be the inevitable outcome of land having been contaminated.

Thus, the best, and indeed the only defensible, advice is that investigators must continually revisit the hazards and pathways being revealed by the work, and ensure that truly representative samples are collected, labelled and "bagged", and then exposed to analytical examination which will be most likely to define whether or not risks could be realised. This places considerable burdens on the site investigator, since all later elements of the risk assessment and risk reduction decisions will rest on the adequacy of the hazard and pathway information which has been collected.

6

Investigation of the Water Environment

6.1 INTRODUCTION

6.1.1 Contamination of the water environment

In contaminated land risk assessments, investigation of the water environment is important because contaminant transfer is largely controlled by water movements and because surface water bodies and groundwaters, particularly aquifers (soil and rocks with pores and fissures capable of holding water), are sensitive to polluting incidents. The capacity of an aquatic system to receive and assimilate pollutants is a function of both the aquatic ecosystem and the nature of the chemical, physical and biological properties of the contaminants.

Generally, the capacity of surface water to absorb biodegradable contaminants is related to the reaeration capacity of the system, i.e. the capacity to absorb atmospheric oxygen. Toxic substances are unlikely to be adequately oxidised and the greatest concern relates to those that are either resistant to degradation or able to accumulate in organisms or both. Heavy metals bioaccumulate, although their toxicity varies according to the chemical characteristics of the water, alkalinity and hardness. Apart from effects on fish and higher predatory life, pollutants have a number of other effects, including destruction of microbial communities so inhibiting decomposition of organic materials and reducing the capacity of water to self-purify; harm to invertebrates by direct exposure and by the deoxygenation of sediments by chronic organic contamination; and damage to algae and higher plants which are important sources of food and oxygen and important habitats. Pollutants are also measured in terms of (i) suspended solids (sand, silt and gravel) which can reduce light in water, inhibiting photosynthesis, and alter species populations through settlement and (ii) heat, which leads to a decrease in dissolved oxygen concentrations. Apart from ecosystem risks, surface water pollutants can affect amenity value in terms of making water unpleasant to look at (for example, as a result of gelatinous microbial growths and staining of the bed), odorous and devoid of plants, and can increase the cost and efficacy of water treatment for supply.

Groundwater (the saturated subsurface soil zone where all of the void spaces are filled with water) has traditionally, in Britain at least, been addressed in terms of contamination sensitivity depending on whether it provides water supply. Although all groundwaters should be protected under the EC Groundwater Directive (CEC, 1980) regardless of whether they currently provide supply, it is inevitable that policy focus (NRA, 1992) provides most risk protection to productive aquifers. The soil column in the unsaturated zone can provide a natural barrier to groundwater pollution as contaminants are purified by processes such as anaerobic decomposition, filtration, ion exchange, adsorption etc. However, reactions between the soil, rock and percolating water may increase the content of dissolved solids. Depending on the characteristics of the soil, the contaminants and the flow system in the zone, movement of a contaminant through and into groundwater may range from being nearly instantaneous (for example in coarse-grained materials such as gravels) to taking hundreds of years (for example in chalk) (Reichard *et al.*, 1990).

Pollutants reaching an aquifer are liable to spread gradually through it in a manner which depends on the nature of the pollutant and of the aquifer. Miscible (soluble) chemicals dissolve and tend to move in the direction of groundwater flow. Immiscible or poorly soluble chemicals (e.g. chlorinated solvents) tend to remain as a separate phase (dense non-aqueous phase liquids (DNAPLs)), either sinking below the water table and flowing separately along low-permeability layers at depth in the aquifer, or being trapped by capillary forces and acting as a long-term contaminant source. A recent suggestion (Grolimund *et al.*, 1996) is that some contaminant transport may be via colloidal particles. This would have implications for the reliability of transport models. The literature on groundwater flow and contamination is large; however, a useful and fundamental text is Freeze & Cherry (1979). Devinny *et al.* (1990) specifically consider groundwater contamination by hazardous wastes.

6.1.2 Water investigation

It is surprising how few contaminated land investigations provide adequate information on surface and underground water qualities. In some cases hundreds of soil samples are taken but only a few groundwater samples and possibly no surface water samples.

Reasons for this imbalance include the fact that not all investigators have yet appreciated the importance of water quality information and the necessity for it, if defensible risk assessments are to be devised, and allowing insufficient time for proper time-sequence monitoring of water quality.

Representative water quality data are difficult to obtain because the water environment is dynamic and subject to externally imposed variations (e.g. seasonal and rainfall effects and the results of such human interferences as the pumping of abstraction boreholes or controlling surface-water levels in canals and rivers). Thus, unless care is taken, surface and groundwater samples which are not truly representative of "worst" conditions will be collected.

The "worst" conditions are those in which the contaminated site has its greatest polluting effects on water resources. In relation to surface waters, these conditions occur at low flow, when dilution potential is minimised. For groundwater, "worst conditions" might occur when groundwater levels are at their annual high point (when deeper groundwater has risen to saturate near-surface contaminated fill layers and dissolve out more soluble chemical substances), or when the pumping out of groundwater has removed the cleaner top water layer and allowed deeper waters, polluted with denser contaminants, to become apparent. Dense non-aqueous phase liquids (DNAPLs), such as chlorinated solvents, often occur lower in shallow aquifers and can give rise to such stratified groundwaters.

An essential aspect of obtaining meaningful water samples is to sample both upstream and downstream of the contaminated site which is being investigated, so as to provide understanding of the potentially "uncontaminated" water relative to contaminant effects downstream. When surface waters are of interest this requirement is more readily satisfied than when groundwater is the medium of concern. In both cases, sampling at off-site locations is likely to be necessary and will call for the approval of adjacent landowners, or in its absence use of statutory powers of access by regulatory authorities. Understanding of "background" water quality will be relevant to the hazard assessment stage (discussed in Chapter 9), and may also be determined by reference to water quality survey data. For example, in England and Wales the Environment Agency takes five-yearly river water quality measurements at 6000 sites. A water quality archive of measurements taken hourly to bimonthly at approximately 50 000 sites covers up to 100 parameters. Groundwater survey data are held for approximately 5000 locations. Unfortunately, this recording network is too coarse to be useful in most contaminated land investigations.

An understanding of the hydrological regime is fundamental to investigations of the water environment. For example, without information on variations in groundwater levels during the year, and on the direction and rate of groundwater flow, it will often be impossible properly to design a strategy for investigation of contamination and migration, complete a risk assessment or develop an appropriate remediation strategy. Information on the geology of the site and surrounding areas will help in the interpretation of the results of groundwater monitoring and prediction of contaminant behaviour in the ground. Information on rainfall patterns (annual, seasonal, peak) may also be a material input to any modelling of actual and potential contaminant movement.

Much of the general guidance in Chapter 5 regarding the sampling and analysis of soil samples is relevant to investigation of the water environment (e.g. quality management; the need to maintain accurate records; use of appropriate analytical methods; handling and transport so as to preserve sample integrity). Sampling the water environment differs in a number of important respects from sampling soils. In particular:

- The water environment is seldom static and sampling over a period of time will often be required.
- Establishment of the nature of the dynamic surface or groundwater regime is usually required.
- Sampling and measurements will often be required outside of site boundaries – however, the ability to do this will often be constrained by difficulties of access.
- There may be a high potential for increasing the dispersion of contaminants, particularly DNAPLs during intrusive investigation.

As part of the overall investigation of a contaminated site it may be necessary to sample:

- groundwater at on- and off-site locations;
- static (or *lentic*) freshwater systems such as ponds, lakes, storage lagoons;
- flowing (or *lotic*) freshwater systems such as rivers, canals, drainage ditches and estuaries;
- docks and enclosed harbours subject to intermittent or occasional changes in water levels;
- sediments;
- wetland areas.

The last comprise a combination of static and flowing water, sediments and land areas. They can present particular difficulties when it comes to devising a suitable sampling programme.

6.2 SAMPLING SURFACE WATERS

The aim of surface-water sampling will usually be to establish concentrations or loads of specified physical, chemical, biological or radiological parameters at selected locations throughout the whole or part of a water course or body of water.

The sampling programme must take into account (Harris, Herbert & Smith, 1995):

(i) Possible variations with location, depth and time of sampling.
(ii) Variations in flow and level arising, for example, from tidal movements, seasonal fluctuations in groundwater levels, and short-term fluctuations associated with rainfall events, and upstream discharges (these should be the subject of separate study).
(iii) Other factors such as temperature, seasonal fauna and flora variations (e.g. algal blooms) and the movement of boats etc.

Guidance on sampling procedures for water quality is given in a series of

International Standards (BSI, 1981b; 1987; 1991a; 1991b; 1993b; 1996a; 1996b).

Possible approaches to sampling are summarised in Table 6.1. Before sampling it is essential to consult the analyst to establish suitable sample preservation and handling arrangements (BSI, 1996b). The types of data required (e.g. concentrations, loads, maximum and minimum values over time or space, arithmetic means, median values etc.) must also be decided, since these determine both the location and method of sampling (e.g. spot versus time-composite samples). Guidance is available on the statistical aspects of sampling in relation to time and frequency (BSI, 1981b).

Table 6.1. Possible approaches to sampling surface waters (Harris, Herbert & Smith, 1995). Reproduced by permission of CIRIA

Method/activity	Applicability/comments
Spot sampling	A single sample taken at a fixed location and depth using hand equipment (e.g. bailer). Time-variable samples may be obtained using automatic equipment. Recommended to detect pollution when unstable parameters are present, e.g. where flow is not uniform, parameters are variable and composite samples would obscure variations
Periodic samples (discontinuous):	
At fixed time intervals	Provides time profile of contamination
At fixed flow intervals (volume dependent)	Taken when quality is not related to flow rates
At fixed flow intervals (flow dependent)	Taken when quality is not related to flow rates
Continuous samples:	
At fixed flow rates	Contain all constituents present during sampling period but do not provide information about variations in concentrations of individual parameters
At variable flow rates	The most precise method of sampling flowing water if both the flow rate and contaminant concentrations vary significantly
Depth integrated sampling	Provides a composite water sample representative of the vertical profile; obtained by lowering and then retrieving a sample bottle at constant velocity
Point sampling at selected depths	Obtained using sample bottles opened at required depths and then retrieved (evacuated or air-filled bottles, flow through samplers or automatic samplers with inlets at specified depths may be used)
Depth profile samples	Series of samples from various depths at a single location
Area profile samples	Series of samples from a particular depth at various locations
Grabs or dredgers	Used to obtain samples of sediment – designed to penetrate under own weight
Cores	Provide information on vertical profile of sediment

To characterise the hydrological regime, and to establish the likely conse-quences of the presence of a contaminated source on users, habitats and inter-connected water systems, additional data must be collected: for example, an estimate (where practicable) of the flow and proportion of surface or rain water in the sampled water body; as appropriate, a description of the depth and nature of the geological stratum (in contact with the water body) from which the sample was collected; rainfall data; and the physical properties of surface-water bodies.

There are established procedures for measuring the physical characteristics (e.g. flow using simple plate weirs, direct gauging etc.) of surface-water bodies, including springs and streams etc. Larger bodies are more likely already to be gauged and the required information more readily available. In the UK, me-teorological data (precipitation rates, long-term monthly extremes etc.) are generally available from the Meteorological Office, the Environment agencies, local authority sources, or from on-site measurements.

Preliminary on-site measurements of a limited range of parameters – e.g. temperature, electrical conductivity and pH – at numerous locations can be used to establish the homogeneity of a surface-water body, and can indicate appropri-ate sampling locations for detailed analysis. Chemical testing parameters will in general be the same as for groundwaters, although a separate and independent measurement of (suspended) solids content may be required. This typically varies markedly across water courses and depth-integrated samples should therefore be taken at several locations.

Representative surface-water sampling of water courses should be:

- at locations where there is good mixing (except if the interest is in the pollution loading from a site) but not excessive turbulence, as this can lead to loss of volatiles, dissolved gases and oxidation of some compounds;
- at locations with sufficient flow rates to avoid stratification due to tempera-ture or density differences (e.g. not close to banks unless of special interest);
- subsurface (from within 0.5 m of surface is generally preferred) rather than from the immediate surface;
- at flow measuring points (if possible) so that flow and concentration data can be combined to provide an estimate of contaminant flux.

Martin, Smoot & White (1992) compared the effectiveness of sampling methods for determining water quality in streams and rivers, specifically surface-grab sampling in which samples are collected from a single point at, or near, the stream surface, and cross-sectionally integrated, flow-weighted sampling using depth-integrating nozzled samples that fill isokinetically (i.e. with no change in stream velocity on entering the intake). They found that, while concentrations of dissolved constituents were not consistently different between the two methods, concentrations of suspended sediment and the "total" forms of some sediment-associated constituents, such as phosphorus, iron and manganese, were signifi-cantly lower in the surface-grab samples.

Hunnes (1995) reported a study comparing the results obtained on filtered (regarded as "good practice" in most guidance) and non-filtered water samples. The study was concerned with the transport of contamination into a nearby surface-water body. It was concluded that, despite the greater analytical uncertainties, measurements on unfiltered samples provided a better estimate of the contaminant load entering the water body. They are also better for detection of hot spots. On the study site filtered samples gave a number of "false negatives" at some sampling locations. Results from filtered samples were best for establishing and calibrating a groundwater model, because these measurements more closely accord with in-ground conditions where there is natural filtration.

Static water bodies can be difficult to sample because of difficulty of access, size and shape (e.g. an irregular shape, such as a bay or inlet, may contain "stagnant" waters) and stratification. Safety always has to be borne in mind considering the depth of the water, and hazardous materials on the surface or in sediments which may be disturbed.

Where area profile sampling is required, the sampling pattern (analogous to soil sampling) may be judgemental, systematic or random (see Figure 6.1).

Where floating layers are present, it must be decided whether to obtain samples of the layer itself or samples that are representative of the water body as a whole; the latter is usually only possible where the water is well mixed (e.g. over a weir), but note the potential for loss of organic compounds at such locations.

Sampling programmes for estuarine waters should take into account tidal currents and the way they are influenced by wind, density, bottom roughness, closeness to the shoreline, the movement of shipping, discharges etc.

The difficulties of obtaining samples truly representative of conditions in water bodies are illustrated by Brick & Moore (1996). They found that concentrations of several metals in water taken from a river in Montana, including copper, zinc, manganese and iron, were up to three times higher in samples taken at night compared to those taken during the day. Over 100 years of mining in the area had left high concentrations of metals in the water and river bed sediments. The river water was also slightly alkaline and saturated with carbonates. All the trace metals were present as particles. The hypothesis to explain the higher night-time concentrations was that the sediments were stirred up by river fauna being more active at night. The particles are acid soluble and thus can dissolve as they move downstream to less alkaline waters. Zinc and manganese were also present in soluble form. It is suggested that at night, as the dissolved oxygen concentration falls, the metals are desorbed from the sediments in a reduction reaction.

6.3 SAMPLING SEDIMENTS

Sediments range from sand to clay particles that are under water. In the context of contaminated land sites, they will lie at the bottom of ponds, lakes, streams or

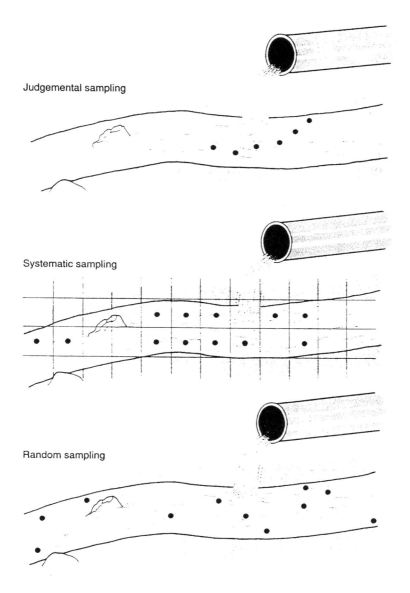

Figure 6.1. Examples of judgemental (top), systematic (centre) and random (bottom) sampling of surface water (Harris, Herbert & Smith, 1995). Reproduced by permission of CIRIA

rivers, canals and docks. Unique sampling problems arise because of the diffi-culty of sampling generally unseen areas under the water. In colder climates, additional sampling problems occur when winter-time sampling requires that holes are cut in ice in order to set up and use sediment sampling equipment.

Access to the sampling area plays an important role in sampling strategy and logistics and in the selection of sampling equipment. There are three basic approaches to the collection of bottom sediment samples: sampling from a platform (e.g. raft, ship, ice, a plane, helicopter etc.), sampling by a diver, and sampling using a remotely operated vehicle. Collection by a diver, although usually more costly and difficult than sampling from a platform, often yields better-quality samples, particularly sediment cores. In areas with sufficient ice cover over the water body, sediment samples can be obtained by drilling a hole through the ice and sampling through this space. The advantages of this tech-nique are a steady platform and a large area for assembling the equipment and processing the samples. Sampling using a remotely controlled machine is not common, but may be the best option when conditions are suspected to be particularly hazardous, either because of contamination or their physical na-ture.

Money spent on the most sophisticated techniques is wasted if samples are collected at inappropriate locations or if an insufficient number of samples are taken to represent the project area. Consequently, the selection of the number and position of sampling stations needs to be carefully designed. There is no one formula applicable to all sediment sampling programmes (BSI, 1996b; CCME 1993a; 1993b).

Sediments are typically layered. Physical properties (e.g. particle size distribu-tion), chemical composition and chemical conditions (e.g. pH and RedoX poten-tial – i.e. an electrochemical potential reflecting the oxidation–reduction status of the water which influences the electrovalency and solubility of chemical spaces) tend to vary with both location and depth. This variation can only be properly addressed by taking a large number of samples.

There are two general types of sediments that may be collected: bottom and suspended. In addition, bottom sediments contain two primary zones of sedi-ment interest in contamination studies: the surficial or upper 100 to 150 mm, and the deeper layers. Sampling of the surface layer provides information on the horizontal distribution of parameters or properties of interest for the most recently deposited material, such as particle size distribution, or geochemical composition. A sediment column, which includes the surface sediment layer (100–150 mm) and the sediment underneath this layer, is collected to study the historical changes in parameters of interest or to define zones of pollution. Hand tools can be used in shallow depths of water to obtain samples of surface sediments. For deeper or fast-flowing waters, or to obtain sediment profiles, grabs or coring devices must be used (see Table 6.1).

6.4 GROUNDWATER INVESTIGATIONS

6.4.1 Introduction

The primary aim in sampling groundwaters is to determine whether or not the groundwater in or beneath a site is contaminated, and/or whether migration off-site is occurring so as to present a risk to the adjacent groundwater or nearby surface-water bodies. Once contamination and/or migration has been established, there will usually be a need to determine the extent of contamination, the pattern of migration and the rate of migration.

The basic requirement is knowledge of the groundwater regime in terms of both horizontal and vertical flow directions within and between aquifers (an aquifer here is defined as any water-bearing stratum with distinct physical characteristics). Information on the flow regime assists in deciding on siting of sampling wells and at what depths to sample. Thus, a phased approach is often required in which flow patterns are first established, and then monitoring wells installed where they are considered most likely to yield the requisite data for the purpose in hand as suggested by simple or more complex modelling. In practice, in many small-scale investigations, boreholes will initially be sited where it is believed that they will both provide information on flow characteristics, and at least initial information on water quality.

Where contamination is suspected the installation of wells itself presents a risk to groundwater, potentially establishing pathways for the dispersion of contamination between levels within an aquifer or between aquifers. This is another reason for taking a phased approach to investigation, and in particular where severe contamination is suspected to begin by monitoring around a suspected source before approaching close to the source itself; the source in this context might be a point source such as a leaking tank or pipe within a site, or the whole of a site such as a waste tip.

Tools other than groundwater-monitoring wells can often be used usefully either before, or in conjunction with, such installations. There is a range of geophysical techniques that can provide information on the disposition of strata and the extent of some contaminant plumes (McCann, 1994). The use of soil-gas analysis can often provide information on the extent of a plume of volatile organic materials such as aromatic hydrocarbons (e.g. toluene) and chlorinated solvents, and sometimes the presence of biodegradable materials such as hydrocarbons or landfill leachate (e.g. by monitoring for methane, carbon dioxide and oxygen depletion), and help in the identification of the most appropriate sampling locations. The employment of such techniques can help to avoid the dangers of spreading contamination and contribute to more economic investigations.

To determine whether migration is occurring requires a comparison of up-gradient and down-gradient (relative to the source) groundwater qualities, which means that the direction of groundwater flow has to be established. While this might appear to be easily determined from local geological and hydrogeological information (e.g. British Geological Survey hydrogeological maps), it has to be remembered that past occupancy of a site will usually have disrupted natural groundwater movements. Even shallow foundations such as road sub-bases can divert groundwater flow, and the covering of "made ground" (which so typifies most contaminated sites) usually gives rise to different, "unique", shallow aquifer conditions.

The definition of up-gradient will depend on the scale of the monitoring network. Local mounding of the groundwater may occur for a variety of reasons, so that locally there can be outflow in all directions from an area. A further complicating factor in determining the results of any quality monitoring regime, certainly in a much industrialised area such as the West Midlands in the UK, is that the water flowing from up-gradient towards the site being monitored may itself be already contaminated. Although the concept of up-gradient and down-gradient monitoring is simple, there is often a strong case for also taking control samples at some distance from the source at right angles to the assumed or derived flow directions.

The design of an appropriate investigation strategy requires an understanding of the sources of groundwater contamination and the way that contaminants behave in the ground. Typical sources of groundwater contaminants are:

- vapours that dissolve in groundwater;
- solids that produce undesirable dissolved constituents (e.g arsenopyrite, a mineral from which arsenic can leach) and that are themselves toxic;
- chemicals sorbed on to solid surfaces that can enter groundwater through various desorption processes;
- pools of NAPLs (non-aqueous phase liquids) that are potentially mobile and are either lighter (less dense) than water (LNAPLs, e.g. gasoline) or heavier than water (DNAPLs, e.g. creosote and trichloroethylene);
- NAPL residuals which are immobile but can leach into groundwater or soil vapour; or dissolved in groundwater and may be capable of sorbing on to the aquifer solids.

Contaminants tend to distribute themselves among these phases or compart-ments to attain equilibrium. By determining this distribution it may be possible to identify one or two phases that contain most of the contaminant (CCME, 1994). The migration of contaminants is largely governed by: (i) the nature of the subsurface materials, for example hydraulic conductivity, mineralogy, physico-chemical interactions with contaminants, and (ii) the physico-chemical proper-ties of the contaminants themselves (e.g. solubility in water and non-aqueous phases, susceptibility to chemical or biodegradation).

Different contaminants may move at markedly different speeds through the ground and may also be chemically changed so that only reaction/degradation products are present more than a certain distance from the source (something that should be taken into account in the analytical strategy). Such products may be less desirable than the original compounds: for example, the formation of vinyl chloride. Different groups of chemicals (say volatile organics and non-volatile organics) impose specific requirements on the subsurface assessment (for example, the choice of sampling method and design of sampling wells).

While much of the remainder of this chapter is concerned with the installation of wells to determine the hydrogeology, water quality or both, there is now a range of real-time techniques that allow direct measurements of certain aspects of groundwater quality. Increasingly sophisticated driven probes are being developed, which, for example, permit direct measurements to be made in the ground as the probe is forced downwards. For example, measurements can be made of pH, conductivity, total petroleum hydrocarbons, or individual hydrocarbons such as benzene. It is often possible to make geotechnical measurements (e.g. resistance to penetration) at the same time so that the measurements of contamination can be correlated with small-scale (perhaps over no more than 100 mm) changes in the physical characteristics of the strata penetrated. Many such devices allow for the injection of a sealing grout as the probe is withdrawn, thus preventing the probehole becoming a pathway for the migration of contamination. Although the mobilisation costs may at first appear high, the rapid nature of the system usually permits many more points to be sampled and the sampling pattern to be modified as the results become available. For example, if a contaminant plume is not found where it is expected, then the search can be extended in a different direction. If conventional techniques relying on laboratory analysis are employed, it could be several days before the information becomes available, although, of course, there are some on-site analytical test methods that can be deployed which may give a general indication of the quality of the water being encountered. But the need properly to develop and purge wells limits the extent to which reliable data can be obtained in this way.

Appropriate computer modelling packages are of great help during most stages in groundwater investigations in helping to analyse and portray data, and derive hypotheses concerning the rate and direction of contaminant movements which can be tested by further investigation and/or used in subsequent risk assessments (predictive models are discussed further in Chapter 10). The limitations of such packages need to be understood, however, as they do not necessarily handle movement in the near-surface zone where there is fill and foundations causing interruptions to the natural flow patterns.

6.4.2 Determination of hydrogeology

The use of insufficient wells and injudicious placements can lead to serious errors in understanding of the hydrogeology. Groundwater flow direction is determined from the head differences in a set of observation boreholes (Figure 6.2). Since this direction is a critical piece of information in understanding the link between source and target, it is important that the water levels measured in the various boreholes are as accurate as possible.

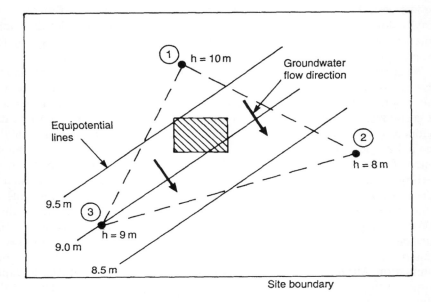

Figure 6.2. Derivation of groundwater flow directions from groundwater level data. Reprinted with permission from Nielsen, 1991. Copyright Lewis Publishers, an imprint of CRC Press, Boca Raton, Florida. © 1991

No groundwater level monitored in a borehole will mirror precisely actual groundwater conditions in undrilled areas of an aquifer, since head losses invariably occur as water flow paths are distorted on entering the borehole. Thus, the aim has to be to have boreholes which reduce this inaccuracy to an absolute minimum (Figure 6.3). In addition, depending on the length of screening relative to the various water-bearing strata, the measured head may represent a combination of the heads within the different layers penetrated. In the case of a confined aquifer (i.e. one overland by a low-permeability or impervious strata such as clay and generally recharged through an unconfined outcrop at a distance), the potentiometric head may even be above the topographic surface of the site, i.e. inserting a well may create an artesian flow. The hydraulic head in a confined aquifer is described by a potentiometric surface, which is the imaginary surface representing the distribution of total hydraulic head in the aquifer, which is higher than the physical top of the aquifer. To achieve the situation shown in Figure 6.3(a) calls for as much hydraulic continuity as possible between the observation

(a) Inevitable (minor) head losses and inaccuracies

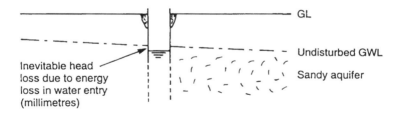

(b) Major head losses and inaccuracies due to poor hydraulic continuity

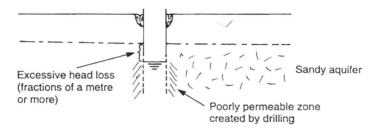

(c) Major inaccuracies created by poor design

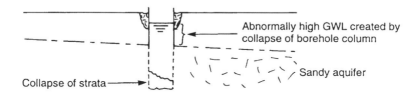

Figure 6.3. Head losses and water level inaccuracies at observation boreholes

borehole and the surrounding aquifer materials. This requires that the borehole be *developed* until water inflow opportunities are maximised (i.e. the hole should be repeatedly pumped and allowed to recover to pull in finer particles which may have been layered along the outside wall of the borehole as a result of the drilling process or the use of drilling muds during its construction). Even when permeable gravel packs have been placed on the outside of the borehole lining, such "development" is essential, as the perimeter of the original drilled hole could be so smeared with clays and muds that water entry is very restricted.

With shallower (and cheaper) water level monitoring points installed into trial pits dug in permeable and open "made ground", development is usually less important. Essentially it is the compacting action of a drilling rig which especially reduces hydraulic continuity between an observation point and the surrounding strata and leads to less accurate groundwater level information.

If care is taken properly to develop groundwater monitoring points and if a reasonable number of such points are installed, than it is possible to move beyond the groundwater contour plan (Figure 6.2) and establish the most likely flow paths of groundwater from various areas under a site by constructing a groundwater equipotential plan (lines of equal hydraulic head which can usually be represented by elevation of the water table in the case of unconfined aquifers and the potentiometric head in confined/aquifers) (Figure 6.4). This is done by dividing the highest groundwater contour line into strips of equal length and constructing flow lines (at precisely right angles to the deduced groundwater contour lines). Within each strip bounded by a pair of flow lines groundwater can be treated (in an open porous, fill layer) as though flowing in a tube and changes in water quality with flow distance can be identified. This process can be

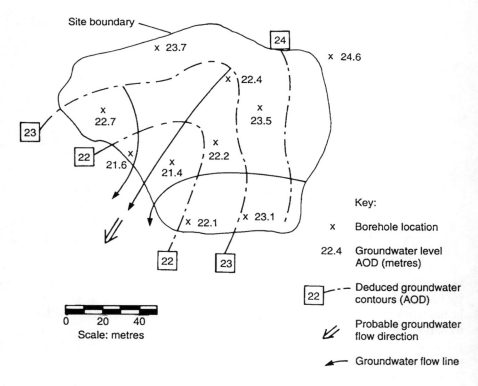

Figure 6.4. Establishment of likely groundwater flow paths

especially useful in cases where groundwater quality differences are obvious between the water entering and leaving a contaminated site and can indicate where more polluting deposits exist. This can guide more targeted subsequent investigation and selection of a remediation strategy.

The relatively simple approach to estimating groundwater flow directions described above is suitable where wells are screened in the same zone and the flow of groundwater is predominantly horizontal. However, with increased emphasis on detecting the subsurface position of contaminated plumes or in predicting possible contaminated migration pathways, it is evident that the assumption of horizontal flow is not always valid. Increasingly, flow lines shown on vertical sections are required to complement planar maps showing horizontal flows, to illustrate how groundwater is flowing either upwards or downwards in the vicinity of a site. The monitoring and analysis of possibly complex three-dimensional flow patterns require careful design of monitoring installations and thorough understanding of site geology (see, for example, Dalton, Hunstman & Bradbury, 1991).

Establishment of the hydrogeological regime is difficult if investigation has to be confined within site boundaries. During exploratory investigations the objective may be to maximise the chances of intercepting a possible plume from a leaking underground tank. Locations will be chosen on the basis of assumptions about groundwater flow directions. Unless these assumptions are confirmed by the investigation, no confidence can be attached to a failure to detect a plume.

For example, in the case of a site immediately adjacent to a landfill, the site assessors relied on regional groundwater flow paths (which were away from, or parallel with, the boundary with the landfill) in their report, but then suggested leachate migration from the landfill as a source of methane found on the site without examining the available data to see if this was a reasonable assumption. In practice, subsequent examination of the groundwater levels suggested that the local groundwater movement was away from the landfill and at roughly right angles to the regional groundwater flow paths.

There is a common tendency to assume that groundwater movement will be towards local rivers. However, although there may be a component in that direction, in large rivers with extensive river deposits, such as the Thames Valley in England, the flow may be predominantly parallel to the river.

An example of how flow patterns can change with time is provided by an industrial site located on river gravels in a broad valley in the Chiltern Hills, northwest of London. The basic assumption would be that groundwater flows had components towards the centre of the valley and parallel to the local stream. However, historical information suggested that this might not always have been the case. A major drinking water abstraction well had at one time been located about 500 metres off-site on the opposite side of the stream and slightly lower down the valley. Subsequently, the industrial site had its own abstraction well. Thus, local groundwater flow directions almost certainly changed on more than one occasion.

During a pre-purchase survey of a site, concern was raised by the potential purchaser about liabilities from possible off-site migration of contaminants. The groundwater appeared to be moving at roughly right angles to the regional flow under the influence of a major nearby up-gradient abstraction well. What was subsequently highlighted was the possibility that if the abstraction was closed and the regional flow direction restored, contamination would be brought on to the site from beneath adjacent land.

6.4.3   Investigation of contamination and migration

Because hydrogeological characteristics and the nature and likely distribution of contamination are all relevant to the design of monitoring wells, an exploratory investigation and sampling programme will usually be necessary before designing any long-term monitoring network. Phasing can provide an economic approach to groundwater investigation (see Table 6.2).

The placing of sampling/monitoring installations must take account of the extent of contamination of the groundwater (which of course is not known at the outset) and the location of known discharge points (into groundwater and from ground to surface waters), where these exist. Information on such discharges may be available from the preliminary site investigation as part of the hazard identification.

As discussed above, it is important to determine the local groundwater flow regime and not to rely on information on regional flow patterns. In urban and industrialised areas the hydrogeology may undergo major changes due, for example, to the opening and closure of major abstractions, urbanisation (increasing runoff), and the construction of deep foundations which impede groundwater flows. In some urban areas groundwater levels are controlled by

Table 6.2. Phasing groundwater investigations (after Harris, Herbert & Smith, 1995)

Phase of investigation	Sampling/monitoring activities
Exploratory investigation	Construction of limited number of simple installations (piezometers or wells) within and around the site based on hazard identification.
	Measurement of water levels
	Preliminary water quality information
Detailed investigation – first phase	Construction of additional monitoring installations on grid basis to give broad cover across area of interest
	Further monitoring of groundwater levels
	Water quality measurements
More detailed phase	Further adjustment of monitoring network where appropriate
	In situ testing (e.g. pump testing, permeability measurements etc. to determine aquifer properties)
	Water level measurements
	Groundwater quality measurements

leakages of groundwater *into* the sewer system. In other areas leakages from sewers or water supply pipes may lead to artificially high groundwater levels. The groundwater table may be elevated locally within a landfill so that local groundwater flow directions may differ from regional patterns (see Figure 6.5).

Monitoring installations should be located both inside and, where possible, around the site or a localised source within a site. Monitoring points should be located both up-gradient (to provide control data) and down-gradient of the source, since most dissolved constituents will move vertically downwards through the unsaturated zone beneath the source area, then horizontally in the direction of flow when encountering the saturated zone (down-gradient installations should be sited in the most permeable strata underlying the source area). In practice, since there will also be a need to determine the spread of contamination, and because some local migration up-gradient (relative to larger-scale flows) cannot be ruled out, monitoring at points laterally away from the source will often be desirable.

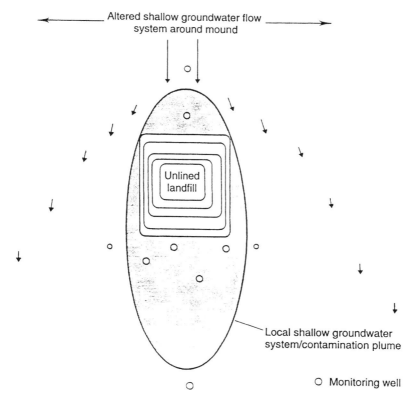

Figure 6.5. Mounding of water within a landfill causing modification of local groundwater flow patterns (Harris, Herbert & Smith, 1995). Reproduced by permission of CIRIA

The selection of optimal sampling points (in terms of both location and depth) must take into account the particular characteristics of the aquifer that is being sampled (e.g. the nature of the groundwater flow (whether intergranular or fissure), the hydraulic gradient and the direction of flow). The use of existing wells or boreholes is only appropriate if they can be shown to be suitable for the purpose of the sampling programme. In many cases, existing wells and boreholes may fully penetrate the aquifer and be open, or screened, throughout their depth, thus making it difficult to examine quality at specific depths.

A sampling borehole to monitor the quality of the groundwater directly beneath the pollution source should be installed, although taking into account the need to minimise the risk of spreading contamination. In addition, at least one borehole should be screened over a narrow depth range immediately below the water table, so that any pollutants which are less dense than water (LNAPL – light non-aqueous phase liquid) will be more easily detected. Further sampling points can then be located at progressive distances down the hydraulic gradient from the source of contamination, and provision should be made for sampling from a range of depths. DNAPLs can also move against the groundwater flow direction (see Figure 6.6). In addition, migration of contaminants may occur not only through advection (the bulk movement of the groundwater) but also by diffusion; if the rate of diffusion is high relative to the rate of flow of the groundwater, contamination may extend up-gradient of the source. A further complicating factor is that flow directions may differ in water-bearing strata at different depths.

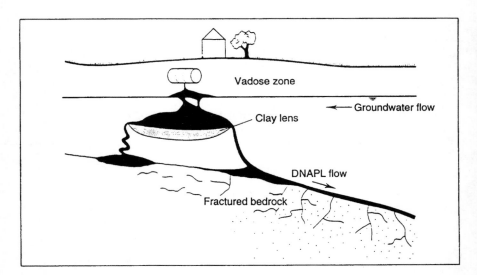

Figure 6.6. Subsurface distribution of dense non-aqueous phase liquid moving contrary to groundwater flow (Harris, Herbert & Smith, 1995). Reproduced by permission of CIRIA

When designing monitoring networks to identify extensive diffuse-source pollution of aquifers, the use of existing sampling points in the form of large-capacity production boreholes is preferable if these exist, as they can provide integrated samples from a large volume of the aquifer. However, in some cases of localised or low-intensity pollution, the use of this type of borehole may dilute the contamination to levels below the analytical detection limit; in these cases smaller-capacity pumped boreholes should be used.

The part of the aquifer which is most sensitive to pollution is that near the boundary between the saturated zone and the unsaturated zone. Other purpose-drilled boreholes should be completed and screened over different depth intervals of the aquifer. Sampling boreholes should be located throughout the area of interest at locations chosen to represent the different hydrogeological and land-use conditions and areas considered to be particularly vulnerable to diffuse pollution.

When the objective is to characterise contamination over a large site or larger area where there might be a number of point sources or diffuse inputs, the number of sampling locations should be proportional to the area under investigation. There is no justification for reducing the number of sampling locations per unit area as the size of the site increases.

6.4.4 Timing and frequency of sampling

A single sampling operation will seldom be sufficient properly to characterise groundwater quality because of the many factors that can result in variations over time. Although changes in aquifers are usually more gradual temporally and spatially than in surface waters, in some aquifers, and particularly where groundwater contamination exists, short-term variations of between several hours and days may occur in the composition of samples.

The most common practice is to take two or three sets of samples quite close together (perhaps separated by a few weeks at most) and then progressively to extend the periods between sampling events if consistent results are obtained. This general approach has to be overlain by consideration of known or expected fluctuations in groundwater levels, flow directions etc.

Continuous monitoring of pH, temperature and electrical conductivity (e.g. using flow-through cells; see, for example, Howsan & Thakoordin, 1996) can assist in identifying the need to increase or decrease the sampling frequency. If continuous monitoring indicates that the rate of quality changes is increasing, the sampling frequency should be increased for any determinands of interest, and vice versa. In cases where there has been a considerable change in quality of any continuously monitored determinand, it is advisable to consider extending the range of determinands to be routinely analysed.

Continuous monitoring is also a useful means of identifying the most appropriate time to sample pumped observation boreholes which are being used to obtain representative samples of aquifer water. Where significant variations are recorded (i.e. $\pm 10\%$ in terms of the concentration (mass/unit volume) within the pumped discharge), this probably indicates local transient conditions within the borehole itself during the early stages of pumping, and samples should not be collected until the monitoring suggests that an equilibrium has been reached. If no significant quality variations occur, the time at which the sample is collected after the commencement of pumping need only be sufficient for the borehole to be purged (see below).

6.4.5 Design, choice and formation of sampling installations

As in all site investigation, the type of installation used depends on the objectives. It is essential to define both short- and long-term objectives (e.g. preliminary sampling as part of an exploratory investigation; long-term monitoring; access for *in situ* testing etc.) at the outset if representative data are to be obtained.

Materials and methods of construction should have minimal effect on water chemistry and be sufficiently robust to withstand the potentially harsh (physically and chemically) environment found on contaminated sites. Table 6.3 summarises information on the types of installation that may be used in groundwater investigations, although in relation to piezometers it should be noted that it is difficult to develop these.

Detailed guidance on the various types of installations used in groundwater-monitoring applications is given in a number of standard texts (Freeze & Cherry, 1979; BSI, 1981b; 1991a; 1993b; Clark, 1988; ASTM, 1990; Nielsen, 1991; USEPA, 1991d; 1991e; WRC, 1992).

It may not be possible to finalise the detailed design of the monitoring wells (e.g. location and length of screened section(s)) until drilling is complete. There will often be a need, therefore, for a degree of flexibility in technical specifications and contract details. In complex strata, or where the nature and predicted behaviour of the contamination requires access to different sampling depths (e.g. stratified systems), a number of installations of varying depth (or nested installations) may be required.

Table 6.3. Types of groundwater-monitoring installations (after Harris, Herbert & Smith, 1995)

Type of installation	Application
Simple standpipes	Measurement of water levels
	Preliminary water quality when installed to correct depth
Large diameter standpipes or observation wells	Water level monitoring
	Water quality sampling (short and long term)
	Permeability testing
	Pumping tests
Piezometers	For accurate monitoring of water pressures, particularly over specific zones
	Measurement of permeability in specific zones

The choice of method depends on the purpose of the borehole, geological conditions and time available. Core extraction is significantly more expensive than open-hole drilling and the need for core samples for testing purposes should be considered at the planning and design stage. If one of the objectives is to provide information on the hydrogeological conditions, including geological structure, secondary permeability etc., core samples should be taken.

Particular care has to be taken in the selection of drilling methods because of the potential to cause contamination. The following precautions should be taken during drilling (Harris, Herbert & Smith, 1995):

(i) Use of appropriate casing, kept coincident with the base of the borehole, at all stages of boring.

(ii) Use of permanent sealing (bentonite or similar) at the top of boreholes to prevent entry of surface waters.

(iii) Use of permanent sealing (bentonite or similar) at intermediate levels where necessary to prevent the creation of preferential paths for the movement of contaminants and leachates between or within aquifers.

(iv) Measures to prevent extraneous matter falling into the borehole (e.g. use of casing during drilling and sealable/lockable covers on completion).

(v) Care in the use of assisted drilling techniques (mud-flush and water-assisted drilling are best avoided where practicable, although they may be necessary in dry strata).

(vi) Provision of means for the collection, treatment or disposal of contaminated liquid and solid wastes arising from drilling operations.

The design of the boreholes (e.g. the open area and length) and the method of construction need to be chosen not only to meet the sampling requirement, but also to minimise contamination or disturbance of the aquifer. The use of lubricants, muds, oils and bentonite during drilling should be avoided if at all possible, particularly when considering sampling for organic compounds. In addition, care is necessary to ensure that boreholes completed with a gravel pack around a solid casing and screens at specific levels are not subject to short-circuiting of aquifer water from different depths via the gravel pack. This can be achieved by sealing the gravel packing in the vicinity of the screens. Attention should also be given to the design of borehole installations at the ground surface, in order to prevent contamination of the borehole by surface water.

Construction materials should not interact with or otherwise modify (e.g. through absorption, leaching or other processes) the contamination present in the ground and groundwater, should be capable of withstanding anticipated ground conditions, particularly where long-term monitoring is intended, and be of sufficient physical strength to withstand installation and development stress. Resistance to chemical attack is important at high contaminant concentrations (10^2 to $10^3 \times$ mg/litre); absorption and desorption are important at lower concentrations (μg/litre).

In view of their low cost, widespread availability and easy handling, polypropylene and high-density polyethylene are recommended for most ground-water-sampling purposes. Groundwater that is highly contaminated with synthetic organic solvents will attack and cause deterioration of PVC (polyvinyl chloride) well casings and screens. In such circumstances, stainless steel or polytetrafluoroethylene (PTFE – Teflon®) are the materials recommended for borehole construction, because of their resistant inert character (BSI, 1993b).

Where composite constructions are used particular consideration must be given to jointing arrangements. Threaded joints on well casings are recommended so that glues and cements do not introduce additional risks of sample modification. Casing, screening and other materials should be carefully prepared before installation. Preparation includes thorough cleaning, and the protection of construction materials against contact with contaminated surface soils.

Equipment or materials introduced into or used for backfilling or sealing the installation may affect the groundwater chemistry. For example, contamination arising from cement grouts (high pH) and filter packs that are not chemically inert are common problems. Packing materials (e.g. sand and gravel packs) should be consistent with the geological properties of the formation. Materials (e.g. quartz sand) imported on to the site for packing purposes should be free of contamination. Although bentonite is considered to have good sealing properties, certain organic substances have been shown to migrate through bentonite with little or no attenuation.

The depth and location of the screen sections should be considered carefully, taking into account the requirements of the investigation, the permeability of the various geological strata and the nature of the contamination, if known. Whenever possible the screen should be fully submerged to avoid groundwater contact with the atmosphere where chemical reactions may occur. The potential for stratification of contaminants within the saturated zone should also be taken into account. For example, low-density organic compounds will float on the groundwater surface and in this case screens should extend above the zone of saturation in order to detect such lighter substances.

The surface finish of groundwater-monitoring installations is important in protecting the installation from unauthorised access and accidental damage. In general, installations should be provided with a durable lockable cover. Measures taken to protect against vandalism (e.g. making the installation as inconspicuous as possible) may increase the risk of damage due to legitimate activities such as vehicle movements and ground maintenance activities. Physical protection measures (e.g. steel casing) will protect against accidental damage.

It is crucial that all wells undergo a process of well development to ensure that they are functional at the time of installation, and capable of providing sediment-free samples for analysis. A wide variety of methods and techniques are available, including bailing, surging and flushing with air and water. The basic principle is to create reverse flow into and out of the installation to break down deposits formed on the surface during installation, and to develop an appropriately graded filter pack within natural or imported layers adjacent to the installation.

Well development may have potential health and environmental impacts, particularly where air-assisted methods are used. These are similar to those for drilling operations.

6.4.6   Sampling methods

Sampling involves collection of data on both physical and chemical properties. It is usually a three-step process: (i) determination of water levels, (ii) purging, and (iii) sampling.

In order to achieve representative sampling within an aquifer, the sampling method needs to be capable of withdrawing samples whose composition reflects the actual spatial and temporal composition of the groundwater under study. Since the majority of sampling points in aquifers are wells or boreholes, they will disturb the natural groundwater system, especially as a result of induced vertical chemical and hydraulic gradients.

In some sampling situations, mineral material (i.e. stagnant water) may accumulate in sampling boreholes between sampling operations. Therefore, the water within the borehole column will be unrepresentative of that in the aquifer under study. Sampling boreholes should therefore be purged before sampling by pumping several times the internal volume of the borehole itself. Recommendations on amount of purging range from three to six times. However, the amount and rate of purging depend on the hydraulic characteristics of the surrounding geology, well construction details and the sampling methods used. Purging should be sufficient to ensure the representativeness of the sample, while minimising disturbance of the regional flow system or the collected sample (note that turbulence may result in the loss of volatile components). Determination of whether sufficient purging has been carried out can be assisted by continuous monitoring of the extracted water for such determinands as pH and electrical conductivity; these should stabilise. In some situations, it may be necessary to clear the borehole, followed by a lower rate of extraction designed to achieve quality stabilisation before sampling.

Vertical stratification in groundwater quality may be natural or a consequence of contamination. For example, diffuse contamination usually results in a more contaminated layer of groundwater at the top of the saturated aquifer, whereas contaminants that are more dense than water tend to accumulate above a less permeable layer at depth or at the base of the aquifer. Sampling methods therefore need to be capable of detecting vertical as well as areal variations in groundwater quality.

The method of sampling also needs to reflect the complexities of groundwater flow, in that it must take account of the aquifer flow mechanism (whether fissure or intergranular), the direction of flow and the hydraulic gradients in the aquifer, which can produce strong natural flows up or down the borehole column itself. Traditionally, two common sampling methods are employed, namely pumped sampling and depth sampling.

Pumped samples

Pumped samples may comprise a mixture of water entering the open or screened length of the borehole from different depths. This sampling method is, therefore, only recommended where groundwater quality is vertically uniform or where a composite vertical sample of approximately average composition is all that is required.

The pumping time required before samples are taken which will be representative of new water drawn directly from the aquifer can be determined by monitoring any changes in dissolved oxygen, pH, temperature or electrical conductivity of the pumped water. In these cases, samples should not be taken until no significant variations ($\pm 10\%$ in terms of quality (mass/unit volume) or $\pm 0.2\,^{\circ}\text{C}$ in terms of temperature) are observed. However, it should be noted that, in addition to measuring surrogates such as temperature or electrical conductivity, it may often be necessary to measure determinands of direct interest, for example complex organic material in cases of groundwater contamination.

The most effective methods for taking samples from an aquifer in which the groundwater quality varies with depth are to sample specific aquifer horizons using specially constructed observation boreholes or, alternatively, to sample from sealed sections of boreholes in relatively close proximity, each completed and screened to enable samples to be drawn from a different depth range of the aquifer. In the latter, samples are pumped from a sealed section of a borehole by means of a packer-pump assembly, thereby providing a means of obtaining a discrete sample of water within a specific depth range of the aquifer. Packer systems consist of one or more sealing devices which can be expanded either hydraulically or pneumatically, once in position down the borehole, to provide a seal. A water sample is obtained from the sealed section by pumping or by gas displacement. This sampling method is only recommended for use in consolidated aquifers; it is not appropriate for use in boreholes completed with a screen and gravel pack.

Depth sampling

Depth sampling consists of lowering a sampling device into the borehole or well allowing it to fill with water at a known depth, and retrieving the sample for transfer to an appropriate container. This method is normally only suitable for use in observation boreholes that are not being pumped. Depth samples should never be collected from within the solid casing of a borehole, since the water cannot have originated at the depth at which the sampling device is activated and, under static conditions, may have altered in quality due to chemical or microbiological action.

Depth-sampling equipment (often known as "thief" or "grab" samplers) consists of devices that can be lowered into a borehole to collect a sample at a specific depth. Designs differ mainly in their closing mechanisms. The design used should be such that no water comes into contact with the sample container until the device is activated at the required depth. Where other methods of sampling are impractical, such as in very deep aquifers (i.e. greater than 100 m), depth sampling is recommended.

Samples may also be collected in a bailer during drilling, to provide crude data on groundwater quality variation with depth. On other occasions, where pumping a borehole is not possible, a simple bailer, such as a weighted bottle or other open container, can be lowered into the borehole to collect a water sample. Bailers should only be used to sample the surface of the aquifer.

Even within the open or screened section of boreholes, depth sampling can be of only limited value because natural or induced flows within the borehole can make the origin of the samples uncertain. Depth sampling is only suitable if the origin of the samples (in terms of depths of water inflow into the borehole) are known. This may be achieved by determining the depths of water inflows to the borehole and flows within the borehole column from interpretation of downhole logs of temperature, conductivity and flow under pumping and static conditions.

Other sampling methods

When the pumping or depth-sampling methods cannot be used or are thought to be inadequate, samples can be taken from discrete points in the aquifer by a variety of *in situ* methods, including porous cups or piezometer points from which water is extracted by vacuum or gas displacement.

Samples from particular depths may be obtained by pore-water sampling. This involves extracting water (usually by centrifuging or by squeezing under high pressure) from soil or rock samples obtained by specialised core drilling. It provides an effective method of quantifying vertical variations in quality and is also effective in the unsaturated zone. It may be the only method of determining contamination within the matrix of a rock such as chalk where flow is predominantly through fissures (borehole water only representing this fissure flow water). However, for periodic monitoring it has the disadvantage of requiring repeated drilling and is therefore an expensive method of sampling. It also has the disadvantage that it can free water that may not normally be removable from the aquifer under natural conditions; this technique should therefore only be used when recommended following specialist hydrogeological advice.

Choice of pumps for purging and sampling

When selecting equipment a distinction should be made between sampling and purging. A variety of devices may be used for purging, but only a few can provide water samples for accurate analysis. In purging the aim is to remove water efficiently; in sampling the primary objective is accuracy, with low flow rates preferred so that sample agitation and aeration are minimised. The performance of a selection of sampling devices with regard to accuracy, flow rates and

Table 6.4. Performance of sampling devices (Young, 1992)

Device	Accuracy	Flow rate	Lift capacity (m)	Available in required materials?	Applicability
Grab samplers	Fair	Low	None	Yes	May affect parameters sensitive to dissolved gas composition, e.g. pH, alkalinity, redox-dependent trace metals
Suction lift	Low	High	5–6	Yes	Not recommended for gas-sensitive parameters
Air displacement	Fair	Low	Up to 75	Yes	Accuracy reduced for gas-sensitive parameters
Electric submersible	Fair	High	Up to 600	Yes	Accuracy fair for gas-sensitive parameters
Piston pumps	Fair	Low	Up to 185	Yes	Accuracy and precision dependent on operator
Bladder pumps	Good	Low	Up to 300	Yes	Highest accuracy and precision

Table 6.5. Performance of sampling devices in relation to contaminant types (USEPA, 1991e)

Device	VOCs, organo-metallics[1]	Dissolved gases	Trace inorganic species[2] Reduced species[3]	Major cations and anions
Grab samplers	May be adequate if well purging is assured	May be adequate if well purging is assured	May be adequate if well purging is assured	Adequate for cations
Suction lift	Not recommended	Not recommended	May be adequate if materials appropriate	Adequate
Gas drive devices	Not recommended	Not recommended	May be adequate	Adequate
Mechanical positive displacement pumps	May be adequate if design and operation controlled	May be adequate if design and operation controlled	Adequate	Adequate
Positive displacement bladder pumps	Superior	Superior	Superior	Superior

[1] e.g. total orgarohalides (TOX), methylmercury (CH_3Hg).
[2] e.g. iron (Fe), copper (Cu).
[3] e.g. nitrate (NO_2^-), sulphide (S^{2-}).
[4] e.g. sodium (Na^+), calcium (Ca^{2+}), chloride (Cl^-), sulphate (SO_4^{2-}).

applicability is summarised in Table 6.4. Table 6.5 considers performance in relation to contaminant types.

In summary, good practice in sampling groundwater requires that:

(i) Sampling commences only when conditions in the well have stabilised as indicated, for example, by pH measurement, electrical conductivity etc.

(ii) Care is taken in the selection of materials used in pump components and other sample delivery equipment (tubing, bailers etc.) to minimise contamination/modification of the sample.

(iii) Sampling procedures minimise the potential for cross-contamination, e.g. sampling upstream before downstream locations, with the order of sampling in the latter dictated by actual or suspected contaminant types and concentrations.

(iv) Sufficient samples are obtained for laboratory analysis – a minimum of 1 litre is usually required.

(v) Test-specific samples are collected in a systematic manner with samples for volatiles, total organic carbon (TOC) and determinands requiring field filtration or determination collected prior to large-volume samples for macroelements, extractable organic compounds and metals etc.

(vi) Some analytical determinations be made on-site, e.g. pH, conductivity, temperature, dissolved oxygen, ammonia.

(vii) Samples are "preserved" on-site as necessary.

(viii) Sample containers are of appropriate materials and prepared for use.

(ix) Post-sampling storage and handling procedures are appropriate to the determinand under consideration.

(x) Potentially hazardous samples are appropriately labelled and protected from physical damage during transport.

(xi) Analysis is undertaken promptly after collection of the samples, and in any event within 24 hours, to prevent deterioration before testing.

(xii) Appearance, odours and location in the field are included in the field description of the sample.

(xiii) Care is taken in the collection of non-water miscible liquids.

Protocols prepared in 1994 for the then National Rivers Authority (now part of the Environment Agency) for those preparing to obtain samples from permanent groundwater quality surveillance boreholes provide useful checklists which can be adapted for use on contaminated sites. Further guidance on sampling and purging is given in a variety of guidance documents and standard texts (e.g. Freeze & Cherry, 1979; Cripps, Bell & Culshaw, 1986; Clark, 1988; ASTM, 1990; USEPA, 1991d; 1991e; Nielsen, 1991). Guidance on the handling, storage and preservation of samples is given in BS 6068: Section 6.3 (BSI, 1996a).

6.5   ANALYTICAL AND TESTING STRATEGIES

6.5.1   Transport, stabilisation and storage

Most groundwater-sampling methods result in the samples undergoing temperature and pressure changes which can alter such variables as pH, electrical conductivity, electrochemical potential, sulphide content and dissolved gas content (particularly oxygen and carbon dioxide). In turn these changes may alter the speciation of some constituents. Atmospheric contact may bring about similar changes and may also result in oxidation, increased microbiological activity, precipitation, volatilisation and changes in appearance (e.g. colour and turbidity). When sampling groundwater, it is important that as many determinations as practicable are carried out on-site, or as soon as possible after sampling. This is particularly important in respect of temperature, pH, electrochemical potential, electrical conductivity, alkalinity and dissolved gases (especially oxygen). A continuous measurement technique is preferable and is best carried out using flow-through cell systems that prevent contact between the sample and the atmosphere (Howsan & Thakoordin, 1996).

On-site filtration of groundwater samples is recommended for stabilising samples, particularly where speciation is under study. No single medium is universally recommended, although glass-fibre filters have some advantages over other media, since they block less easily yet provide similar filtration efficiency in terms of particle size retention. The recommended pore size is 0.4 μm or 0.5 μm, although other pore sizes may be preferable, depending on the particular sampling purpose and the determinands of interest. Whatever medium is used for filtration, the results of subsequent analysis should be reported as "filterable" species (quoting the appropriate pore size of the filter) rather than "dissolved" species. It is particularly important that on-site filtration of anaerobic groundwater should be carried out under anaerobic conditions.

Sample containers should be delivered to the laboratory tightly sealed and protected from the effects of light and excessive heat. If this is not done, sample quality may change rapidly due to gas exchange, chemical reactions and microbial activity. Samples which cannot be analysed within a day should be stabilised or preserved. For storage over short periods, cooling to 4 °C may be applied; for storage over longer periods, freezing to −20 °C is advisable. If the latter technique is used, the sample must be completely thawed before use, as the freezing process may have the effect of concentrating some components in the inner part of the sample, which is the part that freezes last. Samples may be preserved by the addition of chemicals, but care must be taken to ensure that the chosen method of preservation does not interfere with the subsequent laboratory examination (BSI, 1996a).

6.5.2 Chemical analysis

Much of the discussion in Chapter 5 on analytical and testing strategies for soils is equally relevant to water. There is an extensive range of International Standard "methods for the examination of water and associated materials" (these are published as British Standards in the BS 6068 series). The Standing Committee of Analysts has produced a wide range of methods for the examination of "waters and associated materials".

Analytical methods for contaminated soils and measurement of a range of environmentally related parameters have been produced in other countries, including Germany, the Netherlands and the USA. Environment Canada has published guidance (CCME, 1993b) on appropriate analytical methods to be used in connection with the National Contaminated Site Remediation Program.

The UK Water Research Centre, working for the (then) National Rivers Authority, has reviewed leaching tests for contaminated soils with a view to establishing a standard method to assess the potential of contaminated site redevelopment activities to pollute groundwater (WRC/NRA, 1993). A method based on this study is now in common use in the UK.

Non-aqueous phases require careful handling to avoid loss of volatiles and disproportionation (a type of oxidation-reduction process of organic compounds) of the sample etc. Special preparation may be required prior to analysis using instrumental techniques such as infra-red analysis and gas chromatography/mass spectrometry. Sample preparation is often the key issue rather than the analytical method itself.

A discussion of genotoxicity testing can be found in Simmons (1990). The US Environmental Protection Agency has published procedures for short-term genotoxic testing (e.g. for carcinogens and mutagens) (USEPA, 1979).

6.6   FIELD EVALUATION OF AQUIFERS

An understanding of the response of an aquifer to abstraction (which might be required, for example, in a pump-and-treat remediation strategy) is important in understanding groundwater hydrology (see standard texts for a description of the theory of aquifer behaviour). To determine the yield of a groundwater system and to evaluate the movement of groundwater contaminants requires data on the position and thickness of aquifers and confining beds; transmissivity and storage coefficient of the aquifer; hydraulic characteristics of the confining beds; position and nature of the aquifer boundaries; location and volumes of groundwater abstraction; location, types and concentrations of contaminants (Harris, Herbert & Smith, 1995). Aquifer tests to determine temporal changes in water levels induced by abstraction are important tools in the determination of some of these parameters. Multiple well tests provide most information, but useful data can be obtained from a single test well.

6.7 GROUNDWATER MODELLING

Contaminant behaviour in groundwater is influenced by hydraulic gradients, geological factors, abstraction etc., in addition to the physical, chemical and biological processes which result in the dilution, attenuation and degradation of contaminants in the groundwater. It is not generally feasible to provide sufficient monitoring points in the field to obtain a comprehensive analysis of groundwater conditions based on direct observation. Groundwater-modelling techniques have been developed as a way of predicting contaminant behaviour based on fewer monitoring points and interpolation between points using mathematical formulae describing specific physical, chemical and biological processes.

Such models may also be used for risk assessment purposes (see Chapter 10) or to predict likely design requirements, efficiencies and duration of remedial operations in actual contaminated groundwater situations. In these applications, field data describing ground conditions, together with leaching/desorption data obtained through direct laboratory testing on collected samples, should be provided.

Where laboratory and field-based data are not available in sufficient detail, or in those cases where only an approximate indication of remedial requirements is needed, data from similar situations or applications may be used to predict expected outcomes. The greater the input of good-quality monitoring data, the greater the accuracy of the result.

The use of models for contaminant behaviour prediction, for risk assessment or remediation purposes, requires the use of specialist expertise. In all cases, care should be exercised to ensure that the assumptions forming the basis of the model are appropriate and applicable to the situation under consideration.

6.8 A CASE EXAMPLE

A brief example provides the conclusion to this chapter. It serves to illustrate the potential complexity of the water pathway for contaminant movement and impact on sensitive targets (in this case a river).

6.8.1 The site

The site was a 10 ha area of alluvial soil adjacent to the river Derwent (Northumberland, northeast England). Originally a historic mill existed on the site, together with a weir to control the river water level. The mill was demolished in the nineteenth century and was replaced by a coke works which operated until 1970.

Spillages of phenols, toluene, tars, oil and poor disposal of cyanide wastes had visibly contaminated the area around the disused coke works. Initial site investigations (1991) revealed that a near-surface shallow groundwater existed which was polluted with sulphates, tars, oils, phenols and cyanides and that – at low water flows – river gravels, below the weir, could be seen to be cemented in a tarry material.

The soils beneath the former works were sands, invariably contaminated to 0.5 m depths, and heavily contaminated to the levels typically found on gas-works sites. The presence of such contamination in granular soils adjacent to a "clean" river indicated that site remediation might release a considerable pollution loading if excavation work enhanced pathways for contaminant migration.

6.8.2　Hydrological investigation

Initially river water sampling, upstream and downstream of the site, was undertaken. This revealed no water quality deterioration when river flows were at medium or high levels. However, in low-flow conditions a marked reduction in downstream water quality was apparent. This information suggested that the outflow of coke works contaminants was usually low and might imply that concerns over the potential to pollute the river were ill founded.

Later, groundwater-monitoring points were set into shallow trial pits (at 30 m centres) to obtain better information. This work showed that (Figure 6.7):

(i)　the "groundwater" was actually river water which entered the alluvial soils at the northern end of the site and flowed out immediately downstream of the weir, where the tarry cementing was visible;

(ii)　the "groundwater" flows coalesced at the exit point downstream of the weir; and

(iii)　the outflow rate was restricted by a clay bund wall which had been emplaced to depths of up to 2 m into the alluvial sandy soils (presumably built as a flood protection barrier).

It became apparent that the uppermost 2 m of the alluvial soils held about 150 000 m^3 of highly polluted groundwater, plus liqours in toluene and benzene tanks, and that all of this could enter the river very rapidly if the clay bund wall was affected by site remediation works. With low flows in the river such an accident would have significant downstream effects. This demonstration of the interconnection of the river and the site and bund wall forced a treatment of the wastes below the coke works site and the emplacement of cutoff walls before remediation of the soils could be commenced.

This site showed that:

● targeted (judgemental) samples had to be collected where pollutant outflow streams were visible in order to demonstrate the entry of potential pollutants into the river;

● shallow groundwater-monitoring points set in trial pits proved satisfactory and were cheaper than drilled boreholes.

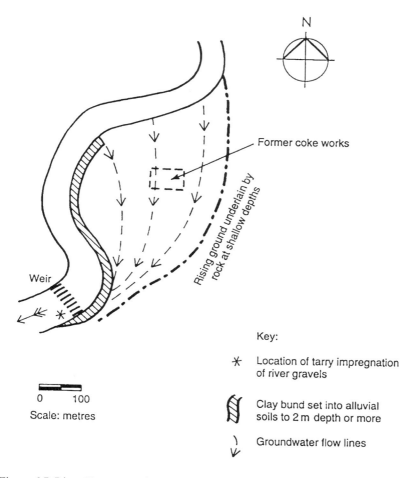

Former coke works

Rising ground underlain by rock at shallow depths

Weir

Key:

✳   Location of tarry impregnation of river gravels

〔 Clay bund set into alluvial soils to 2 m depth or more

〕 Groundwater flow lines

0   100
Scale: metres

Figure 6.7. River Derwent coke works site: conditions revealed by detailed investigation

The site demonstrated the problems and importance of proving the original hypothesis; the importance of revisiting assumptions; the need to revisit the adequacy of investigation data in the light of the developing conceptual model; and the importance of determining the source of "groundwater" and the flow patterns.

7

Gas and Vapour Investigations

7.1 INTRODUCTION

Subsurface occurrences of gases and vapours can pose more immediate hazards than concentrations of solid contaminants in soils. Compared with solid contaminants they present potentially more acute and rapidly acting risks. Gaseous contaminants are highly mobile, and many are lighter than air and so able to migrate up through soil profiles under quite small pressure heads. The consequences should such gases and vapours percolate into enclosed spaces – particularly fire, explosion, asphyxiation and toxicity risks – have been emphasised by a number of accidents (Anon, 1986; Williams & Aitkenhead, 1989; Staff, Sizer & Newson, 1991; Orr *et al.*, 1991; Gendebien *et al.*, 1992).

The heightened awareness that such risks are feasible has made the assessment of gas atmospheres an important dimension of the investigation and remediation of contaminated land in the UK, to an extent which exceeds practice in some other countries (e.g. the USA), particularly in relation to explosion and fire risks.

However, despite a developing guidance on gas surveys, these can still be poorly performed because of a failure to understand the key factors influencing gas generation and flows. No single gas monitoring event will establish worst-case gassing conditions, as a number of controlling variables affect which measurements can be collected.

This chapter addresses the characteristics of gases and their behaviour subsurface, and considers appropriate investigation and monitoring requirements to allow for the characterisation of gases and understanding of risk potential. It concludes with five brief case studies which illustrate the difficulties of obtaining representative soil-gas data and of characterisation.

7.2 PRINCIPAL SUBSURFACE GASES AND THEIR PROPERTIES

7.2.1 Principal gases

While methane and carbon dioxide occurrences are usually emphasised in UK practice, derelict land has often been subject to some degree of tipping and infilling with variable wastes, so that a much wider range of potentially hazardous gaseous compounds needs to be evaluated (Table 7.1). Various industrial wastes and quite commonplace underground conditions can give rise to hazards at least equal to those from landfill gases. The desk study undertaken to establish a conceptual model of a site (Chapter 4) should indicate which site-specific gases and vapours could be important.

Other, less frequently encountered gases (such as hydrogen) could have been included in Table 7.1, as indeed could have been the risks of oxygen deficiency. Oxygen contents in uncontaminated soils are essentially similar to those in the atmosphere (20–22%) and are necessary for the well-being of plants and the safety of people who have to enter trenches and excavations. However, as the concentrations of gaseous contaminants increase, those of oxygen generally decline. When this leads to oxygen levels of 18% or less, health and safety concerns will always arise (HSE, 1991). In many of the contaminated sites being considered for reuse, reduced/lowered oxygen concentrations are apparent (see Tables 7.2 and 7.3) and call for specific assessment and remedial action.

Some of the minor constituents of landfill gas are considered to be toxic, carcinogenic, mutagenic and teratogenic, and have served to focus attention (particularly in the USA) on the air pollution aspects of fugitive gas emissions involving volatilisation processes and gas-producing chemical reactions.

In landfills which have taken industrial wastes, the major contribution to toxicity derives from the presence of benzene, vinylchloride and mercury vapours (Young & Parker, 1983). Trace components in domestic waste usually present more of an odour nuisance problem owing to the presence of thiols, esters and sulphides. The UK view has been that the trace gases are not a health concern, with field studies suggesting that landfill gas should not give rise to toxic concentrations of individual compounds in unconfined localities where moderate dilution is available.

7.2.2 Gas properties and behaviour

Gas and vapour concentrations, pressure and flow rate properties are all relevant to the identification of hazardous subsurface gaseous atmospheres. Concentrations (in percentage terms or in parts per million) indicate how much of a particular gas occurs in a sampled subterranean atmosphere. As some specific concentration values (i.e. 5%, the lower explosive limit (LEL) for methane or the 400 ppm for hydrogen sulphide – Table 7.1) indicate the onset of particularly risky conditions, it is perhaps unsurprising that many regulators and some investigators react with concern when such values are measured in gas observation holes. Such critical concentrations are of concern if they occur in enclosed spaces, houses or structures.

However, for gaseous contaminants to present risks to people or site surface conditions, deeper gas collections have to move up the soil profile and then collect in above- or near-ground enclosed spaces at critical concentrations. For this to happen bulk gas transfer has to occur, and that usually calls for a flow

Table 7.1. Gaseous contaminants found on contaminated sites (listed in order of most frequent occurrence)

Carbon dioxide (CO_2)	Product of the biodegradation of organic wastes. Also produced by coal workings, limestone mines, chalk, sewers and marshes, and chemical reactions (e.g. acids and carbonate).
	Colourless and odourless. Very dense gas (specific gravity 1.53) soluble in groundwater.
	Toxic and asphyxiating to people. Health effects obvious. >3% concentration.
	Toxic to plant species, although the critical concentration varies with species (1% in most sensitive cases to 20% in least susceptible species).
Methane (CH_4)	Product of the biodegradation of organic wastes. Also produced by coal workings, sewers and marshes.
	Colourless, odourless and tasteless.
	Lighter than air (specific gravity 0.55).
	Flammable and explosive in concentrations between 5% (lower explosive level – LEL) and 15% (upper explosive level – UEL) in air.
	Able to expel air from soil voids and so generate asphyxiating atmospheres.
	Harms plant life if methane has expelled oxygen from plant root zones.
Carbon monoxide (CO)	Product of the incomplete combustion which typifies subsurface smouldering of heated carbon-rich wastes (coal, shale, wood, papers etc.).
	Colourless and almost odourless.
	Slightly soluble in groundwater.
	Lighter than air (specific gravity of 0.97).
	Highly toxic to people (>0.1% to 1%).
	Flammable (>12% to 75% in air).
	Toxic to plant life.
Hydrogen sulphide (H_2S)	Product of the degradation of sulphate-rich wastes (e.g. plasterboard) and of oil refineries where sulphur-rich crude oil has been refined.
	Colourless gas. Initially obvious by its "rotten egg" smell, but this becomes less noticeable as the gas is inhaled. Soluble in groundwaters. Heavier than air (specific gravity of 1.19).
	Highly toxic to people (>400 ppm). Flammable at 4.3% to 45.5% concentrations in air.

Lighter "oil" vapours (e.g. benzene, xylene, toluene) (C_6H_6X)	More volatile and diffusible components of spilled fuels. Diffusivity is high enough to permit vapour rise through soil in warmer temperatures. Colourless gases. Marked "petroleum" smells. Lighter than air, although specific gravity varies depending on the gas mixtures. Practically immiscible with water. Highly flammable and explosive in air. Some (e.g. benzene) carcinogenic at low concentrations. (Propane and butane admixtures are often present.)
Sulphur dioxide (SO_2)	Product of the combustion of sulphur-rich wastes (some coals and oils and the spent oxides on gasworks sites). Colourless gas with sharp odour. Denser than air (specific gravity of 1.43). Soluble in groundwater. Toxic to humans. Irritant to eyes and nose at lower concentrations. Toxic to plants.
Hydrogen cyanide (HCN)	Product of the combustion of cyanide-rich wastes, or the acidification of cyanide salts. Colourless gas with faint odour (bitter almonds) soluble in groundwaters. Flammable and explosive in air ($>6\%$). Highly toxic to people (>100 ppm).
Phosphine (PH_3)	Highly toxic to people (listed as First World War poisonous gas). Highly flammable acid, explosive.
Radon	Hazardous due to decay of Ra^{222} to daughter products giving rise to cancer when inhaled. Associated with some rock types (e.g. granites), some limestone mines, and some groundwaters.

Table 7.2. Gas concentrations revealed by standard gas monitoring boreholes drilled to natural strata (former landfill site)

Borehole number	Gas concentrations measured		
	Oxygen %	Carbon dioxide %	Methane %
1	14.1	8.3	0.15
2	1.8	9.6	1.10
3	1.3	8.0	3.96
4	trace	11.9	2.56
5	1.5	7.1	5.35

Note: Gas concentrations usually of concern are oxygen <16%; carbon dioxide >1.5%; methane >1%).

Table 7.3. Gas concentrations found in a multi-point gas observation borehole (former landfill site)

Depth of gas measurement (piezometer response zone)	Gas concentrations measured on specified dates			
	14.4	21.4	2.5	16.5
1.0 m	20/0.3/0	20.4/0/0	21.1/0.6/0	20/0.1/0
1.8 m	20/0.5/0	20.5/0.3/0	21/0.4/0	20/0.3/0
2.5 m	20/0.4/0	20.1/0.2/0	20.8/0.3/0	20.1/0/0
3.5 m	18.6/0.9/0	18.4/0.7/0	18/1.3/0	17.9/1.5/0
5.0 m	0/10.4/1.3	2.1/8.3/0.6	1.3/9.4/1.0	1.0/10/2

Note: Gas concentrations reported in the sequence: oxygen/carbon dioxide/methane in % terms.

(advection) of the type governed by the Darcy flow formula (written as an equation in various forms, all of which include a proportionality constant which is termed the coefficient of permeability). Gases can also move by diffusion, but this is generally of lesser importance – see below. A pressure head (above that of the atmosphere) is needed to drive bulk transfer flows, and positive relative pressure heads can be measured. The flow rates, generated by the existence of higher gaseous pressures deeper in a site, can be quantified with modern monitoring equipment. Deeper gases moving up and out of the ground are exposed to dilution with atmospheric air and to the effects of ventilation in houses, and so concentrations usually fall.

A 5% methane concentration, in soils or fills below land surface, might be a trivial matter if no positive gas pressures and no bulk flow rates occur, or could be serious if high outflow rates can be measured. The gases in the first case will not flow up the soil profile, and so cannot directly harm site users. In the second case, not only will such gas outflow occur, but the flow volumes are likely to be so high that, even when dilution with the atmosphere takes place, high remnant methane concentrations will persist within homes or structures.

Reacting to concentrations of gases and vapours, found to occur at depth in a site, requires knowledge of the gas pressures and flow rates likely to exist in most hazardous conditions. However, it is important that worst-case gassing conditions within the ground are measured and used in the analysis of possible risks. As noted below (Section 7.3), measuring worst-case conditions is not always possible, and calls for an intelligent assessment of when to conduct subsurface gas surveys (Section 7.5).

Barry (1991) notes that "effective ventilation (i.e. dilution) can eliminate all risks from gases, whether toxic, asphyxiant or explosive", adopting the same approach to judging risks from gases and vapours, i.e. that risk only occurs when gases move out of the ground and in such concentrations and flows that dilution will be inadequate to remove risks.

One final, and potentially confusing, type of gas or vapour movement is diffusive flow. This is flow driven by differences in concentrations within a fill or soil profile. It is a slow equalisation process, usually taking weeks or months to become apparent, which will ultimately spread previously localised higher gas concentrations throughout the subsurface profile, once investigation works have pierced gas sealing layers. While this could give rise to quite high concentration results in near-surface layers, these readings would not be accompanied by positive gas pressure heads or measurable flow rates. Thus, migration of meaningfully large outflow rates, to pose risks in homes or other enclosed spaces, would not usually result. Diffusive flows are the usual means by which subsurface gases gradually dissipate to the above-ground atmosphere (Bauer, Gardner & Gardner, 1972).

7.3 ESTABLISHING HAZARDOUS GASSING CONDITIONS

7.3.1 Gas and vapour monitoring equipment

Field-portable gas measuring equipment was originally developed for industrial applications (in mines, oil refineries and factories) and earlier meters were not ideally suited for use in contaminated land investigation. In the early 1970s to early 1980s only a few gas concentration portable meters existed, and these (as well as being limited to providing concentration results) were often of dubious accuracy, particularly if soil oxygen levels were too low ($<14\%$) to allow catalytic-thermal instrument systems to function accurately. A number of older publications (e.g. County Surveyors Society, 1982; Crowhurst, 1987) provide useful details on first-generation equipment and methods for gas monitoring. Such older publications stress the need periodically to confirm the accuracies of field meters, by collecting gas samples in pressurised tubes and then analysing the contents, with more sensitive gas chromatographic equipment, in a laboratory.

The measurement of gas flow rates, or gas pressures, at this time, was something of a specialist activity, and so few gas surveys included information other than concentration results. An obvious consequence of this was a widespread misconception (which influenced earlier official advisory documents) that concentration readings, by themselves, were adequately diagnostic indicators of "safe" or "unsafe" subsurface gas conditions.

This limiting situation has improved more recently, as the contaminated land investigation and reclamation industry has grown in economic importance and become a profitable target for instrument providers. The increased insistence by regulatory bodies that subsurface gaseous atmospheres must be investigated has also been influential.

Modern field-portable gas measuring equipment includes:

(i) Infra-red concentration meters, able to provide accurate measurements of the oxygen, carbon dioxide and methane contents of sampled underground atmospheres. These instruments do not suffer from the inaccuracies created by reduced oxygen concentrations and are easier to use than older meters. Thus, the need for laboratory-based gas chromatographic analyses is reduced.

(ii) Similar concentration measuring equipment for other gases (photon ionisation detectors for organic vapours, and specific monitors for all the gases listed in Table 7.1). Portable gas chromatographic analysers are also increasingly available.

(iii) Electronic pressure-measurement methods.

(iv) Flow rate meters, which can determine higher gas outflow rates.

With such equipment it is possible to establish quite quickly gas concentrations, gas pressures and gas flow rates.

Summaries of the performances of newer monitoring equipment are available (Crowhurst & Manchester, 1993; Raybould, Rowan & Barry, 1996), although the pace of technical innovation is such that for clarification of current gas monitoring capabilities, direct contact with instrument manufacturers is necessary.

7.3.2 Gas monitoring observation works

Several distinct types of gas monitoring holes are available. All have particular advantages, but none is uniquely suitably for all conditions. Thus, a careful evaluation of the required gas or vapour information is needed before a choice of monitoring facility is made.

The commonest are the traditional deeper borehole installations (Figure 7.1), initially devised for monitoring off-site migration of landfill gas. Typically such holes are drilled to large enough diameters (150–200 mm) to allow space for the insertion of a 50 mm internal slotted gas monitoring tube and gravel pack. This, together with the high percentage of open area (typically 6–12%) in the walls of the internal gas tube, allows unimpeded entry of gases from the surrounding ground. The top metre of the installation is unslotted and sealed with a bentonite grout plug to prevent atmospheric air entry (in periods of higher barometric

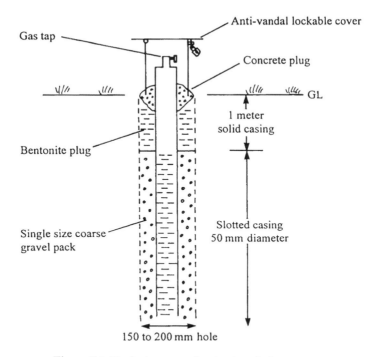

Figure 7.1. Typical gas-monitoring borehole

pressures). These boreholes are usually drilled through the full depths of gas-producing materials, and taken a further metre into underlying natural ground. This, for landfill-monitoring purposes, is a reasonable choice, since landfill gases under high pressures need not rise or move laterally, but can migrate downwards into underlying soils, if these offer the easiest initial outflow pathways.

Advantages of deeper borehole installations include the convenience of being able to make repeated use of the same installation over many months of a gas investigation, and the ease of taking groundwater and soil temperature observations by simply unscrewing the gas tap unit. However, borehole costs are high (about £600 for a 4 m deep hole) and it is seldom economic to install these closer than 30 m apart. This financial limitation can pose real difficulties; not all areas of a contaminated site will usually be gas producing and thus hazardous, but, for the most hazardous conditions to be established, observation boreholes need to be sited in the (unknown) areas of greatest potential concern.

Because of cost constraints, the use of smaller diameter boreholes (drilled at 50 mm diameter and with a 20 mm internal gas collecting tube) has become popular. However, if monitoring works include a mixture of boreholes of different diameters, results will not be directly comparable. Smaller diameter holes will tend to produce higher gas concentration measurements than larger diameter installations: atmospheric leakages into larger diameter holes are more likely to occur, and in larger-borehole volumes where only small inflows of gases occur stabilised concentrations are lower.

For cost reduction reasons small diameter drilled probeholes (usually 20 mm in diameter) and trial pits with gas collection slotted tubes are also used. If trial pit installations are employed, it is necessary to centre a vertical slotted polyethylene pipe, surround it with a granular backfill to within a metre of ground surface, and then cap off the backfill with a clay, or other gas-impermeable, cover. If the voids in the clay are highly water saturated the required impermeability can be assumed, but if the clay (placed close to land surface and exposed to desiccation) dries out, it is likely to be permeable to gases and will provide a poor top seal (Jefferis, 1993). Because of this doubt over effectiveness, a U.P.V.C. sheet, set into the clay cover, is a prudent addition, to ensure that collected gases do not merely migrate to the atmosphere.

The cheapest gas monitoring devices are hand-driven probe holes, to depths of 1.0 m and 1.5 m, which allow closely spaced arrays of gas measuring points. These "spike test" holes are cheap (up to 30 can be installed for the price of a single deep borehole), and so subterranean gas conditions can be mapped (in favourable conditions) and contouring of results can reveal specific locations where the worst gassing is occurring. Sealing the tops of temporary spike test holes to preclude atmospheric air entry is, however, difficult, and it has to be accepted that some dilution from atmospheric air is likely to occur.

One final choice, which seems attractive, is to install several small diameter piezometer tubes in a single large diameter borehole, and separate each piezometer response zone with bentonite grout dividing layers (Figure 7.2). The aim of such an installation is to identify the depth at which a gassing source is located and allow reclamation work to be more precisely focused. An obvious problem is that poor construction and workmanship can give rise to less effective seals, and to gas leakages between piezometer zones.

The use of "flux-box" installations (i.e. containers of known internal volumes set at depths of particular significance, such as the layer of soil just below planned house foundation slabs) is more commonly restricted to proof testing (after a site has been reclaimed) to check that all gas or vapour hazards have been reduced to acceptable levels. The internal atmospheres in these flux boxes gradually come into equilibrium with gas concentrations in the soil, and if three or four samples of flux-box atmospheres are taken at weekly or fortnightly intervals, it is possible to determine the inflow rates of gases (Cairney, 1995). This remains the best method of establishing low rates of gaseous migration.

None of these monitoring facilities is uniquely suitable in all circumstances, and all have their limitations. For deeper boreholes, the primary problem is that the entire internal monitoring tube atmosphere can be influenced by gas inflows from one thin layer within the investigated soil profile. The resulting measurements are representative neither of the strata in the greater part of the borehole nor of the atypical gassing layer. This is not important if boreholes are being used

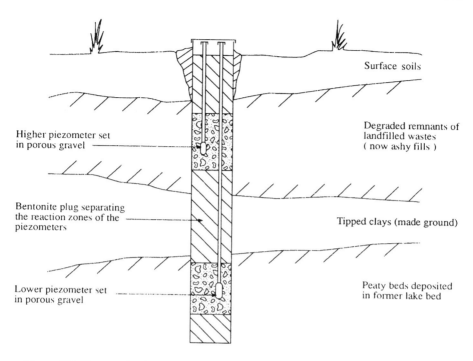

Surface soils

Higher piezometer set
in porous gravel

Degraded remnants of
landfilled wastes
(now ashy fills)

Bentonite plug separating
the reaction zones of the
piezometers

Tipped clays (made ground)

Lower piezometer set
in porous gravel

Peaty beds deposited
in former lake bed

Figure 7.2. Gas monitoring borehole to establish which horizons are gassing sources

for monitoring the perimeters of landfills (where the aim is merely to decide whether gas migration to adjoining land is occurring), but is unhelpful if the reason for monitoring is to identify the required reclamation strategy.

An example illustrates the point. A 1960s domestic waste landfill was investigated to establish whether the land could be used for housing. The landfilled wastes appeared to be an inert mixture of cinders and ashes (from domestic fires) together with broken bottles, tins and stones. However, when boreholes were drilled through the fills (4–5 m) and into the underlying sandy clays, unacceptably high carbon dioxide and methane concentrations were found, together with reduced oxygen contents. On these results (Table 7.2) redeveloping the site was believed to be unacceptably expensive if risks were to be reduced. However, some trial pitting was still in progress, and this allowed spike testing as trial pit depths were advanced (from 1 m to 3 m). This second investigation revealed only uncontaminated soil atmospheres in the ashy fills, and cast doubts on the validity of the deeper borehole results. To resolve the uncertainties, one multi-point borehole (Figure 7.2) was sunk, and revealed that landfill gases only occurred in a thin zone at the top of the sandy clay stratum (Table 7.3). Later excavation proved that this had resulted from the washing down of fine organic matter over some 30 years. This had collected as a 10 to 20 mm thick black accumulation on top of the natural underlying clays, and could readily be removed in site reclamation.

In this example, gases in the thin organic layer were pressurised by the weight of overlying strata, and so moved rapidly into the lower pressure zones offered by the drilling of each gas observation borehole. This demonstrates that all deeper boreholes are low pressure environments which tend to collect gases from considerable distances, and can produce high gas concentration results. Such overestimation of gas concentrations might be acceptable for lighter gases, as the higher than actual gas concentrations might indicate the riskier situations which could occur in future, more extreme, circumstances (Section 7.4), but is less acceptable when denser gases are of concern. In such cases, a deeper borehole can act as no more than a "sump" into which gases such as carbon dioxide flow and collect. Cases where very high carbon dioxide concentrations have been measured in observation boreholes, and where later excavation of sites has revealed no comparably high carbon dioxide contents, are relatively commonplace in areas of old limestone or coal mining (see Section 7.10).

In contrast, shallow spike-tests do not produce such exaggerated gas readings, but tend to give lower concentrations because of dilution with atmospheric air. Additional limitations of shallower observation points are that they can fail to intersect the deeper gas producing layers (see Table 7.3) and can produce results influenced by near-surface phenomena (e.g. the bacterial oxidation of methane to carbon dioxide or the solution of carbon dioxide in shallow groundwaters (Ward, Williams & Hill, 1993)). Despite these limitations, and particularly where sites are underlain by predominantly granular fills, spike-testing at 5 m to 10 m centres can highlight those subareas where gassing potential is greatest, and so allow more expensive boreholes to be sited effectively.

7.3.3 Factors affecting measurement of landfill gas (and other degradation product gas)

The analysis of measured gas and vapour results remains an art form, simply because measurements are affected to a great extent by a wide range of factors. Table 7.4 indicates a commonly encountered range of variation in results, within which some could appear to be entirely safe (i.e. carbon dioxide less than 1.5% and methane less than 1% of the underground sampled atmospheres), while others could be taken as signals for concern.

The factors leading to these variations and interpretation difficulties need to be discussed separately for gases resulting from biodegradation and those resulting from other mechanisms.

Landfill gases dominate UK concerns over subsurface gaseous contamination, simply because so many old landfills exist (in excess of 1000) and because tipped organic wastes occur on most derelict sites. In consequence, professional associations (County Surveyors Society, 1982; IWM, 1990) and government-funded studies (Hooker & Bannon, 1993; Card, 1993) have developed a comprehensive

Table 7.4. The difficulty of measuring worst-case gassing data – variability of gas measurements in a single borehole

Date	Volumetric concentrations of			Observations
	Oxygen	Carbon dioxide	Methane	
05.10.91	20.4	0.3	nil	–
10.11.91	16.1	1.7	0.3	Reading after sudden barometric fall (24 mbars)
04.12.91	20.1	0.4	nil	–
03.04.92	19.7	1.4	0.1	Period of gradually
24.04.92	17.4	2.7	0.6	increasing soil temperatures 8–11 °C
05.05.92	10.4	8.7	2.5	
06.06.92	2.3	12.7	4.8	
01.07.92	0.1	2.3	5.7	Near surface
04.08.92	0.1	1.3	4.1	soils saturated by prolonged rainfall in early July
07.09.92	0.1	16.2	5.5	Near surface soils dried out
04.10.92	12.4	4.9	trace	Soil temperature falling
05.11.92	18.7	1.2	nil	Soil temperature at 8.1 °C
03.12.92	20.4	0.2	nil	Soil surface frozen

Note: Gas source 3 m thick refuse layer at depth of 2.3 m below land surface.

understanding of biodegradation processes. A large scientific literature has existed since the 1950s (see Gendebien *et al.*, 1992 for a review).

Landfill gases are the by-products of bacterial degradation, and so gas emission rates, flow quantities and concentrations depend primarily on the rate of bacterial activity and the expansion of bacterial populations. This dependent link implies that any internal or external factor which affects bacterial populations will inevitably affect the concentrations and types of gases which can be measured in observation holes.

The idealised representation of landfill gas generation over time, depicted in Figure 7.3, is often assumed to represent the actual degradation process from freshly tipped wastes to the final composted residue, suggesting that biodegradation is a fixed and predictable matter. While such idealised conditions might be representative of those of deeper landfills (> 6 m) with an effective cap, even these tend not to progress through the various degradation phases, usually remaining in the unstable Phase III for decades.

In the shallower and less well-engineered organic deposits which typify many contaminated sites, a degradation process suggested by Figure 7.3 can be misleading, as internal and external factors may accelerate or retard degradation, and affect the types, concentrations and emission rates of subsurface gases.

Of the internal controls, the most important are (Rees & Grainger, 1982; Barlaz, Ham & Schaefer, 1990):

- temperatures within the waste;
- moisture conditions;
- waste acidity;
- the presence or absence of toxic compounds;
- the ease with which atmospheric air can intrude into the decomposing wastes;
- the quantity and distribution of degradable carbon.

While these factors are conveniently considered separately, it should be realised that they will usually inter-react in various combinations, which adds a further complexity.

Temperatures of 30 °C to 45 °C encourage bacterial growth and give rise to high gas emission rates. The internal heat of degradation (632 kJ/kg) is enough to support these temperatures, if cover above the waste provides good thermal insulation. However, the 2–3 m of soil cover thickness needed for this seldom occurs over the patches of tipped materials found on contaminated sites. If waste

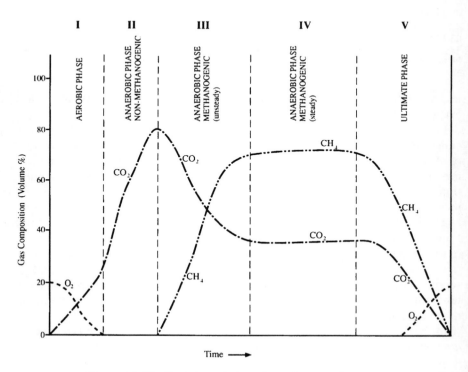

Figure 7.3. Idealised degradation phases for organic wastes

temperatures fall, biodegradation slows and gas emissions decline. Even a small internal temperature drop (of 1 °C) can cause methane contents in sampled underground atmospheres to fall sharply (from 19% to less than 4% in recorded instances). Such an event often results from rainfall entry in winter periods. As chilling continues and waste temperatures fall below 10 °C, gas emission rates effectively halt. Thus, the normal pattern, in thinner waste accumulations, is for methane to "disappear" in winter, and then reappear in hot summers (Table 7.4).

Elevated internal moisture content increases gas production. This is because bacteria obtain their food from the moisture films on waste particles. If the wastes are dry (< 10% saturated), then food availability is essentially zero and the bacteria become dormant. As wastes are wetted, bacterial life and gas emissions pick up, and peak at a waste moisture content of about 60% to 80% of the totally saturated state. At this level, food availability is most widespread and bacterial populations expand. However, if more water enters the wastes, this tends to wash away the organic fine materials, and so reduce food for bacteria. Total water saturation slows biodegradation to the extent that even 30-year-old wastes can appear to be nearly as fresh as the day they were tipped. In one recent land reclamation, this effect was obvious in wastes tipped in the 1950s. Because this landfill had been sited in a shallow clay-floored depression, the basal metre or so of the wastes produced entirely readable newspapers in a matrix of totally saturated but recognisable domestic refuse. Similarly, dry conditions may delay degradation for many years so that it is stimulated only when wetting occurs: for example because of rising groundwater or breach of a low-permeability cap.

Acidity and the presence of toxic compounds (especially metals such as aluminium) can kill bacteria and slow degradation processes. Methanogenesis proceeds optimally in a pH range of 6.5–8.5. Easy oxygen entry (as is the case with smaller volumes of poorly engineered organic wastes found on many contaminated sites) in periods of higher atmospheric pressure prevents methanogenic conditions developing, and precludes the idealised sequence of degradation phases. Together these effects slow biodegradation so that waste accumulations may persist for decades without reaching the high internal temperatures needed for rapid decomposition. This represents perhaps the worst of outcomes, as the wastes pose continued hazards (organic-rich polluting leachates and periodic emissions of carbon dioxide together with small concentrations of methane).

In addition, external climatic and soil factors have their influences on measurements of landfill gases. Abrupt atmospheric high pressures can force oxygen deep into poorly covered wastes so that no landfill gas contents can be sampled. In contrast, a deep barometric low can "pump" out enhanced landfill gas concentrations into an observation hole (Figure 7.4).

Groundwater level changes can act in a similar manner. Sudden rises in shallow groundwaters will force gases into observation boreholes at very high concentrations, while abrupt falls can have the reverse effect (Barry, 1987) and surface climatic effect (i.e. ground freezing or thawing) can close off or open pathways via which gases migrate to observation points. Surface effects can ensure that the gas atmospheres measured in shallow observation holes differ markedly from those which originated at a deeper gas generation location (Ward, Williams & Hill, 1993).

7.3.4 Factors affecting the measurement of other gases

Although the technical literature for gases and vapours not resulting from biodegradation is less comprehensive, enough exists to establish broadly why gas and vapour measurements are subject to marked variability.

The gases created by subterranean heating and smouldering (carbon monoxide, sulphur dioxide and hydrogen cyanide) can be affected by atmospheric pressure effects. Higher barometric pressures force air deeper into the ground and so accelerate smouldering and gas output rates, although this will usually take a day or so to become apparent, and will initially have the reverse effect (i.e. combustion by-products will be less apparent in gas observation holes as a high pressure tracks over a site).

Similarly, variations in the proportion of combustible wastes will have obvious effects. As more abundant concentrations are exhausted, subsurface progress of a smouldering front will slow and possibly decline, although this (and the accompanying marked reduction in gas emission rates) need not mean that the hazard has vanished. In one recently investigated case, where tipped sawdust and newsprint had been ignited by deep migration of a surface fire, output of combustion gases ultimately slowed and then ceased for several months. However, these suddenly reappeared (presumably because remnant soil temperatures had been enough to ignite more distant patches of combustible waste), together with high concentrations of methane. The methane, in this case, was a result of the soil warming increasing the biodegradation rate of garden wastes, which had been tipped within the combustible fills.

Soil temperatures also have marked effects on the concentrations of volatile organic compounds (from spills of fuel oils and petroleum products). Here volatilisation is the main cause of gas emissions, and high levels of, for example, benzenes, xylenes and toluene are most apparent in hotter summer conditions. Atmospheric pressure changes also affect the quantities of these volatile organics which are measured in a manner similar to that noted in Figure 7.4.

These volatile vapours often arise from fuel products floating on shallow groundwaters, and so can be most apparent when groundwater is higher within the ground. Carbon dioxide (from slow biodegradation of heavier oil fractions) is often found to accompany higher volatile organic concentration results. The pattern of occurrence of these vapours and gases can be used to establish the direction of flow of subsurface hydrocarbon spillages (Evans & Thompson, 1986).

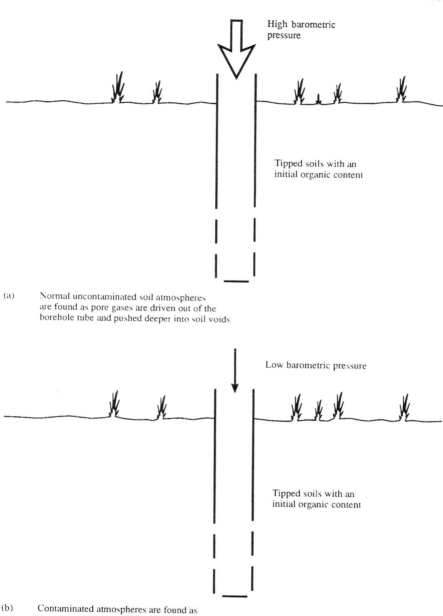

High barometric
pressure

Tipped soils with an
initial organic content

(a)   Normal uncontaminated soil atmospheres
are found as pore gases are driven out of the
borehole tube and pushed deeper into soil voids

Low barometric pressure

Tipped soils with an
initial organic content

(b)   Contaminated atmospheres are found as
pore gases can move into the borehole
tube from the surrounding soils.

Figure 7.4. Atmospheric controls on pore gas concentration measurements

7.4   GAS AND VAPOUR SURVEYS

7.4.1   Observation locations

Two separate problems have to be addressed by well-designed gas and vapour surveys, i.e.

(i) observation holes have to be located in those areas of a site most likely to be emitting gaseous compounds; and

(ii) surveys have to be conducted predominantly in those periods when the variables which affect gaseous emission rates will most probably produce more hazardous results.

If the potential for gaseous emissions were uniform within a site, there would be no need for care in selecting gas observation hole locations. However, this will seldom be the case, as most sites contain subareas free of gassing nuisances and others where much worse gas or vapour results occur. Thus, it is important to try to locate the observation holes in areas of poorest underground atmospheres.

This situation, which typifies many contaminated sites, is quite different from that on larger landfills. These, while not entirely uniform in composition, are more consistent than are many contaminated sites where smaller accumulations of gas producing materials pose risks.

Three methods can be used to locate observation holes in most potentially hazardous gassing areas:

(i) The hazard identification stage should be able to identify where quarry voids once occurred or where small tips were previously sited;

(ii) Gas concentration measurements as investigation trial pits and boreholes are deepened will show which horizons, within a site, produce greatest gas concentrations (Table 7.5).

(iii) Spike-test surveys, together with subsurface temperature readings.

It is seldom prudent to site gas boreholes before preliminary site investigation results have been thoroughly analysed.

7.4.2   Surveys of hazardous gassing conditions

Table 7.4 shows that no single-event gas survey is likely to record worst-case gassing results in boreholes and spike-test holes. As a result, most practitioners try to include between three and five different site visits to measure subsurface gas and vapour information. These, usually at weekly or fortnightly intervals, are a balance between budget restraints (which tend to reduce gas survey events) and the need to measure in different conditions to establish trends and identify more hazardous conditions. However, this practical compromise may not be the technically most effective, since gas concentrations, flow rates and pressures can

Table 7.5. *In situ* gas-monitoring results as a groundwater observation borehole was being drilled

Borehole	Depth (m)	% Oxygen	% Carbon dioxide	% Methane
17(W)	1.0	21.1	0.3	–
	2.0	20.3	0.2	–
	3.0	19.7	0.2	
	4.0	18.4	0.7	<1
	5.0	18.1	1.4	1.2
	6.0	14.2	2.7	1.6
	7.0	19.7	1.3	1.0
	8.0	20.1	0.6	–
	9.0	20.7	0.4	–
	10.0	20.3	0.3	–

Note: Horizon from 4.6 m to 6.4 m depths was an ashy layer (decomposed) landfill material. Materials above and below were clay subsoils, produced as wastes from road construction works in the area. Subsurface conditions elsewhere on the site were found not to be gas producing.

change abruptly and in very short periods (of a day or less). Ideally, continuous gas monitoring would be undertaken. However, equipment limitations and costs usually make this impractical, except on problem sites (e.g. those where gas is threatening existing buildings).

For landfill gas surveys optimum survey periods include:

- times of higher soil temperatures;
- periods when atmospheric low pressure conditions occur immediately after higher barometric highs;
- when near-surface groundwater levels are rising.

If combustion by-product or volatile gases are of interest, higher concentrations and flow rates are likely in times when:

- low atmospheric pressure conditions occur just after a barometric high; and
- soil temperatures are at their annual peaks.

In this way, it is possible to distinguish between the adequacy of gas and vapour surveys and to decide how much reliance should be ascribed to measured results (Table 7.6).

Site investigators, however, can be required to undertake gas and vapour surveys in very unsuitable conditions (for example, cold winter periods when atmospheric high pressure is dominant) and then have the problem of advising clients that, whilst no contaminated soil atmospheres were encountered, this does not preclude the occurrence of much poorer conditions in warmer weather. The potential risks, and hence liabilities, which may be missed by insufficient surveys, combined with the costs of retrofitting gas protection, are high.

Table 7.6. Adequacy of gas and vapour surveys

Reliance possible*	Survey content
Low	• Gas concentrations only measured. Measurements only on a single occasion. No independent checks of measurement accuracy. Readings taken in autumn or winter.
	• As above, but readings taken on at least three occasions and measurement accuracy confirmed. Atmospheric pressure and climatic conditions recorded. Readings extend into periods of warmer soil temperatures.
	• As above, with gas pressure readings included and with changes in barometric conditions over prior 24 hours recorded.
	• As above, with direct flow rate measurements.
	• As above, with soil and gas temperatures measured.
High	• As above. Readings continued to include conditions especially likely to reveal highest subsurface gas emission rates (soil temperature, barometric and other more sensitive conditions adequately covered).

*Increasing reliance from low to high.

7.4.3 Analysis

Measurements are frequently made on-site, although samples for off-site analysis can also be collected using either gas containers or absorption tubes. Substance-specific detection tubes may also be employed down exploratory holes. Samples for analysis may be withdrawn under suction or by flushing the gas from a sample chamber into which it has entered through a gas-permeable membrane of high diffusion impedance (Robitaille, 1992).

For accurate data on the concentration of any bulk gas and all trace gases, samples should be collected for laboratory analysis by gas chromatography/mass spectrometry (GC/MS). It should be noted that the difficulty of getting a real sample in the tube is such that the test results are not necessarily "accurate". Samples should be collected using metal containers, pressurised if necessary (e.g. stainless steel Gresham™ Tubes), which have been flushed through three times with the gas before final filling for analysis. Plastic or rubber containers are not recommended as they are relatively gas permeable and may absorb the gases. Pumps used to extract the sample should be of plastic or rubber composition when sampling at about 100 kPa, or of inert metal composition when higher pressures are involved.

When an integrated measurement is required, an adsorption device is placed in a hole in the ground. This is recovered after a suitable period of time, which may be several weeks, and the adsorbed substances desorbed in the laboratory for analysis using gas chromatography, mass spectrometry and similar instrumental techniques. Adsorption may be onto proprietary devices intended to extract gases from a gas flow, devices intended for other proposes (e.g. monitoring individual personnel exposures), or specially developed (proprietary) devices (Viellenave & Hickey, 1991). Integrating techniques provide for the detection of lower concentrations, and smooth out concentration variations associated with fluctuating water tables and atmospheric pressure that would otherwise necessitate repeated and regular measurement.

Particular consideration should be given to the depth of sampling relative to the location of gas permeable strata, and to the need for long-term monitoring of the gas regime. The installation of "permanent" monitoring positions is invariably required with regular monitoring of *in situ* concentrations and gas emission rates, and laboratory chromatography analysis of gas samples collected from each installation. In addition, ambient and in-ground temperature should be recorded on each monitoring occasion, together with atmospheric pressure, immediately before, during and after the collection of the sample, weather conditions including any recent rainfall, and depth to water table.

7.4.4 Data consistency

Even a less than adequate gas survey will produce a large number of individual measurements. Better surveys may produce hundreds or even thousands of numerical values. It is prudent to check that results are consistent, and that anomalously high values are not a consequence of instrument or human error.

One simple consistency check, when landfill gases are of interest, is to produce a graph of the measured oxygen and carbon dioxide readings. If carbon dioxide is being generated within a site then oxygen will be consumed, and a relationship between the carbon dioxide and oxygen measurements should be apparent. Departures from this relationship could be due to instrument or observer failures or could be a result of other, deeper sources of carbon dioxide. Similarly, the effects of landfill gas migration can be apparent from such graphical plots. For example, if an abrupt rise in CH_4 concentration is accompanied by an O_2 concentration fall below that which would be predicted from the CO_2 concentration from earlier monitoring, it could be postulated that some O_2 is being physically dispelled by landfill gas inflow under pressure.

Consistency checking thus not only gives an indication of data accuracy, but also offers insights into what is occurring within a site (Figure 7.5).

7.4.5 Gassing categories

A well-designed and conducted gas or vapour survey should always reveal which category of subsurface gassing conditions exists below a site. Various categorisations are possible. ICRCL (1990b) advises that a distinction can be made between low gaseous emission rates in older landfills (and most contaminated sites), and the rapid emissions of high volumes of pressurised gases

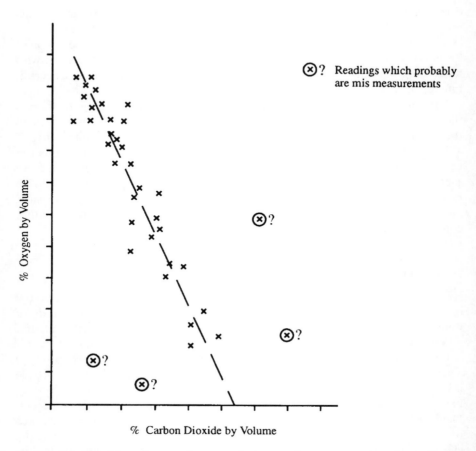

Figure 7.5. Checking the consistency of measured oxygen and carbon dioxide concentrations

more typical of modern landfills. Emberton & Parker (1987) drew distinctions between old deep (5–10 m) landfills and sites more representative of much of contaminated land. This last category was typified by the release of small quantities of methane and the lack of any potential for future large-scale gas emissions. More comprehensive categorisation is possible (Table 7.7) to distinguish between the more or less hazardous conditions which can occur within contaminated ground.

The most hazardous gas category (No. 4) is actually the easiest to identify and the one least open to interpretation and remediation doubts. In contrast, the least hazardous (No. 1) is often more difficult to identify, since gaseous contaminants may not be found in some of the gas surveys.

Table 7.7. Gassing categories on contaminated sites (in increasing order of hazard)

Category	Conditions encountered
1	Gases produced intermittently, often in response to climatic/barometric changes. Flow rates/pressures usually trivial. Concentrations occasionally high.
2	Gases produced continuously, although production rate varies with climatic/barometric factors. Flow rates/pressures usually trivial (although not in highest emissions periods). Gas concentrations often high.
3	Gases produced continuously, although production rate varies with climatic/barometric factors. Flow rates pressures/concentrations often high.
4	Gases produced in large volumes and under significant positive head. Concentrations usually very high, and to the point where dilution with air is still likely to leave fire, explosion and toxicity risks.

7.5 GAS SOURCE IDENTIFICATION

Gas source identification is essential to the choice of remediation strategy. In simpler cases (e.g. Table 7.5) the location, thickness and depth of a gassing source can be directly identified. More commonly, however, all that is known is that gases or vapours are being generated at depth, probably beyond that of gas observation holes. These could be methane and carbon dioxide, possibly from nearer-surface biodegradation, or deeper coal mine gases or emissions from deeper peaty deposits in the soil profile, or arise from off-site sources or leakages from buried gas pipes.

Distinguishing between methane and carbon dioxide atmospheres from such a range of sources is possible. British Gas distribution records, coal mining plans and geological/hydrogeological information can suggest which sources are more or less probable. More detailed gas concentration measurements, using laboratory-based gas chromatography, can reveal the contents of ethane, propane, butane and helium trace gases (Table 7.8). These can prove diagnostic, especially if carbon-dating tests have been conducted. However, since dilution with air will often have occurred before gas concentrations have been measured, it can be difficult to be certain which source is responsible for the measured gases.

If carbon dioxide is the gaseous contaminant a number of sources could be the cause, i.e.:

- pre-methanogenic degradation of organic wastes;
- gases rising from chalk limestone or coal mines;
- the slow degradation of heavier oil residues trapped in soils or contained in shallower groundwaters;
- the oxidation of methane.

Table 7.8. Distinguishing the sources of methane and carbon dioxide measured
in monitoring holes (after IWM, 1990)

Gas source	Properties
Landfills	Young gases (^{14}C dating).
	Methane: carbon dioxide usually 60:40. Carbon dioxide more prevalent than methane on older landfills and on contaminated sites.
	Only minute traces of ethane and propane present.
Peats	Intermediate aged gases (^{14}C dating).
	Carbon dioxide more prevalent than methane. Ethane, propane and butane traces are significant. No odorants, of types found in mains gas, exist.
Marsh gas	As peat gases.
Sewer gas	As peat gases, albeit with younger age (^{14}C dating). Traces of hydrogen sulphide usually occur.
Mains gas	Old gases (^{14}C dating).
	Significant ethane, propane and butane contents occur. Diethyl sulphide has been added to give a noticeable odour and is a diagnostic source identifier.
Mine gas	As mains gas, although without the diethyl sulphide odorant. Helium (in the range of 20 to 200 ppm) commonly occurs.

As with the methane and carbon dioxide associations, it can be difficult to be sure which of the possible types of gas emission sources is actually the cause (see Section 7.10). Other types of gases and vapours generally present fewer source identification difficulties (Section 7.9).

7.6   GAS EMISSION PREDICTION

Whilst the risks from subsurface gases and vapours will usually be demonstrated by the measurement and analysis of information from gas survey facilities, some workers have attempted to calculate the ultimate volumes of gases which might be produced. This approach depends on estimating the volume of total carbon in a site and then assuming a realistic decay rate. Lord (1991) supported this approach because of the difficulty of measuring very low rates of gas flows.

Although theoretically sound, the method does call for a large number of assumptions (e.g. the actual amount of degradable waste in a variable tipped site, the ultimate methane emission volumes per tonne of dry waste, etc.) which will be difficult to make, or subject to technical uncertainties.

Given the present improvements in gas monitoring equipment, it is difficult to advise that emission predictions will offer additional useful insights to risk assessment.

7.7   CASE STUDY 1 – GLASGOW GREENBELT

A particularly simple and solvable landfill gas problem was encountered when the greenbelt around the city of Glasgow was extended to release land for housing. In one area of some 4 ha the land was gorse- and heather-covered rough grazing, underlain by a thin (1–2 m) layer of sandy glacial clays with peaty horizons. Below this are massive Coal Measures sandstone rocks, which are usually highly weathered (to gravely sandy rubble) in their uppermost beds.

The presence of peat layers suggested that subsurface gaseous nuisances might occur, and a spike-test survey was undertaken to confirm or deny this. Spike-tests at 15 m centres were driven to 1.5 m depths, or to rock head if this occurred at shallow depth. In addition to the concentrations of methane, carbon dioxide and oxygen in each spike hole, soil temperatures were also measured by an electronic thermocouple device.

This survey was necessarily carried out in an especially cold January, since the arrangements for the legal transfer of land ownership had been more time consuming than had been anticipated. No significant biodegradation gaseous products were expected, and the client was advised that the survey ought to be delayed until warmer soil temperatures developed. However, given the low probability of thin peat layers being able to generate gassing problems, the client's wish for an immediate gas survey was accepted.

This assumption (of no measurable biodegradation) proved to be correct except in two quite large areas, where carbon dioxide concentrations reached levels of up to 6%. Elsewhere only trace levels of this gas (<0.5%) were found. At the two anomalous localities, soil temperatures were found to be at, or near, 10 °C, in sharp contrast to the 1–1.5 °C subsurface conditions recorded in the rest of the site.

Shallow excavations revealed the presence of bagged food wastes, of recent age, obviously derived from a food-processing source, and hidden below a 0.5 m thick cover of local soil. These bagged wastes represented illegal waste disposal in shallow quarries; no official records of disposal existed.

Redevelopment of the land was delayed until the following summer, because of a downturn in the sales of private housing, and this allowed more detailed monitoring of the anomalous areas in warmer soil conditions and as atmospheric and climatic conditions varied. This showed that, while no other gassing areas existed, landfill gases (methane) from the tipped wastes could and did migrate outwards (through the decomposed granular rubble at the top of the coal measures sandstones) for distances of up to 300 m.

In this case, shallow spike-testing fitted the site conditions, and was able to identify the buried food wastes only because active degradation was still in progress. Had the wastes been older and more fully degraded, it is probable that the initial gas survey would have revealed little useful information. Excavation and removal of the gas source were economical because of its especially shallow location.

7.8   CASE STUDY 2 – PAPER MILL SITE, CENTRAL SCOTLAND

This proved to be a more complicated site, on which gas surveys were conducted on four occasions (April, May, July and October) to include the effects of warmer soil conditions, and also periods when barometric low pressures followed rapidly after atmospheric highs.

The site had housed a paper mill (now demolished) and included several areas where localised tipping of paper-ash and woodpulp wastes had taken place. Below the surface, silty alluvial soils with thin peaty horizons were encountered over massive and open-jointed granite rock. South of the site is a deep railway cutting (15 m below site surface) and a modern landfill lies some 300 m south of the railway line. Records exist to show that the landfill is within a granite quarry whose base is 12 m lower than the level of the railway cutting.

Initial gas surveys (April and May) made use of spike-test networks (10 m centres between spike holes) supplemented by 12 deeper boreholes in areas whose spike-testing had revealed poorer subsurface atmospheres. These deeper boreholes were terminated at the upper surface of the granite rock.

The carbon dioxide and oxygen relationships revealed by these earlier surveys (Figure 7.6) followed the predictable pattern. The April survey showed little oxygen depletion and only slightly enhanced carbon dioxide concentrations. No methane or other flammable gas occurrences were recorded, and pressures in observation boreholes proved to be identical to those of the atmosphere. By May, oxygen depletion and carbon dioxide increases were more obvious, although the graphical relationship established in April persisted. Methane (up to 1%) did occur in several observation holes, although gas pressures still remained low.

This change was interpreted as a consequence of degradation of the tipped woodpulp and incompletely combusted paper-ash, and appeared to indicate that only a single shallow gas source was affecting the site.

In July and more especially in October, however, matters became more complicated. A hot summer had occurred, and it had been assumed that this would only enhance the oxygen/carbon dioxide relationship already established. In fact (Figure 7.6) a quite different graphical pattern was obvious, with increased carbon dioxide concentrations co-existing with only slightly reduced oxygen contents. In addition, positive and high gas pressures were found (particularly in the southern-most boreholes) and explosive concentrations of methane (up to 12%) occurred. During the October gas survey, even more anomalous results were obtained.

The pattern displayed in Figure 7.6 suggests that two quite distinct gas sources affect this site:

(i) a shallow source of poorly degradable wastes (woodpulp and other tipped matter); and

(ii) a deeper source of pressured gases with much higher methane and carbon dioxide contents.

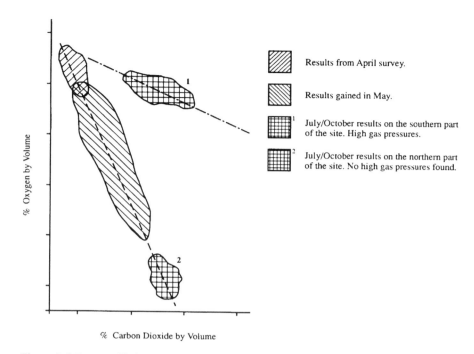

Figure 7.6. Paper mill site: consistency checks which indicate the presence of two separate gassing sources

Further work confirmed this view, and suggested that the deeper gases originated from the local authority's active landfill, and that these gases – when pressure conditions increased as soil temperatures rose – could force their way through joints and fissures in the granite rock to reach the paper mill site.

Resolving the problem, in principle, should not have been difficult. Shallower woodpulp and similar wastes could be identified and excavated. The migration of pressurised landfill gases through discontinuities could be intersected by a suitably located gas vent/barrier trench. Together these measures would have been adequate to permit a safe reuse for the paper mill land. However, on practical and economic grounds, the feasible solutions could not be entertained. The vent/barrier trench would have to be excavated to depths of about 27 m, and the bulk of this would be in especially hard rock. Therefore, the risk reduction costs would have been out of all proportion to the final value of the land.

This case demonstrates that gassing problems can prove to be too expensive to solve in some more complex cases, and also illustrates the advantages of well-conducted and timed gas surveys. If only the April gas information had been available and the site redeveloped, substantial problems and risks would have been found at later dates.

7.9 CASE STUDY 3 – FORMER STEEL MILL SITE, WEST MIDLANDS

Gases and vapours, other than degradation by-products, can prove to be a major concern on sites where past industrial use has included spillages of hydrocarbon fluids. One such case occurred on land previously used as the by-products manufacturing area of a steelworks, where tanks and vats of lighter fuel oils, phenols and ammonium salts had existed until the site was demolished in the 1970s.

Prior to its use for manufacturing, the area of land had been a convenient tip for excavated soils, combustion ashes, slags, scrap timber and other refuse from the adjacent steel mill. In consequence, land levels were raised by 4–5 m and a shallow perched groundwater table had developed at the base of these fills. Natural deeper strata are mined-out coal measures, mining having been carried out at shallow depths (< 20 m).

When the land came to be redeveloped an initial gas survey (of 10 deeper boreholes located on areas of deeper and shallower fills) was undertaken. This, however, revealed a very confusing body of information:

(i) Some boreholes showed only traces of carbon dioxide without any measurable gas pressures. These instances occurred where timber and other poorly degradable materials were most obvious in the fills.

(ii) Other boreholes revealed carbon dioxide at consistently higher than atmospheric pressures, which could have originated from coalmine workings or from subsurface geochemical effects (possibly either the decay of limestone bands or decomposition of heavy oils which had collected in the mine voids).

(iii) A few boreholes in the areas where lighter oil storage tanks had existed gave rise to measurable concentrations of volatile organic compounds (mainly benzene). These VOC contents could occur at high pressures, although when pressure readings were continued for periods of up to four hours, significant measurement variations were apparent.

This lack of consistency forced a more detailed review of the available site investigation information and the siting of a large number of small diameter probeholes in areas thought to be likely to typify the different conditions.

In a manner similar to that described for the paper mill site, it was possible to establish that two sources of carbon dioxide (without the presence of methane even in hotter soil conditions) did exist. The most widespread was the occurrence of low CO_2 concentrations (up to 4%) of unpressured gas, and these measured values varied in response to barometric conditions. No sizeable response to changes in soil temperatures was apparent. Wherever such conditions were encountered, shallow deposits of tipped timber, cloth and similar materials were obvious which could easily be excavated.

The different occurrences of pressurised carbon dioxide (at concentrations of up to 10% in worst cases) were found to exist only in small areas, where phenolic liquor tanks had stood. Around these, surface vegetation was absent (despite the land being unused for more than 20 years) and both soils and shallow groundwaters proved to contain phenol contents of more than 200 ppm. These anomalous unvegetated areas seemed not to correspond to the recorded locations of coal mining. Thus, it was concluded that microbial break-down of phenol-rich liquors was the probable cause of the measured gaseous contaminants. The depths to which phenol spillages had extended (up to 10 m), and the volumes involved if excavation were to be undertaken, were too great for an economic source-removal solution. Fortunately, the land areas were not significant, and it would be possible to omit these from planned redevelop-ments.

The volatile organic compounds, however, proved to present a greater risk and were found to occur whenever topographic variations brought the perched groundwater table nearer (to within 3 m of) the surface. Photon ionisation detector surveys revealed a wide variation in measured VOC concentrations, and these, when plotted on a site plan, outlined the probable flow paths of lighter hydrocarbon liquids. Trial pitting was able to confirm that, where the worst VOC contents were found, the local perched groundwater had a visible floating "oil" layer.

More detailed concentration and pressure measurements, conducted over several hours at a time and on subsequent days, revealed that VOC concentra-tions at a locality rapidly altered in amount (0.3 to 151.6 ppm in one quite typical instance). This suggested that the VOC emissions were especially sensitive to quite small variations in ambient barometric conditions and also indicated that even less acceptable vapour concentrations would occur in warmer soil condi-tions.

Laboratory gas chromatographic analyses revealed that these vapours consis-ted of a varying mixture of benzene, toluene, xylene and naphthalene, all poten-tially carcinogenic and/or toxic. While it was economically possible to deal with the two types of carbon dioxide contamination, full resolution of the VOC hazard would have called for pumping and chemical treatment of the shallow groundwater and would have been a long and excessively expensive solution.

Thus, the sole reasonably affordable solution was to advise that full gas-proofing measures (and the provision of vapour alarms in basements and smaller enclosed spaces) would be necessary. This precluded a safe domestic housing reuse for the land, given the difficulties in managing gas-proofing effectiveness in hundreds of individually owned properties, and forced development plans to be altered.

7.10   CASE STUDY 4 – MIDLANDS FOUNDRY SITE

Gas sources entirely unconnected with prior uses of a particular piece of land can occur and will often pose especially difficult interpretation problems. One such case (Cairney, 1995) was proved on an old iron foundry, where casting sands and other solid residues had been used to raise land levels by some 4 m. A few scraps of poorly biodegradable materials (mainly plywood and cloth) were visible in these fills, and this led to an initial spike-test gas survey being carried out. As this revealed elevated (<4.2%) carbon dioxide concentrations, and minute traces (<0.2%) of flammable gases, more permanent monitoring boreholes were installed.

Initially, four shallow (4 m deep) boreholes were sunk and revealed fairly widespread occurrences of carbon dioxide at concentrations of up to 3.5%. A later set of six rather deeper boreholes (8 m depths) was then drilled to check if gassing conditions in the lower sands and gravels (below the iron works fills) were acceptable. These showed a widespread and unexpected oxygen depletion (down to a near zero content) associated with high carbon dioxide levels (up to 19%). Given the concern which these contaminant concentrations generated, five deeper (15 m to 25 m deep) boreholes were sunk into the underlying rocks. Table 7.9 shows that these produced high (up to 27%) carbon dioxide results.

The pattern of deeper observation holes producing poorer subsurface atmospheres suggests that the more worrying carbon dioxide results originated at considerable depths below the site. More detailed monitoring (including measurements of gas pressures and flow rates) was undertaken over several months, and in various climatic and barometric conditions. Additionally, flux boxes were installed at depths of 100 mm below the levels planned for domestic house foundations. This work revealed that:

Table 7.9. Gas survey results for Midlands foundry site

Borehole	Week 1	Week 4	Week 8
1	18/2/0.1	16/3/0.1	17/2.5/0.12
2	19/1.5/–	18/1.5/0.08	16/3/0.16
3	14/3.5/0.1	12/4.2/0.18	10/2.4/0.2
4	16/2.5/0.1	13/3.5/0.2	12/4/0.2
5	–	3.5/10/–	2.1/12/–
6	–	2.4/12/–	1.4/15/–
7	–	2.8/13/–	1.0/18/–
8	–	1.2/16/–	0.6/20/–
9	–	1.1/17/–	0.6/19/–
10	–		0.2/23/–
11	–		<0.1/27/–
12	–		<0.1/25/–
13	–		0.15/23/–
14	–		<0.1/23/–
15	–		0.12/26/–

Notes: Results reported in the sequence: oxygen/carbon dioxide/methane.
Boreholes 1 to 4:   4 m deep
Boreholes 5 to 10:   8 m deep
Boreholes 11 to 15: >15 to 25 m deep

(i) In times of higher barometric pressures, sampling of shallower borehole atmospheres produced no significant carbon dioxide contents (all were less than 2.5%) and no measurable gas pressures.

(ii) In low atmospheric pressures, all the boreholes gave rise to depleted oxygen and enhanced carbon dioxide concentrations, and small positive gas flows were measurable. These flow rates, however, declined as soon as the barometric pressure began to rise, and positive gas pressures then vanished. The flux-box installations never showed any increase in carbon dioxide contents, even when abrupt barometric lows tracked over the area.

As the rocks beneath the site are the unmined shales of the Lower Carboniferous Series, coal workings could not be the gassing source. However, below these shales, and at depths in excess of 80 m, are massive limestone horizons, which had formerly been mined (1780 to 1830). Plans of the limestone mines showed not only that mining did take place at depths below the iron foundry site, but also that the mine entrances, in an adjacent valley, are still open. Thus, a possible gas source existed in continuity with the atmosphere, and so affected by barometric pressure changes.

Further investigation confirmed the hypothesis, and indicated that if no stress-relief holes were driven into the solid rock substratum, then only trivial carbon dioxide contents could be measured in the top 4 m of the site's profile. These carbon dioxide traces originated from poorly decomposed materials mixed in the site's fill capping layer.

As discussed earlier, investigation boreholes alter subsurface conditions, and provide stress-release zones into which gases can concentrate. With heavy gases, this effect is especially misleading, and the use of deeper boreholes on this site increased investigation costs and duration. Passive ventilation of house foundations was the solution to gassing conditions, as toxicity risks from deeper carbon dioxide occurrences were unlikely to affect near-surface soil layers.

7.11 CASE STUDY 5 – LEISURE DEVELOPMENT, LONDON DOCKLANDS

In less easily interpretable cases, it may be necessary to carry out gas monitoring for many months before gassing conditions become understandable. A planned leisure centre in the London Docklands proved to be a good example. This site is underlain by a thin concrete and made-ground cap over 8.5 m of alluvial clays with subordinate bands of peats and silts.

Subsurface gas investigations were necessarily comprehensive, given the size of the leisure centre and the commercial investment involved, and included numbers of large diameter observation boreholes supplemented by multi-point boreholes with piezometer response zones against the soil layers most likely to produce gaseous contaminants. In all some 36 boreholes were drilled over the 2 ha site.

Monitoring showed that, at the eastern end of the site, up to 50% concentrations of methane could occur (although without any enhanced carbon dioxide content) particularly in warmer soil conditions. In contrast, boreholes at the western end of the site generally showed only uncontaminated soil atmospheres.

Since gas readings had indicated that measurable methane concentrations occurred to depths of up to 5 to 6 m below site surface, initial views were that perhaps peaty bands or more organic clays could be the gas sources. Because of this, multi-point boreholes were added to the gas monitoring network. Several months of monitoring at weekly intervals took place to confirm this hypothesis, and concentration measurements were supplemented by gas flow-rate estimations. In 1989 when this work commenced direct measurement of smaller flow rates was difficult, so use was made of the indirect nitrogen purging technique. In this, monitoring borehole internal volumes are flushed out with pressurised nitrogen, and the recovery rates of gaseous contaminants with time are measured (Figure 7.7a). From the steepest gradient of the gas (methane) recovery rate, the inflow rate of gaseous contaminant can be estimated (Smith, 1993a), although it can be difficult to purge entirely borehole volumes and then recovery curves (Figure 7.7b) can be unhelpful. Additional difficulties can arise if only small, localised volumes of gases are contained in the strata around purged boreholes, and Smith (1993a) cites a situation where an entirely uncontaminated borehole atmosphere existed 33 minutes after the completion of nitrogen flushing, only to change to a 2% methane and 0.5% carbon dioxide atmosphere 25 minutes later, but then to revert to uncontaminated conditions a little later. In such circumstances, nitrogen purging flow estimates are meaningless.

After a considerable period of monitoring, no unique stratum in the soil profile could be identified with certainty as the main source of the methane problem. Attention was then directed to the geographic pattern of methane results and particularly to the fact that high methane concentrations occur only in the eastern portion of the site. Historical research in the local archives revealed that this area had been an open stream, in which untreated sewage effluent had been disposed until the 1890s. With the stream's outflow to the river Thames controlled by the flap valve, which closed in high tide conditions, it became clear that the former stream bed would have become silted up with coarser sewage, and that this might be the prime gassing source affecting the site. Further local enquiries showed that methane emissions from this (now infilled) stream were known to affect the inland area for distances of up to a kilometre from the river Thames junction.

With this information, the results of gas monitoring over the site's area became explicable. Exploration pits were dug, to identify if organic wastes which could be excavated could be found in the infilled stream bed, and these rapidly caused methane concentrations in adjacent gas monitoring boreholes to decline. From this, it was concluded that gas venting works could resolve the situation and two trenches were excavated and found to remove the methane nuisance. While it

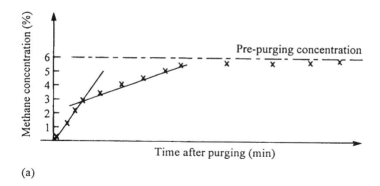

Figure 7.7a. Effective recovery of gas concentration after purging

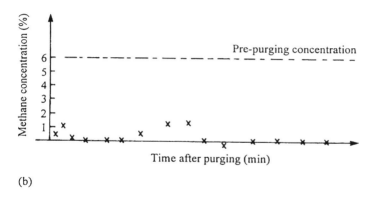

Figure 7.7b. Ineffective gas concentration recovery (non analysable)

took many months at considerable cost before experienced investigators understood the cause of the gassing problem, later discussions with the local authority revealed that the existence of a linear gassing zone along the infilled stream channel was well known to local residents. The basic error had been at the hazard identification stage in a failure to carry out adequate desk studies.

8

Qualitative Risk Assessment

8.1 INTRODUCTION

The principles of risk assessment in the context of managing the risks of contaminated land have been described in Chapter 2. Both qualitative and semi-quantitative (discussed in Chapter 9) risk assessment embrace the steps of hazard identification and hazard assessment as defined in Chapter 2. If carried no further (i.e. no need is found for site-specific estimates of risk), they must also include a risk evaluation stage in which the uncertainties underlying the assessment and the significance of the findings are evaluated. The assessment will be based on the results of preliminary, exploratory and other investigations (Chapters 3–7), the robustness of the latter, and other relevant information (e.g. plans for the site, attitudes of regulatory bodies, concerns of local residents etc.).

The term qualitative is used here to refer to an assessment which uses generic guidelines and/or standards to make judgements about the significance of the risk information held about a site. Although a great deal of quantitative information in the form primarily of site investigation data will underpin this assessment, only a descriptive or narrative statement about the risk to sensitive targets is made. Generic guidelines or standards indicate nationally, or sometimes regionally, acceptable levels of risk to certain targets assuming defined (often worst-case) exposure scenarios. In a qualitative risk assessment these guideline values are used directly in relation to the site and targets of concern. No site-specific estimate of the risk is calculated, i.e. no site-specific estimate of the potential and extent of exposure and of the resulting dose received. The assumptions inherent in the generic guidelines are used as a surrogate for the conditions relevant, or potentially relevant, to the site of concern. Fundamentally, qualitative risk assessment requires a detailed understanding of the basis of the generic values and the assumptions inherent within them if robust decisions are to be made. Qualitative risk assessments form the most numerous and common assessments made about contaminated sites, although the required understanding of the basis of generic guidelines is the most frequently observed deficiency.

Throughout the preceding chapters the iterative nature of risk assessment has been stressed. An objective of the preliminary investigation is to enable preliminary conclusions to be drawn about potential hazards, geological and hydrological (including hydrogeological) settings, current and future uses of the site itself and of neighbouring land, so that a conceptual model of the site can be derived in terms of potential, plausible, source–pathway–target scenarios. This information is used as a basis for the design of the exploratory investigation to test hypotheses, and the detailed investigations (including analytical strategy) to understand the contaminant and environmental conditions.

All of this information will need to be reappraised when the formal assessment of site investigation information is carried out, i.e. the process of hazard identification using all available information is the first step of the formal assessment, and the results of the preliminary investigation will be relevant to all subsequent stages of the assessment. At each stage, judgements must be made about the sufficiency of information in terms of type, quality and quantity. For example, the analytical strategy will have been based on the preliminary investigation. However, during the on-site works it is possible that evidence will be found that suggests that other contaminants should have been sought. If the investigation strategy has been sufficiently flexible for an adjustment to be made to the analytical strategy, then the risk assessment should not be comprised.

The design of the site investigation should have taken into account the assessment process that is to be followed. However, sometimes it is necessary to make an assessment based on someone else's data. A formal evaluation of the data for sufficiency in terms of quantity, quality, type and temporal span should always precede any assessment. For example, the quantity of data is important, since it dictates what confidence can be attached to the assessment of the results and is fundamental to the application of generic guidelines and standards. The data must be analytically reliable and the procedures used to obtain and handle samples should have avoided such pitfalls as cross-contamination. The data must be of the right type (e.g. do the generic guidelines require the use of "total", "leachable" or "plant available" concentrations?) and the methods specified for use with the guidelines used, or alternatives, technically justified.

The testing of data sufficiency also needs to extend into the geological and hydrological data, the desk study and, on a development site, the confidence that can be attached to the developer's plans for the site (changing commercial conditions or attitudes by planning authorities can result in a change of use of the land).

Any assessment will rely on the manipulation and presentation of the data so that, for example, locational patterns, statistical characteristics and relationships between contaminants can be determined and appreciated; not only by the assessor, but also by other interested parties. Chapter 3 draws attention to the need for those preparing reports to keep the potential wide readership in mind, and also the need to maintain a clear distinction between factual information and matters of opinion. Assessment reports must always state clearly the assumptions on which they are based and provide sufficient detail to enable any calculations etc. to be checked.

Qualitative risk assessment as discussed here focuses on soil and water contamination. Analogous processes are followed for other hazards, such as those
arising from bulk gases (see Chapter 7) and biological hazards. The assessment of
ecological risk is not discussed here specifically. General guidance has been
provided by Harris, Herbert & Smith (1995) and readers are referred to
specialised texts and guidance (e.g. Bartell, Gardner & O'Neill, 1992; Calabrese
& Baldwin, 1993; Suter, 1993; CCME, 1996a). For a recent view of the progress
being made in developing knowledge and guidance in this still emerging science
see Renner (1996). Little has been written concerning the assessment of biological
hazards arising on contaminated sites. However, an exception is a paper by
Turnball (1996) on potential risks from anthrax spores, which discusses sampling, interpretation of the results of examination and possible action to be taken
in regard to contamination with the agent of anthrax.

8.2   HAZARD IDENTIFICATION

Chapter 4 discusses preliminary investigation as part of hazard identification.
Stress is laid on achieving an appropriate balance between generic sources of
information which list contaminants known to be linked to particular industries
(e.g. Barry, 1985; Bridges, 1987; Department of the Environment 1995/96; Steeds,
Shepherd & Barry, 1996); available site-specific information (for example, on
activities known to have taken place), and information on background contamination in the area.

The objective of the preliminary investigation is to enable the assessor to
determine which of the identified properties of substances, operation or process
could lead to an adverse impact. If the preliminary assessment produces insufficient information to assess whether a specific hazard is present, the assumptions
for the purposes of the remaining components of the assessment should be that a
hazard is present (Department of the Environment, 1995b). This equates with the
"precautionary principle" which underpins UK government policy on sustainable development (Anon, 1994). The principle as adopted at the United Nations
Conference on Environment and Development in 1992 means that, where there
are potential threats of serious damage to the environment, a lack of scientific
certainty should not be used as a reason for not taking action.

The output from the hazard identification will be an understanding of the
historical use of the site, the potential hazardous substances which may be
present, and the targets existing and future which may be at risk. Table 8.1
summarises information resulting from a hazard identification of a large-scale
plating works attached to an engineering works. Whilst all buildings remain, all
plant except for the effluent treatment plant has been removed. The intention is
to develop the site for commercial purposes.

The hazard identification needs to be able to identify those source–pathway–target scenarios which are plausible and those which are likely to be critical.
Examples of plausible scenarios for the plating works (Table 8.1) in relation to

Table 8.1. Hazard identification of a plating works

Sources/ contaminants	Pathways	Targets current	Targets future
Cadmium	For humans:	Casual visitors to	Structures, floors and
Chromium	ingestion,	the site	walls
Copper	inhalation, direct	People working on	Site investigation team
Cyanide	contact	underground	Demolition and
Nickel	Runoff	services	refurbishment
Sulphate	Subsurface	Soil	workers
Low pHs	migration	Local stream	Remediation team
Mineral oils	Aerial deposition	Groundwater	Construction workers
Chlorinated		Surrounding	Occupants of buildings
solvents		farmland	Grounds staff
Hydraulic fluids			Maintenance staff
Cutting oils			Visitors
Lead and zinc from			Plants used for
paints and aerial			landscaping
deposition			

the stream as the target of any contaminants are (i) via runoff from the roof where there is dust accumulation, and (ii) from the surface of the site via infiltration through the ground and subsequently through the banks via the granular fill around surface drains. Once site investigation data are available it may be possible to eliminate some of the plausible scenarios if no evidence is found of suspected contaminants above "natural" background concentrations, or above what can be regarded as negligible concentrations. Table 8.2 summarises a hazard identification for an industrial site in existing light engineering use but with historical use as a metal finishing works.

Compilations of background concentrations are available for most elements (Smith, 1991 and Chapter 2), although these often refer to cultivated agricultural soils where they may have been input via fertilisers (cadmium, for example). In addition, background concentrations will vary depending on the bedrock in the area and the extent to which there has been large-scale transportation of soil-forming materials. "Naturally" high concentrations should not be considered as contamination (see discussion in Chapter 1). However, this does not mean that they will not cause "harm". Plant communities are frequently adapted to soil chemistry where there are high concentrations of otherwise phytotoxic elements.

Differentiation between background concentrations of, and contamination by, organic compounds tends to present greater difficulties than inorganics because, although they are generally derived from human activity, a number of the compounds of concern (e.g. polycyclic aromatic hydrocarbons – PAHs, some halogenated hydrocarbons) do arise "naturally" and have "always" been present in the environment to some degree. Some organic compounds have become

Table 8.2. Plausibility of source–pathway–targets scenarios for a site in use

Target	Examples of sources	Possible pathway	Plausibility	Justification
Human health – site users and occupiers*	Chemical contamination, e.g. fuel hydrocarbons, solvents, metals	Ingestion of contaminated dust/soil	×	No exposed soil
		Inhalation of contaminated dust/soil vapours	×/✓	No exposed soil but check subsurface vapours entering buildings
		Skin contact with contaminated dust/soil	×	No exposed soil
		Consumption of contaminated food	×	Not relevant to existing use
		Consumption of contaminated water	×/✓	Mains supply, but hydrocarbon fuels and other organic substances may be present as a result of local industrial use
	Methane, carbon dioxide	Migration into confined spaces	×	No known on- or off-site source, relatively impermeable strata
Water	Chemical contamination	Migration through soil and in water	×/✓	No local surface-water features or abstractions, but minor aquifer classification
Flora and fauna	Chemical contamination	Direct contact or via off-site migration	×	No landscape plants or locally important or protected species or ecosystems
Building materials and structures	Chemical contamination	Direct contact	✓	Possibility of effect on water supply pipes, concrete structures etc.
	Methane, carbon dioxide	Migration into confined spaces	×	No known on- or off-site sources

* Note that workers may need to be protected during subsurface maintenance works.

widely distributed in the environment. However, there is a general shortage of data on background concentrations of organic compounds (but see Jones *et al.,* 1996). The simple presence of many organic chemicals will have to be regarded as sufficient evidence to suspect contamination.

In practice, there may be no need to establish that contamination exists provided that guideline values are available which define the concentration of a substance which presents no additional risk. Examples of such criteria include:

- Dutch "target" values (MHSPE, 1994);
- USEPA screening values (USEPA, 1996a);
- Canadian interim criteria (CCME, 1991);
- Australian and New Zealand investigation levels (ANZECC, 1992).

The hazard identification represents the first (and compulsory) stage of the hazard assessment.

8.3 HAZARD ASSESSMENT FOR SOILS

8.3.1 The hazard assessment process

The hazard assessment followed by risk evaluation is intended to answer the question: do the plausible, and more particularly, the critical source–pathway–target scenarios matter?

Hazard assessment and the accompanying risk evaluation make use of:

- dedicated generic guidelines and standards, including those designed to identify "negligible risk" levels;
- non-dedicated guidelines and standards;

where:

- standards are mandatory in the legislative and regulatory framework in which they are set;
- guidelines are intended to be applied using professional judgement.

The UK ICRCL trigger values (ICRCL, 1987; being replaced in 1997 by Department of the Environment, 1997e) are examples of dedicated generic guidelines. Occupational exposure standards and water quality standards are examples of non-dedicated standards. As discussed in Chapter 1, many jurisdictions have now produced generic guidelines or standards relating to soil and/or groundwater contamination. Most were originally set using "professional judgement", but increasingly they are being set following application of a risk estimation model that takes into account a number of routes of exposure for specified targets (in the UK, human health). Some take both human health and ecological risks (ecorisks) into account; either a single consolidated set of numbers may be provided, as in the Netherlands (MHSPE, 1994), or separate numbers provided, as in Sweden (SEPA, 1997) and as is the intention in Canada.

As discussed in Chapter 2, where guidelines have been derived by a risk assessment, it could be argued that they allow for an estimation of risk to be derived for a site. However, this is only the case if the exposure scenarios (activity patterns, sensitivity of exposed individuals, age, gender etc.) relevant to the site and the potential adverse effects (death, cancer, chronic disease etc.) of concern are the same as those considered within the assessment which derived the guidelines. However, it must be stressed that it is not possible to make a direct statement about an estimated site risk to a defined target through the use of generic guidelines, and the latter may not provide fully for the source–pathway–target scenarios relevant to a site. This is why this text has preferred to draw a distinction between this approach and the quantitative site-specific estimate discussed in Chapter 10.

The guidelines developed by authorities have to be able to balance the importance of protecting humans and the environment from harm in a cost-effective way: for example, not requiring overly expensive remediation. The costs of risk reduction are rarely linear, as suggested by Figure 8.1. If soil clean-up guidelines are set at levels which make achievement prohibitively expensive, it is possible that remediation will not go ahead.

The principal stages of the assessment are:

(i) Evaluation of the hazard identification and site investigation data and other information available for sufficiency in terms of quantity, quality and appropriateness. The validity of a conclusion that a contaminant is not present or is only present at negligible concentrations must be carefully checked. For example, did the sampling strategy in terms of numbers, types and analytical methods correspond with those specified in the accompanying text to the screening values? Was the site investigation suitably targeted based on the site history and subsequent on-site observations? What probability is there that the site investigation will have missed a hot spot (at this stage this would be defined as any area outside of the range of natural background, e.g. mean plus 2 or 3 standard deviations if local data are available)? Does experience lead you as the assessor to "trust" the conclusion?

(ii) Manipulation of the data to obtain basic statistical parameters such as maxima and means, and possible use of graphical methods to look at the statistical distribution of data; the data for pre-determined zones can be compared and contrasted by this means and if necessary a new zoning derived.

(iii) Confirmation of dedicated and non-dedicated guidelines or standards for use in the assessment and review of their relevance to the task in hand: for example, what source–pathway–target scenarios were taken into account in their derivation (for a commentary on the ICRCL trigger values see Smith,

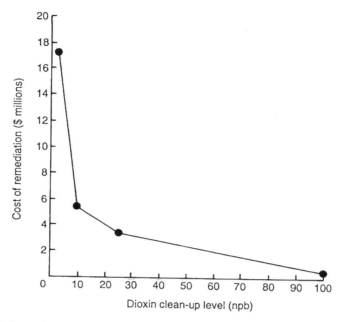

Figure 8.1. Estimated costs of soil removal and incineration for different target clean-up levels for TCDD at a site in Missouri, USA (Paustenbach *et al.*, 1986 quoted in Ferguson & Denner, 1994). Reproduced by permission of Academic Press Inc.

1993b), or has sample analysis been undertaken in accordance with the guideline requirements?

(iv) Screening of the data against "threshold" values which define an "acceptable" level of risk.

(v) Screening of data against guidelines or standards which mandate some form of action, which might be a presumption that remediation is required or that further assessment is required (the Dutch "intervention" values which represent a significant level of risk (i.e. 1×10^{-4} excess cancer risk) mandate a site-specific risk assessment).

(vi) Evaluation of the results of these screening processes, preparation of statements about the significance of the contamination, and recommendations concerning any need for urgent action to deal with immediate hazards, further (supplementary) investigation, or preliminary recommendations as to the nature of remediation required.

An example of a formalised assessment process which builds a staged approach into decision making is the Risk-Based Corrective Action (RBCA) standard introduced by the American Society for Testing and Materials (ASTM, 1995; Begley, 1996). RBCA applies a tiered approach to the specific problem of petroleum-contaminated sites. The first tier of the approach equates with the hazard assessment process described above. This "tier" includes a site investigation and identification of principal chemicals of concern, the extent of contamination and potential migration pathways and receptors. The site is classified according to immediacy of threat or risk and site conditions are compared with conservative risk-based screening levels (i.e. only relatively low levels of contamination are required to make the assessor move to the next stage). If the risk evaluation at this point suggests that remediation may be appropriate, the assessment moves to tier 2, which replaces use of generic screening levels with site-specific target levels. A tier 3 site-specific quantified risk assessment only becomes necessary for a site where active remediation is likely.

8.3.2   Screening against guidelines and standards

Risk assessment is concerned with deciding whether observed concentrations of contaminants actually matter. In qualitative and semi-quantitative assessments this is done using generic guidelines and standards. Certain concepts underlie these guidelines. For example, for soils the following situations can be identified:

(i)   Observed concentrations of a contaminant are below, or are consistent with, typical ranges to be found in uncontaminated soils: the soil is *uncontaminated*.

(ii) Observed concentrations are above background or "negligible risk" values but are at a level which do not increase the risks to a specified target by an unacceptable amount: the soil is *contaminated but presents no unacceptable additional risks* to the specified target (note: other targets may be at risk). This value is sometimes termed the *threshold* value.

(iii) Observed concentrations are above a level at which some increase in risk is anticipated and an assessment is needed to determine the extent of these risks and whether any action is required (such action may take the form of additional site investigation and some form of estimation and evaluation of risks): the soil is *contaminated and may present additional risks* to specified targets.

(iv) Observed concentrations are such that some form of action is essential either to overcome immediate problems or to avoid future problems arising from a change of use: the site is *contaminated and presents an unacceptable risk* to defined targets (e.g. to children playing on the site or to underlying groundwater or adjacent surface water because the soil contains soluble contaminants). The site is almost certainly *polluted* according to the definition given in Chapter 1.

Figure 8.2 describes the above concepts. Any such simple categorisation is complicated when naturally occurring background concentrations of a substance are already above a concentration at which harm is likely, or has already occurred, to specified targets.

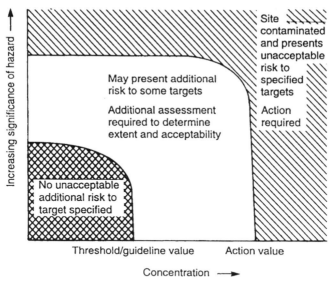

Figure 8.2. The threshold/action concept

The concept of the threshold or guideline value applies only to those contaminants for which some threshold can be identified below which it would be considered that the level at which contaminants are present would not normally require action, even though they may be higher than the background concentrations. However, there will be some substances where thresholds cannot be set because they present some risk at very small concentrations; for example asbestos. The objective should be to keep these risks as low as is reasonably possible.

The intermediate area between the threshold or guideline and the concentration at which a contaminant is considered to present a risk and action is required presents a significant decision-making challenge. It should indicate that there appears to be a potential risk but that further information is required, often additional site investigation and consideration of the critical source–pathway–target scenarios. It does not necessarily indicate that action is automatically required; merely that consideration should be given as to whether action is justified.

Although most of the guidelines produced in different countries have used an approach based on risk assessment (often qualitative rather than quantitative in the past) in their derivation so as to provide for the protection of targets of concern, these different guidelines differ markedly in purpose and hence in design. Since the risk characteristics of a contaminated medium vary according to its use, different assumptions and uncertainty factors can be used to develop a range of generic values for different uses. The concept of variable end-use has been applied to the development of dedicated generic guidelines in a number of cases: UK trigger concentrations were defined for different uses of the site (ICRCL, 1987); national Canadian criteria are available to cater for the different uses of soils and groundwaters associated with the site (CCME, 1991; 1995; 1996b). In Ontario different criteria are available for soils located at different depths (> 1.5 m or < 1.5 m) (OMEE, 1994; 1996). In the Netherlands only one set of values is available, because the policy context of multifunctionality states that soils and groundwater should be assessed according to their ability to perform any function (MHSPE, 1994).

The following are examples of how different jurisdictions have provided different solutions to the screening problem (it is relevant to note at this point that none of the systems deals with combinations of contaminants or possible synergistic effects):

(i) Region III of the US EPA has produced "screening values" representing negligible risks (e.g. 1×10^{-6} increased individual lifetime cancer risk) based on application of a standardised USEPA-type risk estimation model to defined sensitive targets (USEPA, 1989a; 1989b). If the screening values are not exceeded the risks are negligible. No higher set of values is provided. The USEPA has more recently published a list for wider application (USEPA, 1996).

(ii) In the Netherlands "target" and "intervention" values have been set to support specific legislative requirements for the identification, assessment and, if necessary, remediation of sites which if they meet certain criteria are "registered". Target values represent negligible human and ecological risks. They are targets to be achieved as a result of remediation; although practice indicates that cost–benefit decisions also come into play in consideration of the remediation of specific sites to the target values. Intervention values have been set to represent a significant level of risk to human health or to the soil ecology. In the case of human cancer risks the intervention value has been set at an excess cancer risk of 1×10^{-4} (Van den Berg, Denneman & Roels, 1993). However, it is not relevant to say that these are 100 times less safe than the Region III USEPA guidelines. Different assumptions in the exposure assessment for both sets of guidelines will have contributed different safety factors (Ferguson, 1996). If the intervention values are exceeded a site-specific risk assessment is required in order to judge the urgency of remediation. There is no equivalent of the ICRCL threshold value representing an "acceptable" level of risk, nor any help with what to do about the many situations where concentrations are above the target value, but below the intervention value.

(iii) The interim CCME criteria (CCME, 1991) used low (i.e. stringent) screening values to identify the presence of contamination and land-use-based remediation values to serve as targets for remediation; the values being a starting point from which site-specific values may be derived if required. The assessment criteria related to approximate background concentrations or approximate analytical detection limits for contaminants in soil or water. The remediation criteria related to three land uses: agricultural, residential/parkland, and commercial/industrial. Those for water were based on the existing water quality guidelines and for drinking-water quality (HWC, 1989). A new protocol for derivation of risk-based soil quality criteria (CCME, 1995; 1996b) describes the derivation of criteria for both ecological and human receptors and the remediation criteria represent levels to "clean down to". The guiding principles are that the criteria will not result in an appreciable risk to humans interacting with a remediated site; will be based on defined, representative exposure scenarios; and will be derived from a consideration of exposure through all relevant pathways. In relation to non-threshold contaminants (i.e. mutagens and carcinogens – see Chapter 10 for further explanation) the remediation level should be at a minimum within the range 10^{-4} to 10^{-6} increased risk of cancer from soil. The protocol will be backed by a series of criteria documents summarising available human and ecological health information (e.g. Environment Canada, 1995).

Experience in the UK is of a tendency among site assessors to look for any guidelines which include the contaminant of concern, not least where the national trigger values have been deficient. However, this tendency to favour use of other guidelines has often been despite a failure to understand their basis.

For example, the Dutch criteria have been used in risk assessments for UK contaminated sites. However, the Dutch criteria are based on ecotoxicological risks as well as human risks. This results in guidelines for some contaminants which are more stringent than if they were only protecting human health. Careful attention would have to be paid to the basis of the specific guidelines and to the UK policy context in which they were to be used. For example, it could be costly to apply a Dutch intervention value based on ecological risks to a site covered by concrete and where there is no potential for leaching to an adjacent water course. The Dutch criteria would need to be adjusted if used in the UK to take account of differing soil characteristics (organic matter content, clay content etc.). Furthermore, in the UK policy context of "suitable for use" criteria which do not relate to land uses may be inappropriate. Fundamental problems can arise if assessors use criteria set in other jurisdictions without understanding that they have been updated. This problem has arisen in the UK where assessors have continued to refer to the old Dutch "ABC" values, unaware that the new intervention values are now in use in the Netherlands.

Many jurisdictions have published associated guidance with their guidelines in relation to sampling strategies and analytical methods which it is assumed will be followed (e.g. CCME, 1994). The Dutch intervention values are considered to be exceeded when "the mean concentration of a soil volume of at least $7 \times 7 \times 0.5$ m^3 is compared with the value . . ." (Van den Berg, Denneman & Roels, 1993), although no specific information appears to be given on appropriate sample sizes for determining the mean concentration. The guidance appears

Table 8.3. ICRCL trigger values (ICRCL, 1987)

Contaminants	Use code	Reference value trigger concentrations (mg/kg air-dried soil)	
		Threshold	Action
Group A: Selected inorganic contaminants that may pose hazards to health			
Arsenic	1	10	**
	2	40	**
Cadmium	1	3	**
	2	15	**
Chromium total	1	600	**
	2	1000	**
Chromium (hexavalent)[1]	1, 2	25	**
Lead	1	500	**
	2	2000	**
Mercury	1	1	**
	2	20	**
Selenium	1	3	**
	2	6	**
Group B: Contaminants that are phytotoxic, but not normally hazards to health			
Boron (water soluble)[2]	4	3	**
Copper[3,4]	4	130	**
Nickel[3,4]	4	70	**
Zinc[3,4]	4	300	**
Contaminants associated with former coal carbonisation sites			
Polyaromatic hydrocarbons[5,6,7]	1	50	500
	3, 5, 6	1000	10 000
Phenols[5]	1	5	200
	3, 5, 6	5	1 000
Free cyanide[5]	1, 3	25	500
	5, 6	25	500
Complex cyanides[5]	1	250	1 000
	3	250	5 000
	5, 6	250	NL
Thiocyanate[5,7]	All	50	NL
Sulphate[5]	1, 3	2000	10 000
	5	2000[8]	50 000[8]
Sulphide	All	250	1 000
Sulphur	All	500	20 000
Acidity (pH less than)	1, 3	pH 5	pH 3
	5, 6	NL	NL

Use codes
1 Domestic gardens and allotments.
2 Parks, playing fields, open space.
3 Landscaped areas.
4 Any use where plants are grown (applies to contaminants that are phytotoxic, but not normally hazards to health).
5 Buildings.
6 Hard cover.

Conditions
1 Tables are invalid if reproduced without the conditions and footnotes.
2 All values are for concentrations determined on "spot" samples based on adequate site investigation carried out prior to development. They do not apply to analysis of averaged, bulked or composited samples, nor to sites that have already been developed.
3 Many of these values are preliminary and will require regular updating. For contaminants associated with former coal carbonisation sites, the values should not be applied without reference to the current edition of the report, *Problems Arising from the Redevelopment of Gas Works and Similar Sites*.
4 If all sample values are below the threshold concentrations then the site may be regarded as uncontaminated as far as the hazard from these contaminants is concerned, and development may proceed. Above these concentrations, remedial action may be needed, especially if the contamination is still continuing. Above the action concentration, remedial action will be required or the form of development changed.

Footnotes
** Not specified.
NL No limit set as the contamination does not pose a particular hazard for this use.
[1] Soluble hexavalent chromium extracted by 0.1 M HCl at 37 °C; solution adjusted to pH 1.0 if alkaline substances present.
[2] Determined by standard ADAS method (soluble in hot water).
[3] Total concentrations (extraction by $HNO_3/HClO_4$).
[4] Total phytotoxic effects of copper, nickel and zinc may be additive. The trigger values given here are those applicable to the worst-case phytotoxic effects that may occur at these concentrations in acid, sandy soils. In neutral or alkaline soils phytotoxic effects are unlikely at these concentrations. The soils pH value is assessed to be about 6.5 and should be maintained at this value. If the pH falls, the toxic effects and uptake of these elements will be increased.
  Grass is more resistant to phytotoxic effects than most other plants and its growth may be adversely affected at these concentrations.
[5] Many of these values are preliminary and will require regular updating. They should not be applied without reference to the current edition of the report, *Problems Arising from the Redevelopment of Gas Works and Similar Sites*.
[6] Used here as a marker for coal tar, for analytical reasons. See *Problems Arising from the Redevelopment of Gas Works and Similar Sites*, Annex A1.
[7] See *Problems Arising from the Redevelopment of Gas Works and Similar Sites* for details of analytical methods.
[8] See also BRE Digest 250: Concrete in sulphate-bearing soils and groundwater.

to relate to near-surface exposure only. Deeper contamination is presumably considered to present less risk. Such guidance would need to be followed in detail if the guidelines are used, or any deviations from the recommended sampling strategies etc. would need to be explained in the assessment report.

One further example of the misuse of guidelines arises in the UK. In the 1970s a set of values indicating different degrees of soil contamination across the Greater London area were derived: the Kelly values. Soils were classified as "uncontaminated", "contaminated", "heavily contaminated", etc. (Kelly, 1980). This was a purely arbitrary classification based on the observed variations in soil quality across the London area. The classification only reflected the frequency of occurrence of contaminants and their distribution, and gave no indication of "risk": for example, concentrations for lead which were classified as uncontaminated were in many instances considerably above natural background concentrations. The Kelly values, or GLC guidelines, were produced to assist the council in making judgements about a large number of redevelopment sites in the capital and to identify potential and relevant waste disposal sites for the soils removed. The GLC was itself developing "trigger values" for use in site assessment (GLC, 1976). The accessibility and apparent ease of use of the Kelly values have led to their misuse as a basis for assessing contaminated site risk. Indeed, such is the extent to which these values have become part of UK practice that even the Health and Safety Executive incorrectly reproduced them in guidance on occupational hazards and contaminated land (HSE, 1991).

8.3.3 UK risk-based values

New risk-based guideline values designed to replace the ICRCL trigger values (Table 8.3) have been developed in the UK following criticism (e.g. House of Commons Select Committee on the Environment, 1990) of the restricted nature of the ICRCL values and the lack of transparency as to their basis. Indeed, this has been an international criticism of generic guidelines. The lack of transparency, combined with the danger that they convey the impression of completeness of knowledge about the underlying processes, has led to a search for more robust means of deriving guidelines (Sheppard *et al.*, 1992).

The new guideline values are derived using the Contaminated Land Exposure Assessment (CLEA) model (Ferguson & Denner, 1993b; 1994; 1995; Ferguson, 1996). CLEA does not cover water pollution or ecotoxicity, although it is expected that further development of guidelines relating to the latter will be forthcoming.

CLEA considers human exposure risks via 10 pathways:

  (i) Outdoor ingestion of soil.
 (ii) Indoor ingestion of dust.
(iii) Consumption of home-grown vegetables.
 (iv) Ingestion of soil attached to vegetables.

   (v) Skin contact with outdoor soil.
  (vi) Skin contact with indoor dust.
 (vii) Outdoor inhalation of fugitive dust.
(viii) Indoor inhalation of dust.
  (ix) Outdoor inhalation of soil vapour.
   (x) Indoor inhalation of soil vapour.

CLEA derives a realistic rather than maximum exposure level (as in the MEI – discussed in Chapters 2 and 10) among an exposed population appropriate to the different end-uses of a site (residential with gardens; residential without gardens; allotments; parks, playing fields etc.; commercial/industrial). Background exposure can be taken into account, as can the effect of time on the degradation of contaminants and the different age intervals of the exposed population from 0 years through childhood and working life to 60–70 years. Organic matter content and soil pH are taken into account where appropriate.

CLEA appears to be the only model developed for deriving national regulatory guidelines that is based on Monte Carlo simulation (Ferguson, 1996). The essence of Monte Carlo modelling (discussed further in Chapter 10) is that single-point values of key variables are replaced by probability density functions reflecting the uncertainties or variabilities associated with exposure parameters. CLEA produces a distribution of values for a range, and likelihood, of possible exposures rather than deterministic single values. For example, a range of ages and bodyweights, exposure frequencies etc. would be accounted for. The key to using guidelines derived by such methods is to ensure that the characteristics of the actual exposed population are reflected in the ranges incorporated in the guideline probability functions. For example, if the guidelines are designed to protect the general population with a natural range of ages and typical exposure patterns, but the exposed targets on a particular site are all children under the age of 5, and the primary exposure scenarios would present most acute risk in this age range, the assessor would need to know whether the guideline values would be sufficiently protective.

The development of the guidelines has emphasised the need for action levels for human health protection. The ICRCL trigger levels had primarily included only threshold values not action levels, resulting in uncertainty about the level at which a regulatory body might require investigation and action. When the action values are exceeded, either a further phase of site investigation should be carried out *or* remediation. If the former is chosen the emphasis is on establishing whether exposure conditions differ significantly from the general conditions used in CLEA. This is expected to be more cost-effective in general than site-specific risk assessment procedures. It does, however, require that the exposure assumptions and data values used in deriving action values are understood by the risk assessor so that the necessary comparisons with site-specific values can be made. Action values may differ depending on site use, pH, soil organic matter etc. reflecting the influence of such parameters on contaminant partitioning between soil solids and soil solution. The values are being published with documentation explaining their basis (Department of the Environment, 1997e).

Adoption of CLEA into national guidelines for regulatory purposes has required a number of key decisions, such as:

- What percentage of the population should be protected (e.g. 90%, 95% etc.)?
- What proportion of a tolerable daily intake should come from a contaminated site given that compared to other potential exposures (e.g. in the workplace, through the diet, from polluted urban air etc.)?
- What is an "acceptable" or "negligible" additional cancer risk from a contaminated site given that there will be background risks from other sources?

To answer these questions requires high-level policy decisions that should be equitable across media and fair to all parties. At the time of this book going to press understanding of the new guidelines is limited awaiting their publication.

8.4   HAZARD ASSESSMENT FOR WATER

Chapter 6 discusses site investigation for the assessment of risks to water. Consideration of the risks to groundwater has three aspects:

- the concentrations of contaminants at a point in the ground which may be on- or off-site;
- the potential for migration vertically or horizontally from one location to another;
- the potential for contaminants associated with soil (or other solids) to leach.

Similarly, assessment in respect of surface waters is concerned with the quality of water *in situ* in a body of water such as a stream, pond, lake, canal or dock; and with the potential for contaminants to be translocated, either already dissolved or in the form of contaminated solids, to such a surface-water body.

As for soils, concentrations in groundwater may be compared with generic guidance values published for the purpose. For example, values taking into account both human health and ecological risks are available in the Netherlands (MHSPE, 1994): both "intervention" and "target" values are available. While if the intervention value is exceeded, there is a presumption that some form of action is required, the urgency of action is to be decided on the basis of site-specific factors. Thus, if there is no extraction of water from the location concerned, or likelihood of the contaminated water migrating to a sensitive surface-water body (without attenuation), the urgency for action could be judged to be low.

Whereas a single set of values has been produced in the Netherlands, in Canada a series of alternative interim criteria has been published (CCME, 1991) for different water uses (e.g. drinking-water abstraction, agricultural irrigation). These have been largely adopted from already established water quality criteria.

In the UK, the Department of the Environment (1993) suggested "default" criteria to be adopted when making judgements about the closure of landfill sites. These were largely based on drinking-water standards multiplied by a factor of 10 on the assumption that some dilution/attenuation will always occur between the site and any abstraction point or receiving water body. These criteria could provide a starting point for the assessment of other types of site presenting a risk to the water environment.

In the absence of guidelines intended for the purpose, comparisons can be made with those intended for other purposes (e.g. for drinking-water abstraction, fisheries, recreational use), with or without a suitable factor to allow for attenuation. In England and Wales, the Environment Agency (National Rivers Authority) has sometimes set control limits at what it judged would be suitable as a "discharge consent" for the at-risk receiving surface-water body, taking into account any aspirations concerning improvement of the quality of the receiving waters in future years. It has then been for the "problem owner" to show via modelling etc. that the agency should be willing to accept higher concentrations in the site under investigation because attenuation was certain to occur.

A qualitative approach to assessment of the potential, for example, of contamination at the surface to penetrate to underlying strata and then to migrate laterally may be possible based on published geological and hydrogeological information and the results of limited site investigation. For example, a report for the Department of the Environment (1994b) proposed a framework for the assessment of potential impacts of contamination on ground and surface waters, the first stage of which is a desk study and site visit (a preliminary investigation) – if contamination has already occurred then full quantification (Stage/Step 2) will always be required.

The National Rivers Authority (NRA) (which became part of the Environment Agency in 1996) has provided a framework for the designation of "groundwater protection zones" (NRA, 1995a) and commissioned a series of groundwater vulnerability maps for England and Wales (NRA, 1995b). These are based on a form of qualitative risk assessment which might be adaptable for application on specific sites should reference to the published maps and accompanying information be insufficient.

The groundwater vulnerability maps are an aid to developers planning *new* activities, and to planners assessing *new* proposals or drawing up strategic planning documents such as statutory local plans. They provide a "first pass" screening tool; site-specific studies will always be required for detailed proposals. The maps are also intended to be of use in the consideration of existing activities which may give rise to (presently undetected) pollution. They allow owners of multiple sites to prioritise their investigative and subsequent remedial actions where historical practice may have given rise to land and groundwater contamination.

The groundwater vulnerability maps show the degree of risk of a pollutant, which originates at the surface, moving down into an underlying aquifer by assessing the:

- soil type – classified by soil leaching potential;
- likely presence/absence of low-permeability surface drift deposits;
- permeability of aquifer material.

The soil classification is based on the physical and chemical properties of undisturbed soils. It can be applied to all soils, but only those soils overlying "major" or "minor" aquifers have been classified on the maps. Wherever geological investigations identify low-permeability surface drifts overlying major or minor aquifers, their presence is indicated by a stipple ornament on the maps. A stipple is used to highlight the inherent variability of drift deposits, both laterally and vertically, to emphasis *the need for detailed site-specific information to assess individual situations.*

The primary use of groundwater protection zones (NRA, 1995a) is to signal that within specified areas there *are likely to be* particular risks associated with the quality and possibly the quantity of water abstracted from the related pumping stations, should certain activities take place on or within the land surface. They are, first and foremost, a screening tool to be used with caution when assessing specific activities.

The delineation of groundwater *source* protection zones adopts a similar concept, but takes additional factors into account in order to delineate three protection zones:

(i) *Inner Zone I* is defined by a 50-day travel time from any point below the water table to the source. It is located immediately adjacent to the well and the selection of the 50-day isochron is based principally on accepted biological (mainly bacteriological) decay criteria. It is designed to protect against the products of human activity which might have an immediate effect on the source.

(ii) *Outer Zone II* is defined by the 400-day travel time or 25% of the source catchment area, based on the protected yield of the source, whichever is larger. The travel time is derived from consideration of the minimum time required to provide delay, dilution and attenuation of slowly degrading pollutants. The zone is generally not delineated for confined aquifers.

(iii) *Source Catchment Zone III* is defined as the area needed to support the protected yield from long-term groundwater recharge (effective rainfall).

The NRA framework allows for zoning using available geological and hydrogeological information, to be followed by progressively more source-specific assessments using increasingly sophisticated modelling techniques.

The framework report for the Department of the Environment (1994c) stressed the importance of all stages of the source–pathway–target scenario in an assessment. It identified key inorganic and organic contaminants that might be of concern, relevant pathways and receptors, and reviewed the value of existing models, information systems and information sources, in particular the ground vulnerability maps referred to above. In respect of the latter it was concluded, "they provide little information on minor aquifers and the relationship of groundwaters to surface waters except by inference. They therefore provide a useful information source for an assessment but deal only with one of many aspects which define impact on a site-specific basis."

The specific questions that it is suggested should be addressed in respect of potential surface-water impacts once it is suspected or established that contamination can be mobilised (by assessment of basic physical and chemical information such as physical state and solubility) are:

● Does the site have a discharge?
● Does the precipitation infiltrate or run off?
● Does the site flood?

These can be compared with the more detailed checklist proposed by Cairney (1995) – discussed in Chapter 9.

The comparable questions in respect of groundwater are:

● Does the site have a direct discharge to, for example, a soakaway?
● Does infiltration from the site penetrate the soil horizon and the unsaturated zone to the saturated zone?
● Is groundwater protected by a thick, low-permeability soil horizon, drift horizon or soil formation?

For the Step 2 assessments, a series of guidance sheets are provided to aid in checking that all factors have been considered and possible interaction identified. These cover site activities and history; nature of contaminant; rainfall/infiltration/runoff; land use; leachability; attenuation processes; soils; drift geology; solid geology and hydrogeology; surface water and groundwater. The framework document provides a number of 'assessment plans' which match investigation data requirements to the source–pathway–receiving water body chain for a range of scenarios relating different contaminant characteristics (soluble, immiscible, non-persistent, volatile etc.) to the nature of the receiving waters (groundwater, surface watercourses, open-surface water bodies). It then discusses methods of assessing potential impacts, including simple calculations and the use of more complex models.

Litz & Blume (1992) describe a system for predicting the vulnerability of soils to organic chemicals. Typical soil characterisation information, such as organic matter and cation exchange capacity; physico-chemical properties of individual chemicals; climatic information (e.g. mean temperature, rainfall); and geological information (e.g. depth to groundwater), are used to predict, for example, volatility from soil, degradation under aerobic or anaerobic conditions, and the potential to pollute groundwater. A similar system is likely to be included as an informative annex to a proposed International Standard on the characterisation of soils in relation to groundwater protection (ISO, 1996b).

The potential for already contaminated soils to release contaminants is difficult to assess. A theoretical approach of the type described above may be possible. Measured total concentrations could be combined with soil characteristic data to indicate a potential for release. For example, factors likely to limit the release of metals such as lead, copper, zinc and cadmium would be a high pH (i.e. high alkalinity), high organic matter content, and high clay content (an indirect measure of the cation exchange capacity). A high pH might, however, indicate a high potential for mobility of arsenic (at very high pHs zinc could also be mobile). It is important to remember in all such assessments that organic matter content and pH are not fixed properties of soil: acid deposition and poor husbandry can lead to acidification of soils and loss of organic matter.

Direct measurement of the water-soluble fraction may be more relevant and standard (or near-standard) tests are available (e.g. the NRA test (WRC/NRA, 1993)) using either water as an extractant or a weak acid approach. Once such results are available a judgement must, of course, be made as to what they mean. As discussed above in relation to groundwater, comparison is possible against a range of dedicated and non-dedicated criteria. The concept of leachability "trigger levels" for certain water catchments with known hydrological characteristics has started to be considered in the UK, although at the time of writing only work in relation to the Upper Tame catchment of the Severn Trent Region in the Midlands is known. The values are intended to indicate the concentrations of contaminants that would be acceptable in a direct discharge to surface water.

8.5 RISK EVALUATION

The evaluation of the results of the assessment depends heavily on professional judgement and experience. Factors to be taken into account following screening against the relevant guideline values (or other appropriate values), and deciding whether exceeding these matters, rely extensively on experience and professional judgement. Such factors will include:

- the relevance of the source–pathway–target scenario underlying the value to the site-specific situation;
- the amount of data available;
- the number of values exceeding the guidelines, their location and how far they are above the thresholds, including whether action trigger concentrations are exceeded;
- the number of contaminants for which there are exceedances;
- the chemical and physical form of the contaminants;

- experience on when "false positives" may occur from the analytical method used;
- information on the likely "safety margin" offered by particular values.

The conclusions from this evaluation may include the following:

- The database is insufficient for firm judgements to be made: for example, although no concentrations exceed the relevant guideline values, there is so little data that the investigation was a waste of time and money.
- The database is sufficient (in practice there will often be some limitations that will need to be mentioned).
- Although guideline values are exceeded for a number of samples, it is judged that this does not matter.
- Action values are exceeded – remediation is required.
- Concentrations are so high that immediate action should be taken to prevent access to the site.
- Further investigations should be carried out.
- A site-specific quantified risk assessment should be carried out (see Chapter 10).

8.5.1   Case examples of deficiencies in qualitative risk assessments

The following four brief cases provide some examples of common deficiencies in assessments.

Development site

A 0.3 ha site, previously a bus garage, was to be developed for 12 blocks of 3–4-storey flats around a central car park. Small (9 m^2) partly paved gardens were to be included for amenity use only, and it was not intended that residents would be able to grow vegetables etc. for consumption. Only narrow (6 m wide) grassed strips would exist on the borders of the land.

The site investigation had been restricted to six boreholes (to 6 m depths) with five samples. Thus, the investigation could not be regarded as more than a limited exploratory survey, whose findings would need to be supplemented. This limited survey provided information to show the following:

 (i) The site was underlain by a shallow (1.1–1.9 m) cover of made-ground. Below this were alluvial clays over thick boulder clays.
 (ii) The made-ground was predominantly composed of sandstone debris and clays, although ashy soils existed in layers 0.4–1.0 m in thickness.
(iii) The made-ground – from past OS map data – appeared to have been imported to raise the surface level of what was low and wet land.

(iv) The limited chemical information (one sample) suggested that the alluvial strata were clean; that non-ashy made-ground (one sample) was clean other than a slight excess in the mercury content (at 1.1 mg/kg); and that the ashy sands were surprisingly uncontaminated (three samples). Two of these ashy samples contained copper concentrations of 312 mg/kg and slightly enhanced sulphate levels. All of the samples were distinctly alkaline.

On the basis of this information, which excluded any gas surveys or tests on the combustion susceptibility of ashier bands, it was concluded that the sole risks to residents were those from contaminant inhalation/ingestion (from fine granular ashes) and that these could adequately be reduced by the installation of a 500 mm thick clean clay cap in all areas not having a hard cover.

A dense housing development would be especially at risk from gas emissions or localised land subsidence (should partly burned ashes be heated sufficiently by buried power cables to be ignited). Likewise, if soluble contaminants (the existence of which had not been considered in the sampling) were present, polluting seepage could affect an adjacent surface stream. It was clear that the site investigation was inadequate and had not been appropriately focused on the potential source–pathway–target scenarios. Conclusions about the site's safe development should not have been drawn. However, neither the relevant planning authority nor the regulatory agency made any objections and was willing to accept the assessment that had been offered. This case illustrates the limited scope of many qualitative risk assessments, but also unfortunately the uncritical response which can be forthcoming when officials (as well as investigators) without adequate training are involved.

Former gasworks

A former gasworks adjacent (300 m distance) to an estuary where tourism was a major activity was to be sold for retail and warehousing use. The site owners had agreed to reclaim the site (simply by excavating hot spots which lay within the top metre of fills) to produce a "safe site". Since such a reclamation would leave *in situ* deeper contamination, the prospective purchaser requested an assessment which would demonstrate that no future liabilities would arise.

The site's location and predictable contamination condition indicated that off-site contaminant migration and water pollution were the critical risks to address in detail. However, the risk assessment failed to identify:

 (i) the groundwater flow direction;
 (ii) groundwater quality;
(iii) whether deeper strata contained soluble/mobile contaminants; and
(iv) whether the adjacent estuary quality had been affected by pollutant loading.

However, the risk assessment concluded that no future liabilities would arise.

This qualitative assessment was merely a personal statement of opinion unsupported by any appropriate evidence.

Development of munitions waste site

The site was an area, formerly waste land, adjacent to a Second World War armaments factory. Wastes had been tipped on the site and combustible materials had been burnt over a period of about 10 years. As a result of these activities a surface (75 cm to 1.5 m deep) of rubble, ashes, metal scrap etc. existed, and within it residues of propellants (cordite and nitroglycerine) were apparent. These munition wastes presented a significant toxic risk. Below the made-ground capping were alluvial gravels with a subordinate sand content. The local groundwater was at about 2.5–3 m depths.

The proposal was to develop the site for housing. The planning permission which had been granted for this development contained conditions which specified that "the removal of toxic metals and other deleterious substances in the *surface layer* [emphasis added] is required. Soils have to contain chemical concentrations lower than the ICRCL threshold trigger values". It should be noted that the potential for contamination at depth was known and the developer had taken the view that the planning condition in effect agreed with his view that this would be dealt with by remediation of the surface layer.

All of the made-ground was excavated, screened into various size factions, and chemical proof testing demonstrated that all of the obvious toxic metals, oils and tars were removed. Clean fractions of the made-ground were crushed and then reused as fill. The upper 50 mm of the underlying alluvial beds was also removed, as oils and silt heavily contaminated with metals had collected there.

Proof testing of the exposed surface of the alluvial gravels demonstrated that at depths below the original land surface of 1–1.6 m, arsenic, copper and chromium (hexavalent) concentrations exceeded the threshold trigger levels (ICRCL, 1987), although by less than an order of magnitude. Further investigation showed that these enhanced soil concentrations extended down to the groundwater. The response of the local authority to these data was that the alluvial beds should be completely removed.

The developer rejected the local authority's demand, with the following reasoning:

(i) The ICRCL threshold values referred to surface soils, not to deeper soils;
(ii) The chemical results were obtained on the fine fraction between individual pebbles in the gravel matrix. As pebbles accounted for about 80% of the alluvial beds, a true chemical analysis (i.e. if the quartz pebbles had been crushed and milled before analysis) would have resulted in concentrations below the threshold values.

The local authority accepted point (ii), but argued that it was the fine fraction between pebbles which could give rise to inhalation and ingestion risks should residents dig in their gardens to depths below 1 m. After considerable debate the developer's view prevailed, largely because it was demonstrated that clean replacement soils (to restore land levels) would make deep excavation by the residents difficult. The fine material between the alluvial pebbles constituted a small volume and so limited quantities would be available via ingestion or inhalation.

Whether the reader agrees with the developer's arguments, this case illustrates the commonplace problem of applying guidelines inappropriately (in this case irrespective of depth of contaminants in the soil profile) and with little attention to the plausible source–pathway–target scenarios.

Development of a landfill

A 4 ha area of land, previously a local authority landfill (1930s–1960s) and later a sportsfield with club facilities, had been acquired for development as a retail park. Site investigation had involved boreholes at 25 m centres and had shown the following:

- "Ashy" deposits occurred below a 0.5 m thick clay cap.
- These ashes varied from 0.5 m to >7 m deep.
- Natural clays, gravels and bedrock underlay the ashes. These materials were described in the investigation reports as "clean" and "ashy".
- The ashes had relatively high arsenic, copper and zinc concentrations (although only 13 samples from the 160 collected were regarded as representative). Some concentrations of cadmium, nickel, lead and mercury were enhanced.

Consultants acting for the landowner specified the following contaminant values above which soils would not be "suitable" and would have to be removed:

	mg/kg air dried soil
Arsenic	40
Cadmium	15
Chromium	1000
Lead	2000
Mercury	20
Boron	3
Copper	200
Nickel	70
Zinc	500
Cyanide	25
Phenols	1000
Sulphates	5%
Toluene extract	10 000
pH	>3

These values were mainly the ICRCL (1987) threshold values for non-domestic land use.

The selected reclamation contractor had the site investigation data checked. Aware that the limited number of analyses which had been undertaken would make it difficult to estimate the volumes of "unacceptable" soils which would have to be removed, he conducted a limited investigation. This suggested that about 100 000 m³ of soils would have to be removed for disposal at a cost of about £1.5 million. The landowner's consultants had estimated that only about 15 000–20 000 m³ of soils would need to be removed, with attendant costs of no more than £200 000.

A risk assessment approach was taken to consideration of the soils which would have to be removed and to establish which could remain. The assessment considered:

- depths at which "unacceptable" materials occurred (i.e. below buried services and plant rooting depths would be regarded as "acceptable");
- the risks which individual contaminants would present;
- whether contaminants were soluble/mobile;
- whether soil treatment could reduce contaminant concentrations.

The "ashy" samples revealed the following:

	mg/kg
Arsenic	31–76
Copper	240–560
Nickel	55–70
Zinc	420–920
Toluene extract	6000–12 000
pH	5.5–7.5

Therefore almost all of the 123 000 m³ of ashes would have failed to meet the contract specification.

The ashes invariably produced CO_2 concentrations $> 1.5\%$ by volume, although measurable CH_4 was not encountered in the winter months when the reclamation commenced. Gas emission rates were always low. The ashes proved to be susceptible to heating (from, for example, buried power cables) and ultimately would burn and give rise to 30 to 35% volume losses, which could produce unacceptable levels of ground subsidence.

The worst conditions were in the the finer (< 5 mm) fraction of the ashes. Soil screening might be able to reduce reclamation difficulties (but had not been costed for in the original contract), however in the wet winter conditions when the soils were frozen screening would not be a viable option.

The work proceeded. The contractor's agent on site had to justify the reuse of soils which had failed to meet contract specification. His usual justification was that the higher concentrations were at depths which would not create unacceptable risks. Screening did remove the more hazardous gassing and potential fire-risk materials which were then encapsulated below parking areas. However, during adverse weather screening failed to separate particle sizes adequately.

The project was unduly complicated and over-ran budget and time constraints. These problems could have been avoided had an adequate site investigation been conducted and if more defensible site-specific criteria had been properly defined.

8.6 CONCLUSION – PROBLEMS OF QUALITATIVE RISK ASSESSMENTS

Any risk assessment has to be (Cairney, 1995):

- consistent – i.e. replicable among different assessors;
- formal – i.e. follows accepted protocols;
- flexible – i.e. transparent as to the data and assumptions used so that changing risk protection priorities and/or developments in guidelines can be incorporated into a re-evaluation;
- comprehensive – i.e. all plausible scenarios should be considered;
- able to identify information deficiencies;
- cost-effective.

The UK approach to site assessment has traditionally been based on qualitative methods. However, reflecting the late (compared with some other countries) promotion of risk assessment as an environmental tool; a pragmatic reliance on generic guidelines; the focusing of responsibilities for identifying contaminated sites on the developers of sites, and a relatively low level of professional expertise in contaminated land assessment, there is no doubt that the use of a qualitative approach has sometimes been "more by good luck than by good judgement". Assessors have failed to follow consistent approaches; there has been a wide scope for personal bias, and the judgemental processes which have underpinned decisions have often not been clear (Cairney, 1995). Cairney provides a number of examples of qualitative evaluations of risk which have either overestimated the potential risk through a failure to understand the availability of contaminants to the target, or underestimated the risk through a failure to collect sufficient site data to characterise the hazard source and understand the pathways. It is hoped that publication of more specific risk assessment guidance, and of guideline values derived by a risk assessment approach together with the model underpinning them, will provide for more robust qualitative assessments. However, it has to be accepted that the contaminants, and the targets at risk, covered by the guidelines are limited.

9

Risk Ranking and Semi-quantified Assessment

9.1 RISK RANKING

Semi-quantified or semi-numerical risk assessment can be relevant to consideration of the risk from a single site, or for comparing a large number of sites. It is for the latter application that most systems have been produced (see Table 9.1 for examples). Most use a numerical approach which puts all information and situations onto a common (albeit arbitrary) scale by assigning scores or subscores to risk factors relevant to the source–pathway–target scenarios of concern. As a minimum these factors usually reflect the nature of the hazards, the existence of pathways and the proximity of targets. Once each risk factor is scored, the subscores are combined into a final score which provides a semi-quantitative assessment of the risks related to a site or group of sites. At the most simplistic the scores are descriptive (high, medium, low etc.), as presented for a number of criteria in Table 9.2.

Semi-quantified approaches provide an indicator, rather than an estimate, of risk and allow primarily for decisions to be made about priorities. Such approaches are a practicable and efficient means of:

(i) a regulatory authority determining priorities for further investigation and assessment across a number of sites in its area;
(ii) landowners determining priorities for remediating sites or taking relevant risk-reducing actions in relation to future liabilities;
(iii) financial institutions and insurance companies determining the potential for future liabilities prior to investment or the provision of environmental liability insurance;
(iv) screening out insignificant potential risks so as to help focus a full (qualitative or quantitative) risk assessment.

The terminology used in different countries in relation to different tools varies. Some countries refer to "hazard ranking", others to a "qualitative risk assess-

Table 9.1. Examples of risk ranking tools discussed in the literature

Name or organisation	Purpose	Source
Birmingham City Council	Prioritisation of closed gassing landfills using actual gas levels	Harget & Miller, 1996
DPM	Defense Priority Model ranks US Dept. of Defense hazardous waste sites based on potential threats to human health and ecology	Hushon, 1990b
DRASTIC	Rating scheme for assessing potential groundwater pollution. Adapted for specific sources, e.g. pesticides	Canter, 1991 Close, 1993 Aller *et al.*, 1987
ERPS	Environmental Restoration Priority Scheme for determining fund allocations for site remediation for US Dept. of Energy; shelved	Jenni Merkhofer & Williams, 1995
Gaz de France	Prioritisation of gasworks in France for investigation	Costes *et al.*, 1995
HALO	British system for assessing operating and closed landfills; never published for use	Gerrard & Kemp, 1993
HRS	Ranking of sites for US National Priority List	USEPA, 1990b; 1990c
Idaho State Water Quality Division	Decision analysis tool for prioritising groundwater pollution sources	Shook & Grantham, 1993 Droppo *et al.*, 1993
National Classification System	Canadian system for use by municipal authorities for prioritising federal sites	CCME, 1992
New South Wales Environment Protection Authority	Ranking system for prioritising potential contaminated sites for further investigation	McFarland, 1992
RISC-URGENCY	Prioritisation of industrial sites for Province of Friesland, the Netherlands	Hemel *et al.*, 1992 Goldsborough & Smit, 1995

ment" or "preliminary risk assessment". Some systems require no site-specific measurements (for example of soil contaminants) to be made and require only the same information as would be forthcoming from the desk study (Chapter 4). Examples of such systems are that developed by New South Wales Environmental Protection Authority to allow local authorities to rank potential contaminated sites so as to identify those where full investigation may be required (McFarland, 1992), and the hazard ranking system developed for preliminary assessment of gassing landfills by Warwickshire Environmental Protection Council (1995). Such systems provide for the impracticability of carrying out

Table 9.2. Simple hazard ranking criteria (after Wales, Myers & Vogt, 1993)

Risk factor Site sensitivity	Details	Score
Surface-water proximity	On site	High
	At boundary	Medium
	Beyond boundary	Low
Surface-water dilution	Small river	High
	Major river	Medium
	Tidal/lock	Low
	Coastal water	Very low
Geology (solid)	High permeability	High
	Low permeability	Low
Geology (drift)	High permeability	High
	Low permeability	Low
Groundwater	Usable	High
	Non-usable	Low
	No aquifer	Very low
Degree of contaminative use	High	High
	Medium	Medium
	Slight	Low
	No contaminative use	Very low

more than a desk study where a large number (perhaps hundreds or thousands) of sites have to be assessed and so provide for a relatively rapid, although coarse, screening. For example, in Table 9.3, translated from Warwickshire Environmental Protection Council (1995), the grading through from low to high risk is gradual and not clearly definable. It merely provides a means of identifying priorities for investigation: all of those sites graded in the high category being the first priority.

Other systems use measured concentrations of contaminants in soil, air or water and in essence are a semi-quantitative hazard assessment translating the information available from the detailed site investigation and the comparison with generic guidelines (as discussed in Chapter 8) into a set of scores to provide a common basis for comparison of potential risk across a number of sites. Harget & Miller (1996) report such a system developed by Birmingham City Council to allow it to identify closed landfills which present a priority for remedial action, using measured gas concentrations (but not pressures or flows) available from regular gas monitoring. Systems using actual site data can form a good second-tier assessment after an initial screening based on historical and desktop information and a visual site reconnaissance. In the first instance some of the scores might have only two points: e.g. is there, or is there not, housing on the site? Is there, or is there not, surface water within or close to (within 100 metres of) the site?

Semi-quantified assessment does not require that the environmental effects of contamination be predicted, or that the relative magnitudes of the effects be

Table 9.3. Example hazard ranking scheme for gassing sites (adapted from Warwickshire Environmental Protection Council, 1995)

Source/ pathway	Target 1	2	3	4
A	M/H	H	H	H
B	M	H	H	H
C	L	L/M	M	M/H
D	L	L/M	L/M	M
E	L	L	L	L/M

Key:
L = Low hazard; M = Medium hazard; H = high hazard
A High gassing potential, migration known
B High gassing potential, ground conditions favourable for gas migration
C High gassing potential, ground conditions not favourable for gas migration
D Low gassing potential, ground conditions favourable for gas migration
E Low gassing potential, ground conditions not favourable for gas migration
1 All buildings and/or services > 250 m from site
2 Buildings and/or services > 50 m–< 250 m from site
3 Buildings and/or services < 50 m from site
4 Buildings and/or services on the site

determined, or that decisions on acceptable risk levels are taken. It can be sufficient that they are correctly ordered in terms of the magnitude of effects. Therefore semi-quantitative assessments have fewer computational requirements, but have a greater degree of potential uncertainty attached to them in terms of understanding the site-specific risk than a quantified assessment, as suggested conceptually in Figure 9.1 (after Droppo *et al.*, 1993). It must be stressed that the term semi-quantified does not imply any greater degree of understanding of site-specific risks than a qualitative assessment. The two approaches are not comparable in this manner. A semi-quantified approach is different in that it seeks to put information about sites (or a single site) onto a common basis. If it uses actual site data then uncertainty is decreasing (Figure 9.1). Uncertainty as depicted in Figure 9.1 assumes that a source–pathway–target chain potentially exists.

Semi-quantified systems are frequently derived for use by non-experts to allow decisions to be made as an expert would in a robust and consistent manner. Systems derived for authorities to exercise regulatory responsibilities in relation to contaminated land provide examples (USEPA, 1982; 1990b; 1990c; CCME, 1992). The use of "expert systems" or "knowledge-based systems" in environmental decision making has been growing, particularly since the late 1980s (Hushon, 1990a). However, few "genuine" expert systems have been developed as full computer-based tools for use in relation to risk management for contaminated land. All knowledge-based systems require that the relevant knowledge, rules, inferences and heuristics which an expert uses to consider and assess the source–pathway–target scenarios form the basis of the system and are

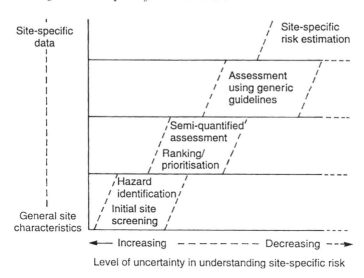

Figure 9.1. Uncertainty versus information requirements in site risk assessment

explained and transparent. Significant problems can arise when the person undertaking the scoring of a site fails to understand the basis on which different scores might be applied for a specific factor. Yet the reader who turns to many of the papers and reports explaining specific hazard ranking schemes will frequently encounter difficulties in understanding why certain factors were included, the logic behind the rating scheme used or why certain factors were weighted as being more important than others. Scoring systems must be tested for their sensitivity to assumptions and ability to be used correctly to classify sites before they are generally applied (Suter, 1993), as was undertaken in a testing programme for the Canadian National Classification System (CCME, 1992). Scoring systems should always be updated as more information becomes available.

Semi-quantified systems are based on the primary risk assessment assumption that risk is related to exposure pathways, target sensitivity and contamination potential:

Risk = Target sensitivity × exposure pathway × contamination characteristics

It is possible, and often desirable, to build in factors which relate not only to direct physical characteristics for each of the source–pathway–target components (e.g. nature of contamination; depth to aquifer; proximity of residential population etc.), but also provide for mitigation, site working practices, management factors and societal sensitivities to be scored. For example, a system which attempts to prioritise petroleum storage sites in terms of potential risk may include factors relating to the management controls, such as existence of an environmental monitoring programme; the existence of such a programme being taken as having the potential to mitigate risks by providing an early warning device. A system used by local authorities to prioritise all potentially contaminated sites in its jurisdiction to identify those requiring immediate investigation might include a factor which provides for social concerns to be recognised. Examples may include the presence of housing on a site where residents would expect investigation, even if the physical characteristics of the site may suggest a low potential risk, or a valuable site where development opportunities may be important. Here the system is being used to take risk management decisions, not merely risk assessment decisions focused on the physical risks.

Shook & Grantham (1993) report a tool for ranking sources of groundwater pollution. This uses three rating factors: the public health risk factor; the aquifer vulnerability factor; and the regulatory adequacy factor. The latter is used as a measure of the ability of any regulatory programme to prevent or remedy contamination. The rationale for this factor is that an adequately regulated contaminant source is less likely to contaminate groundwater than a poorly regulated, or unregulated, source. Such a role for regulation to "lower" risk scores on polluting industrial processes is also underpinning risk assessment application in the UK, as evidenced in the OPRA (Operator and Pollution Risk Appraisal) system being developed by the Environment Agency for regulating operational processes (HMIP, 1995).

Such systems emphasise the difficulty of separating technical factors from social, political and economic factors in risk assessment: the primary debate which has underpinned risk assessment development in general (as discussed in Chapter 2). Some technical experts are reluctant to devise tools which appear to "breach" the technical–socio-political divide. However, tools which do this more explicitly account for socio-political objectives and priorities and therefore are arguably more useful in the decision process.

Systems where risk factors are scored involve technical issues which do not appear in other types of assessment schemes. Examples include the treatment of uncertainty caused by a lack of data or missing values (are they allocated the highest, median or lowest possible score?); weighting to indicate relative importance (is a human carcinogen twice as important (risky) as a fish toxin?); scaling (arithmetic or logarithmic scales?); and procedures for combining subscores (additive, multiplicative, or root square mean?) (Suter, 1993). Decisions on some of these issues are illustrated in the following example systems.

9.2   MULTI-SITE RISK RANKING AND PRIORITISATION

9.2.1   The USEPA HRS

The USEPA's Hazard Ranking Scheme (HRS) is perhaps the best-known prioritisation tool (USEPA, 1982; 1990b; 1990c). HRS is the principal method used to rank the potential risks posed by sites reported to the EPA under CERCLA. Once an identified site is placed on the CERCLA information system it is subject to a preliminary assessment (primarily desktop). This is followed by an initial site investigation and the data gathered are used as the source for applying HRS. The scheme is a typical ranking system in that it relies on expert judgement to allocate scores related to defined criteria of high to low risk. Sites with high scores (see below) are then considered for placement on the National Priorities List for more detailed risk assessment and potential remediation.

HRS assigns numerical scores to a variety of factors that affect the risk potential, including characteristics of waste toxicity, exposure potential, and potentially exposed populations. The three original pathways were groundwater, surface water and direct exposure by air. Revisions have added to these, including the potential for contaminated groundwater to reach surface water and addition of a soil exposure pathway. Each of the pathways consists of a set of factors that represent the potential to cause harm:

(i) the manner in which the contaminants are contained, actual releases and the potential of a contaminant being released;
(ii) the characteristics and amount of the materials, soils etc.; and
(iii) the nature of the likely targets.

A score for each of the pathways is obtained by allocating a numerical value or "no score" (to denote not significant) for each of the characteristics or factors within each pathway. Each value is multiplied by a specified weighting factor and then the factor scores are multiplied together to give an overall pathway score:

$$\frac{\text{Pathway}}{\text{score}} = \frac{\text{Likelihood}}{\text{of release}} \times \frac{\text{Material}}{\text{character}} \times \frac{\text{Exposed}}{\text{targets}}$$

The pathway scores are normalised and combined using a root square mean approach rather than a simple average. This has the effect of emphasising a facility recording a high score for one particular pathway which can then be used to discriminate between sites with severe problems in one respect compared with sites with a combination of less severe problems. Sites are ranked on the basis of the total score and the score for each pathway; a score of 28.5 (on a scale 0–100) warrants inclusion on the National Priorities List. The score of 28.5 does not imply a specific level of risk, merely a screening level to allow priorities for further investigation and assessment to be set.

Table 9.4 provides, as examples, the factor categories and individual factors which are scored in relation to the groundwater and air migration pathways. Scores are assigned according to prescribed guidelines. For example, when considering the likelihood of release for a pathway the maximum score possible is 550 for an observed release. Where there is no observed release then a score not exceeding 500 is assigned to the potential for release. This score is composed of a

Table 9.4. Example pathway factors and score derivation within the hazard ranking system (USEPA, 1990c)

(a) Groundwater pathway		
Likelihood of release ×	Waste characteristics ×	Targets

1 Observed release
  or
2 Potential to release
  2a Containment
  2b Net precipitation
  2c Depth to aquifer
  2d Travel time
	4 Toxicity/mobility
	5 Hazardous waste quantity
	6 Waste characteristics
		7 Nearest well
		8 Population
		9 Resources
		10 Wellhead
		   protection

Groundwater pathway score =

[Likelihood of release (higher of 1 or (2a × (2b + 2c + 2d))) × (6) × (7 + 8 + 9 + 10)]

(b) Air migration pathway		
Likelihood of release ×	Waste characteristics ×	Targets

1 Observed release
  or
2 Potential to release
  Route characteristics
  2a Gas potential to release
  2b Particulate potential to
    release
	4 Toxicity/mobility
	5 Hazardous waste quantity
	6 Waste characteristics
		7 Nearest individual
		8 Population
		9 Resources
		10 Sensitive
		   environments

Air migration pathway score =

[Likelihood of release (higher of 1 or (2a or 2b)) × (6) × (7 + 8 + 9 + 10)]

series of pathway-specific factors. Taking as an example from Table 9.4 the groundwater pathway factor relating to the potential for release, the following maximum values can be assigned: factor 2a, containment – 10; factor 2b, precipitation – 10; factor 2c, depth to aquifer – 5; factor 2d, travel time – 35.

Reviews of the HRS have revealed characteristics that unintentionally biased its results (Suter, 1993). For example, waste type was scored on the basis of the most hazardous component of the waste, whilst waste quantity was scored on the basis of all components. Therefore, a large amount of a relatively low hazard material could receive the same final score as a small quantity of a highly toxic material. However, like all such systems, the HRS is not designed to distinguish between the risks presented by two sites where the scores are similar, but rather to provide a meaningful indicator of the relative risks between sites with very different scores.

9.2.2 Other international examples

A similar prioritisation or ranking system was developed in Canada under the National Classification System (CCME, 1992) for application to federal sites.

The NCS allocates a maximum of 100 points per site: 33 relating to contaminant characteristics; 33 to exposure pathways; and 34 to targets. Each of the evaluation factors in the classification is assigned a score ranging from 0 to 18. These scores are designed to weight factors according to the potential or actual relevance in contributing to the site risk.

For each factor, several possible scenarios are presented. For example, the physical state of contaminants could be liquid, sludge or solid; the topography of the site could be steep, moderate or flat. Scoring guidelines are suggested for each scenario presented. The scoring guidelines can be adapted by the assessor as long as a score does not exceed the maximum allotted. Table 9.5 provides some examples of the scoring guidelines for different evaluation factors.

The NCS handles information deficiencies for a particular evaluation factor by assigning a score that is half of the allowable score. This suggests a plausible rather than conservative approach to uncertainty which would not be preferred by some. However, where "estimated" scores are allocated the system provides for them to be added separately so that the degree of uncertainty is transparent and understood. If information is unavailable to assign a score to factors valued to a total of 30% of the maximum possible score, it is considered that there is insufficient information to assess the particular site, and the assessor is required to obtain further information. This approach has potential advantages over systems which simply allocate the maximum score where information is unavailable and adds these values into the scores derived from actual data and information. Under the latter approach some sites actually presenting low risks may be added to the list requiring further investigation, so wasting resources.

The Canadian system also (as with HRS) provides for a distinction between contamination or impacts that are known to have occurred and those that have the potential to occur. Although known impacts are not rated any higher (i.e. the maximum allowable score is the same for the known and potential sections of each category), their differentiation provides for a clearer appreciation of site conditions. In Table 9.5 the known contamination of groundwater is provided as an example.

Sites are classified on their individual characteristics and placed into classes (CCME, 1992):

- Class 1 (Score 70–100): Action required. Site shows a propensity to high concern for several factors and measured or observed impacts have been documented.
- Class 2 (Score 50–69.9): Action likely to be required. There is probably no indication of off-site contamination, however the potential for this is high.
- Class 3 (Score 37–49.9): Action may be required. Site is not a high concern, however additional investigation may be carried out to confirm the classification.
- Class N (Score < 37): Action not likely to be required.
- Class I (Score ≥ 15): Insufficient information to classify the site.

Table 9.5. Example evaluation factors from NCS (CCME, 1992)

Category	Evaluation factor	Scoring guideline
Contaminant characteristic	Degree of hazard	
	High concern-high concentration	14
	High concern-low concentration	11
	Medium concern-high concentration	8
	Medium concern-low concentration	5
	Low concern	3
Physical state of contaminants	Liquid/gas	9
	Sludge	7
	Solid	3
Exposure pathways	Groundwater	
	Known contamination at or beyond property	
	Groundwater significantly exceeds	11
	Canadian Drinking Water Guidelines by $>2\times$ or known contact of contaminants	
	Between 1 and $2\times$ guidelines or probable contact	6
	Meets drinking water guidelines	0
	Potential for groundwater contamination	
	Engineered subsurface containment	
	No containment	4
	Partial containment	2
	Full containment	0
	Thickness of confining layer over aquifer	
	3 m or less	1.5
	3–10 m	1
	<10 m	0
	Hydraulic conductivity of the confining layer	
	$>10^{-4}$ cm/sec	1.5
	10^{-4} to 10^{-6} cm/sec	1
	$<10^{-6}$ cm/sec	0.5
	Annual rainfall	
	>1000 mm	1
	600 mm	0.6
	400 mm	0.4
	200 mm	0.2
	Hydraulic conductivity of aquifer of concern	
	10^{-2} cm/sec	3
	10^{-4} to 10^{-6} cm/sec	1.5
	$<10^{-6}$ cm/sec	0.5
Receptors	Known adverse impact on humans or domestic animals	18
	Strongly suspected adverse impact	15
	Potential human exposure through land use; use of land and surrounding site	

Land use (current or future)	0–300 m	300 m–1 km	1–5 km
Residential	5	4.5	3
Agricultural	5	4	2.5
Parkland/school	4	3	1.5
Commercial/industrial	3	1	0.5

The NCS is accompanied by detailed evaluation forms and a user's guide which provides the rationale for each evaluation factor and scoring, together with the method of evaluation to be used and the sources of information.

McFarland (1992) reported a scheme devised by the Environmental Protection Agency, New South Wales, Australia, for the ranking of sites by local authorities to enable priorities for detailed site investigation to be identified. The latter, in contrast to HRS and NCS, used a simple scoring system, where each of 10 factors was scored on the same basis of 1 to 5, indicating low to high risk respectively. Simplicity, and hence potential assessor variance, was weighed against the expense of collecting site-specific data and the need for expert skills. The 10 factors considered were:

- Proximity to sensitive environments.
- Proposed land use.
- Current and prior land uses.
- Types of contaminants.
- Characteristics of contaminants.
- Distribution of contaminants.
- Potential for site disturbance.
- Potential for groundwater resource exposure.
- Significance of the site, particularly for development.
- Emotiveness of the site and/or contaminants.

In this system the scores for each factor for a site were simply added together, with one site having the potential to score a maximum of 50 if presenting a significant risk. By sorting all sites in descending numerical score order each site would be prioritised on a relative scale.

9.2.3 UK systems

In the UK, work in the late 1980s was started on the Hazard Assessment of Landfill Operations (HALO) package (Gerrard & Kemp, 1993). HALO was designed to rank the operations and environmental impact of existing operational waste disposal sites. The methodology was loosely based on the USEPA HRS, with scores developed for material pollution potential; landfill operations; groundwater pathway; surface-water pathway; landfill gas and also public perception. As with HRS each component and its related factors was assigned a numerical value which increased as the degree of hazard increased. These scores were then normalised and subsequently weighted according to their relative importance.

HALO has never been published for general use and considerable debate has focused on the scoring and weighting developed, its relevance to non-operational sites, and the extent to which the relatively complex evaluation system would be able to be applied by individuals in the field. An adapted HALO methodology was developed for the hazard assessment of methane-generating sites (Department of the Environment, 1990).

In 1995, a new prioritisation tool for use in the UK was published (Department of the Environment, 1995b). The tool has two parts. Part I leads to a preliminary prioritisation of a site based on assessment of proximity of a target assessed under three headings: development (humans, plants and the built environment), surface water, and groundwater. The assessment uses simple flow charts which take the user through a series of "yes–no" questions. For example, in relation to development the scheme asks whether there is any residential development, school, playground or allotment on the site or within 50 m. If the answer is "yes" the site is determined a priority for a Part II assessment. If there is any commercial or industrial development on site or within 50 m of the site or residential development within 250 m, then the site is determined as the next priority for a Part II assessment. Unfortunately, as discussed in Chapter 4 the user is presented with no rationale for the distances and there is little assistance to the non-expert user to take appropriate risk decisions outside of the simplistic flowcharts.

The Part II assessments use more detailed information about the sources and pathways as well as the targets. A desk study is required following the type discussed in Chapter 4, together with a site reconnaissance. Again, the user follows a number of flowcharts relating to the three types of targets of concern. For example, in relation to development, the chart starts by asking whether there are any contaminants in the soil which are toxic by ingestion, inhalation or skin contact in concentrations which exceed relevant action values and/or present an unacceptable risk? As the assessor is unlikely to have site investigation data, a "no data" answer follows the same path as a positive answer. If there are also areas of exposed soil (e.g. gardens or landscaped areas) or there is no data, the site is prioritised as a Category 1 site suggesting that the site is probably not suitable for it present use and urgent action is needed in the short term. This system takes a highly conservative approach to dealing with uncertainty which appears to have been necessary because of the simplistic checklist approach with no use of a score. At the time of writing there appears to be little experience of use of the scheme which, although published by the Department of the Environment, is not official guidance in terms of required use. It is not clear whether it was extensively tested with potential users before its publication.

9.3   SINGLE SITE ASSESSMENT

Cairney (1995) describes an approach to the semi-quantified assessment of various risks presented by single sites. The approach uses comparisons with published guidelines, criteria which the author has devised from experience, and qualitative site data. This should be seen as an example of possible viable assessment systems; it should not be used directly without careful consideration because published guidance may have changed, and the criteria listed involve a number of personal judgements by the author. As the author says, "Use is necessarily made of suites of contaminants and trigger concentrations to identify potential land contamination hazards. These have not been based on any new theoretical or research findings, but instead are the parameters and values which experience suggests will be practically appropriate to use."

Seven environmental risk situations are considered:

Group I
(1) Risks of polluting surface waters.
(2) Risks of polluting groundwaters.
(3) Risks of producing area-wide air pollution.

Group II
(4) Risks of gases and vapours entering dwellings.
(5) Risks of attack on construction materials.
(6) Risks to plant populations.
(7) Risks to human health by contaminant contact, ingestion or inhalation.

The Group I risks are those of interest to regulatory bodies charged with minimising wider environmental degradation. Group II risks are those of direct concern to individuals or companies owning or occupying land which is, or might be, contaminated.

For each risk situation a three-part assessment is proposed:

(i) Part [A] – The potential for future risk is identified on the basis of a desk study. Certain limitations are placed on the assessor and there is a requirement that any decision to terminate the risk assessment process at this stage must be justified. No use is made of numerical assessment values.

(ii) Part [B] – The quantification of the probable magnitude of future risk. In this, information from the site investigation is used, and any limitations of the information are highlighted and penalised. Scoring of "liabilities" (as negative values) allows the relative magnitudes of the different risk situations to be established. This in turn indicates the particular emphasis which should be observed when remediation methods are being selected, and prevents the importance of a specific risk being overlooked.

(iii) Part [C] – The focus here is on how far site remediation has removed or reduced risks. Results (as positive scores) are subtracted from those obtained in Part [B] to give as precise as possible indicators of the scale of remnant liabilities. Particular emphasis is placed on the availability of proof that the remediation has been effective.

Simple checklists are provided for use in the assessment. For example, for surface-water risk assessment the following are asked:

[A] *Potential for future liability*

  (i) Is the site located adjacent to a surface-water body, or do ditches from the site drain to a surface-water body?
 (ii) Would the site topography encourage runoff to a surface-water body (within 300 m of the site boundary)?
(iii) Would any runoff flow easily to the surface-water body?
 (iv) Does the site surface appear to have the potential to produce polluting runoff (see list of contaminants and trigger concentrations)?
  (v) Does the site appear to be creating surface-water pollution currently?
 (vi) Is the quality of any nearby surface-water body (likely to be affected by the site; i.e. within 300 m upstream of the site) currently high?
(vii) Does the National Rivers Authority (or equivalent control bodies) zoning of any surface water, within 300 m, show the water quality to be high?
(viii) Does the control body monitor surface-water quality conditions within 300 m of the site (either upstream or downstream)?
 (ix) Is there any evidence to suggest that a near-surface polluted groundwater could exist in the site, and that this could flow into nearby surface waters? Confirm by reference to groundwater pollution risk analysis.

If the answers to all the above questions are "no", then the assessor terminates the surface-water pollution liability assessment and accepts that the liability risk is likely to be negligible. Clearly, the reader might wish to add other questions to the checklist: for example, no questions relate to the current or planned use of the water, there is no reference to the presence of discharge pipes (e.g. for surface-water drainage, unknown purposes) or of old land drains, or of the potential of the site to flood (it might be in the floodplain or lie in a flood control channel). The framework developed for the Department of the Environment (Department of the Environment, 1994b) stresses the importance of taking into account weather and seasonal factors in whether an actual discharge will be observed. The limit of 300 m is arbitrary and might need review in relation to the potential magnitude of any possible discharge relative to the size of the receiving water body.

The Part [B] assessments of the magnitude of future liability for surface-water pollution combine numerical scores based on observed or predicted (from the site history) concentrations of listed substances which have the potential to pollute water with scores relating to the potential for polluted groundwater or polluted runoff to affect the surface water. The highest score is given to highly mobile or leachable substances such as free oils, tarry liquors and phenols which are present in excessive concentrations and throughout the site. A lower score is allocated to hot spots of groups of contaminants, and the lowest score to individual contaminants which occur with negligible mobility. The assessment allows for "no data" and data deficiencies to be scored high and requires them to be noted specifically.

The scheme presented by Cairney is intended to:

> offer the rigour of the quantified approach, whilst still retaining the flexibility needed to overcome scientific information deficiencies. In this method, assessor freedom of choice is deliberately restricted, gaps in the necessary site information are highlighted and penalised, and assessors must inform colleagues, and the individual in charge of the assessment group, of any abnormal information or conditions. This permits an agreed evolution of the assessment approach, as experience proves necessary. (Cairney, 1995 p. 57)

The system goes beyond the schemes discussed earlier in that it is intended to provide for judgements about the risk reduction potential or remediation alternatives to be assessed, thus making a link through from the risk assessment to risk management.

Probably semi-quantified assessment's greatest value is in the requirement to make explicit judgements about the components of the source–pathway–target chain. It is not a replacement for either a qualitative or a quantitative assessment of a specific site.

10

Site-specific, Quantified Risk Assessment

10.1 INTRODUCTION

This chapter deals with quantified, site-specific risk assessment. It is not a manual on how to carry out such assessments. Rather, it is intended to highlight some of the important areas where professional judgement is required in order to make the reader aware of the need for great care and attention to detail if reliable answers are to be obtained.

As discussed in Chapters 2 and 8, such approaches should not be viewed either as "better" than an approach using comparisons with generic guidelines etc., or as being diametrically opposed. In practice, it will often be convenient, and sometimes essential, to use a combination of the two approaches. Site-specific risk assessment applied to all sites is not a cost-effective approach to contaminated land risk management, and experience is that the majority of sites can be dealt with by a qualitative assessment (if conducted rigorously), as discussed in Chapter 8. As introduced in Chapter 1, a site-specific risk assessment is likely to be required where:

- generic guidelines do not cover the contaminant(s) of concern and/or are insufficiently protective relative to the target(s) of concern in terms of their sensitivity, activity patterns, etc.;
- generic guidelines would be overly protective as a result of conservative assumptions used in their derivation;
- observed concentrations of contaminants exceed the generic guidelines to an extent which indicates the potential for risk to be realised;
- local background levels are high compared to the generic guidelines;
- a site is causing considerable public concern (for example, about threats to child health arising from particular contaminants) and there is demand for a fuller understanding of the risks presented by the site (a rare situation).

Generic guidelines are (now) derived using the same approach as is inherent in a site-specific risk assessment, except that exposure assumptions have to be made which will reflect political choices as to the degree to which the most sensitive target is to be protected. In the site-specific risk assessment estimates are made of risks to targets in light of the actual (observed, measured or predicted) chemical (source), pathway and target characteristics. Assumptions about all of the latter have to be made in the derivation and use of generic guidelines and standards, but in most cases generic criteria are intended to be protective.

Site-specific assessments suffer both the disadvantages and the advantages of allowing individuals to exercise professional judgement. There may be significant opportunity for personal bias or for pressure to obtain a desired outcome (usually in the direction of a lower rather than a higher estimate of risk). It also leaves open to judgement the question of "What constitutes an acceptable (or tolerable) risk?" in a particular situation. However, problems can also be created by overly prescriptive official guidance which specifies how the risk assessment should be performed at each step. Whilst consistency and quality control in risk management activities form the basic reasons for the development of procedural guidance, standardisation can lead to insufficient attention being paid to the different types of risk that need to be addressed and the different objectives of risk assessment (Covello & Merkhofer, 1993). "Manuals" which provide detailed flow and decision sheets, but which rely on the user's environmental and scientific understanding, can be counter-productive.

In quantitative risk estimation, a number of uncertainty factors or assumptions will be used to trace the contaminant from the source to the target which, collectively, may outweigh the variations in analytical data obtained. This does not, however, obviate the need to know how reliable the data are, since even small errors in determined values, or failure to detect a contaminant at the target location, can have profound effects on the estimation of risk. At the time of writing there is a move to the development of probabilistic as opposed to deterministic risk assessment. The latter computes a single value for the level of adverse health or environmental consequences, with occasionally the statistical uncertainty inherent in this value estimated. The degree of uncertainty and potential conservatism inherent in the latter has been a source of expert concern (Cullen, 1994). The use of Monte Carlo analysis to produce distributions of risk reflecting uncertainty and/or variability has traditionally formed an important component of the risk assessment of radioactive waste disposal. Monte Carlo techniques are computationally more complex than point estimate approaches, and there is potential for mistakes and abuse (Morgan & Henrion, 1990; Burmaster & Anderson, 1994). However, in the UK it has already been built into a risk assessment tool for the assessment of new landfills (Hall *et al.*, 1995) and, as introduced in Chapter 8, is inherent in the CLEA model being used for the derivation of guideline criteria for contaminated soils.

In Chapter 2 the caution often attached to the use of site-specific quantitative risk assessment compared to the use of generic guidelines was illustrated by the Ontario guidance (OMEE, 1996). The latter places administrative controls on the application of quantitative assessments. Issues relating to uncertainty, bias and the need to justify and explain (to the public) the derivation of a clean-up programme based on a site-specific assessment are covered.

The discussion that follows concentrates on human health risks from soils, but the general principles are similar for other targets and routes of exposure. The chapter identifies key components of the assessment in terms of the exposure and effects assessment and the determination of acceptable risks. It includes a case example of an assessment applied to a contaminated site occupied by a school. The chapter concludes with a discussion of how site-specific risk-based criteria can be derived using risk assessment.

10.2 BASIC ISSUES

It is unlikely that a site-specific assessment will be started unless some form of qualitative or semi-quantitative assessment, along the lines described in Chapters 8 and 9, has first been carried out. Plausible, and probably critical, pathway–target combinations will have been identified, and possibly also a limited number of contaminants of concern. In the move to a site-specific assessment there are number of questions to address, including:

- Which source–pathway–target scenarios should be taken into account?
- What assumptions regarding exposure duration etc. should be made?
- Which reference data on relative toxicity and cancer potency should be used?
- Which risk estimation models/protocols should be used?
- Which site data should be used in the calculations?
- What are the acceptable, tolerable and/or unacceptable levels of risk?

There are two key areas which are important in the understanding of risk assessment and particularly health risk assessment: the use and limitations of models, and the approach to dose-response assessment. These two issues are introduced here prior to further discussion within the following sections which elaborates on the risk assessment process: i.e. exposure assessment, effects assessment and risk evaluation.

10.2.1 Models

Models used in risk assessment are of two types: (i) those focused on understanding the movement of contaminants in the pathways of concern and the transfer between media, and (ii) those which relate exposure of a target to an effect. In any site-specific assessment a combination of such models may need to be used. For example:

(i) prediction of the concentration of a volatile organic compound dissolved in water that would arise due to migration from a source allowing for processes of attenuation such as dilution, adsorption, biodegradation and abiotic degradation;

(ii) prediction of concentrations in the breathing zone of water extracted and used for showering;

(iii) prediction of the dose adsorbed as a result of breathing contaminated air;

(iv) calculation of the individual excess cancer risk (see below) resulting from exposure to that dose.

The use of environmental fate and transport models places considerable demands on the site investigation stage. For example, if we consider potential movements of a chemical from a landfill through the unsaturated zone to groundwater and transport to a drinking-water source, data required for modelling purposes would include: soil concentrations for each chemical; average annual precipitation; runoff coefficients; hydraulic conductivity; hydraulic gradient; suction/conductivity variations with moisture content; the area of groundwater flow, and the transverse dispersivity of the aquifer material. Major data requirements for fate analysis often involve parameters such as water solubility, vapour pressure, octanol–water partition coefficient, bioconcentration factors, soil sorption constant, water–air ratio, and degradation rate constants in water, air, soil and biota. These provide a clear warning as to the information demands on site investigation (addressed in Chapters 5–7) and where such parameters cannot be measured, or are not known, the need to understand the behaviour of the chosen model if default values have to be used.

The key to the exposure assessment component of the risk estimation is the appropriate choice of transport and fate models. Selection of a particular model will be dependent on many factors including: (i) relevance to the site conditions, (ii) degree of model validation, and (iii) availability of required input data (USEPA, 1988a). Many hundreds of models exist. Reviews and explanations of the use of various models are found in Jorgenson, 1984; Covello & Merkhofer, 1993; USEPA, 1988a; OECD, 1989; Bartell, Gardner & O'Neill, 1992; Calabrese & Baldwin, 1993; Suter, 1993; Department of the Environment, 1994c.

No environmental fate model produces entirely reliable results and yet without these models the risk assessment cannot proceed. The objective of these models is to determine the average concentrations of a pollutant for a particular time frame and target group for one or more exposure pathways. Air pollutant transport and fate modelling has reached a relatively high degree of sophistication, addressing the processes of pollutant transport, diffusion and deposition. The two factors which are important in determining dispersion are wind direction and speed, and atmospheric stability. The primary outputs of the models are atmospheric concentrations of pollutants and deposition rate, both "wet" (i.e. during rain or snow) or "dry". Modelling of dispersion and deposition from continuous emission sources (such as chimneys) is easier than modelling "puffs" which might occur over short periods, for example from a spillage. Modelling of emissions from landfills poses particular problems as the emission height is at ground level and wind speed is often at zero. Gaussian models which form the common basis of air models cannot process zero wind speeds, therefore predicted ambient air concentrations have a wider margin of error (Petts & Eduljee, 1994).

Atmospheric models have been developed for the indoor air environment and are relevant to consideration of transport of dust or vapours indoors and as a result of volatilisation of chemicals during use of contaminated water. In the outdoor air mixing results in large dilutions of gaseous contaminants. However, soil vapours migrating into living spaces of houses may reach concentrations harmful to human health. Where subsurface soil gas or groundwater seepage enters indoors, compartmental models have been developed to account for the multiple sources and sinks of pollutants and the ventilation rates among discrete volumes of air within a building. Some models of soil-vapour ingress are specific to particular contaminants and also to style of house construction (e.g. models for radon transport). Ferguson, Krylov & McGrath (1995) report a simple generic model relevant to the construction style of houses in the UK. However, in general there have been few studies to validate models of this type (Ferguson, 1996).

Surface-water transport and fate models are similar to atmospheric, although there are key differences in pollutant behaviour: for example, dissolution of acids; volatilisation of contaminants such as benzene and toluene; precipitation of metals to the bottom; adsorption onto sediments; and dissolution of metals resulting in possible uptake by plants and animals all require attention. Models designed for particular types of water bodies (streams, rivers, estuaries etc.) can be used for estimating average concentrations as a function of distance from continuous or discontinuous point-source discharges. Simple models assume steady-state conditions over time, dynamic models simulate contaminant concentration in space and time.

Simple models which estimate dilution factors and contaminant concentrations downstream of a contaminated discharge, or which estimate travel time for a contaminant travelling through the saturated zone, provide some scope for varying input values according to known or assumed site conditions and will avoid the inherent conservatism of a risk assessment based on comparison of a measured concentration with a water quality standard (as discussed in Chapter 8). However, simple models do not allow for a high degree of site specificity, and a note of caution is required relative to their common-place use. The Department of the Environment framework report on water assessment (1994c) has an appendix on water models based on UK reviews of commercially available analytical and numerical models, including one carried out in 1991 for Her Majesty's Inspectorate of Pollution (Department of the Environment, 1992b).

Groundwater models have largely been developed (particularly in the USA) to analyse the transport of chemicals leached from waste disposal sites. They calculate the distribution of contaminants in the unsaturated soil and in the saturated zone and typically include two components: groundwater flow which provides estimates of water velocities or flow paths and travel times, and contaminant transport which accounts for degradation and transformation of the pollutants. Hundreds of groundwater models exist and the USEPA published a guide to selection (USEPA, 1988e).

Food-chain models are essential to the assessment of bioconcentration and bioaccumulation of chemicals up the foodchain via plants and animals. Foods grown in soil will be affected by aerial deposition and uptake through the soil system, while fish and shellfish will be in contact with surface water and sediments, which may themselves be affected by surface runoff, ingress of contaminated groundwater etc.

In general, the transfer of pollutants via the foodchain is poorly understood. The uptake of soil contaminants by plants occurs via root uptake and subsequent translocation by transpiration; uptake from particulate matter deposited on the leaf surface; uptake of vapour via leaf pores; and in certain oil-containing vegetables uptake through oil cells. Uptake behaviour differs markedly depending on soil type, pH, organic matter content, plant species, individual cultivars, and by the presence of other contaminants (particularly in the case of metals) (Ferguson, 1996). Models simulating the pathway uptakes are rare. The US Terrestrial Foodchain Model (TFC) (Travis & Cook, 1989) has been used in the UK, although primarily in relation to risk assessment of incineration risks from continuous stack emissions (Petts & Eduljee, 1994). Although simple regression models have been built into contaminated land risk assessment models such as AERIS (1991), these are considered to be limited, as key site-specific values for parameters such as K_d (the soil solid/soil liquid partition coefficient) are usually not available, leading to recourse to default values which cannot be appropriate to all plant-soil conditions.

The use of environmental fate modelling results in the development of behavioural profiles of chemicals in a specific environmental setting. The prediction of behaviour may relate to the persistence of the chemical, the dominant fate processes, the magnitude of chemical exposure and the relative distribution of residual concentrations. Of particular relevance to environmental fate modelling is the work of Mackay and colleagues (see Mackay, 1991 for a review) who derived a series of general evaluative models to characterise the behaviour of chemicals in the biosphere. The principal concept centred on the capacity of chemicals to accumulate in different environmental compartments (air, water, soils, fauna etc.). This "escaping" from one medium to another is termed "fugacity" and can be related to environmental concentrations for model predictions. Fugacity models are the most widely accepted for multimedia fate modelling.

Multimedia models simulate the transport and transformation of contaminants in multiple environmental media: for example, solvents in landfills moving from soil to groundwater and surface water. Moskowitz *et al.* (1996) review two models used to support remediation of hazardous waste sites under Superfund. These simulate transport, fate and effects. However, a USEPA Science Advisory Board review has expressed concern about the lack of validated models for regulatory purposes, including site-specific risk assessments (USEPA, 1995f).

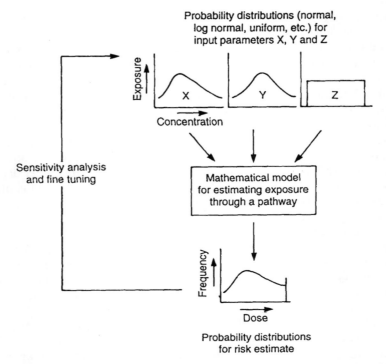

Figure 10.1. Dealing with parameter uncertainty using probability analysis (after Tyler *et al.*, 1991; and Moore & Elliott, 1996)

Exposure-dose models convert the output of the environmental fate models into the dose received by the individual target. These models (discussed further in Section 10.3.4) estimate pollutant intake by multiplying the concentration in the medium of concern by an estimated intake rate by a duration or period over which the target is exposed. If the pollutant is present in multiple media or multiple exposure routes exist (e.g. direct ingestion of soil, ingestion via drinking water, inhalation of dust), then it is normal to sum the exposures received by each route of exposure. Soil ingestion and skin contact with soil and dust present two significant areas of uncertainty arising from lack of knowledge about key parameters such as the nature, frequency and duration of events, the amount of soil ingestion and the area of skin exposed and the bioavailability of soil-borne contaminants (see Ferguson, 1996 for a review of problems). Models have to provide a means of estimating exposure in the absence of comprehensive monitoring and population behaviour data. The problems arise not from inherent deficiencies in modelling technique but from a lack of understanding and of data on transport and fate processes and from difficulties in predicting target behaviours (Covello & Merkhofer, 1993).

Many of the models in use for calculating dose from, say, soil concentrations, are linear in form, producing a single, deterministic calculation. However, a number of models have been used, using stochastic or probabilistic approaches in which discrete values for variable are replaced by probability distributions for each parameter, which are then sampled many times to build up an output in the form of a probability curve, for example for dose received. Figure 10.1 illustrates the concept. Tyler *et al.* (1991) applied the technique to an analysis of arsenic ingestion from soil exposure; the resulting estimates of intake ranged across seven orders of magnitude depending on the assumptions made. Whereas the simple linear model is amenable to hand calculation, stochastic models are reliant on computers to make the necessary calculations.

Computer programs are available which can be used to assist in site-specific risk assessments. Table 10.1 lists a few of the commercially available models. Some integrate fate and transport and exposure assessment models, others combine these with dose-response estimation to provide a final risk estimate. If these are to be used, it is essential that the user understands the basis of the model and is able to modify in-built assumptions when necessary. Easily usable packages may generate a sense of achieving "the right answer". However, no model is a "correct" representation of the real world. In an examination of three models, including AERIS, applied to the derivation of clean-up criteria for a site in Canada, it was found that different quantitative estimates of exposure for most of the 24 measured contaminants were obtained and no model gave consistently higher or lower estimates of total exposure. Suggested soil clean-up criteria varied widely, in one case (in relation to PCBs) by a factor of 350 between the lowest and highest suggested value (Jessiman *et al.*, 1992 discussed in Richardson, 1996).

10.2.2 Dose-response approaches

In general the underlying concept is that toxic effects (such as adverse health effects) occur only after exposure exceeds some threshold level. At sufficiently low exposures no effects will be detected, regardless of duration of exposure. However, as the dose increases a greater proportion of the population will respond with subtle effects in those most sensitive through to more severe adverse effects. At a high enough dose of sufficient duration nearly the whole population will experience severe effects, ultimately death. This concept is presented in a generalised manner in Figure 10.2 (Covello & Merkhofer, 1993).

It is usual to assume that all non-carcinogenic (non-cancer) toxic effects have a threshold, and that below that threshold no adverse effect will occur: this is referred to in the US literature as the reference dose (Barnes & Dourson, 1988), that is the maximum amount of a chemical (in mg/kg body weight/day) that the human body can absorb on a continuous basis without experiencing adverse health effects. In the UK literature the term tolerable intake (usually tolerable daily intake) is used. While the assumption underlying the reference dose greatly oversimplifies the relevant biology and has been demonstrated to

Table 10.1. Examples of commercially available risk assessment packages

Name	Source of information	Features
AERIS (Aid for Redevelopment of Industrial sites)	AERIS, 1991 Decommissioning Steering Committee Canadian Council of Resource for Environment Ministers, Canada Environment, Canada	Multi-media risk assessment Generates site-specific clean-up guidelines related to human health Designed for industrial sites available for redevelopment to defined land uses Considers five exposure pathways Not suitable for evaluating spills
DSS (Decision Support System)	American Petroleum Institute (API), Washington, DC Geraghty Miller Inc, Maryland, USA (and in UK)	Exposure and risk assessment decision support system Petroleum contaminated sites Works in deterministic and Monte Carlo mode Estimates of site-specific exposures and risks
RBCA (Risk-Based Corrective Action System)	Groundwater Services Inc, Houston, Texas, USA Environmental Systems and Technology, Inc, Virginia, USA ASTM, 1995	Calculates site-specific risk-based soil and groundwater remediation goals Tiered decision system, using spreadsheets, with increasingly site-specific levels of assessment as appropriate Focused on hydrocarbon-contaminated soil and groundwater
Risk* Assistant	Hampshire Research Institute, Virginia, USA USEPA Research Triangle Park, North Carolina, USA Thistle Publishing, Alexandria, Virginia, USA (distributor)	Estimates exposure and human health risks from any contaminated site based on concentrations in air, water, soil, sediment, biota Holds USEPA IRIS and HEAST databases so can be used as a reference source Includes international soil guidelines Can adapt exposure to site-specific conditions
RISC-HUMAN	Van Hall Institute, Groningen, the Netherlands	Estimates exposures (using CSOIL-model) and site-specific human health risks based on Dutch target and intervention values

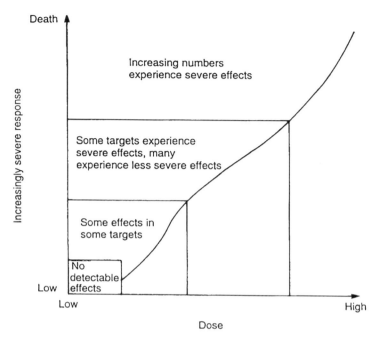

Figure 10.2. A general dose-response relationship (after Covello & Merkhofer, 1993)

be inappropriate for some chemicals and toxic effects, its analytical simplicity is such that it is the model usually applied. When sufficient data are available a dose-response relationship can sometimes be derived directly from statistical analysis. However, most data for dose-response models are derived from animal studies with extrapolation to humans. Where exposure is by ingestion and inhalation, this extrapolation is typically achieved by conversion factors based on body weight or surface area.

It is important to stress that the tolerable intake does not represent a clear dividing line between a "safe" and an "unsafe" dose, not least because of the large uncertainty over the application of safety factors to the extrapolation of laboratory animal data to humans. Certainly, there is no significant difference between an estimated risk which is 0.9 of the tolerable intake and one which is 1.1.

There are various categories of dose-response model, ranging from simple dose-response curves which relate a single measure of dose to a measure of response (e.g. number of fatalities) to complex pharmacokinetic models. The latter are based on the study of absorption, distribution, metabolism and elimination of chemicals in humans and animals. Pharmacokinetic modelling recognises that the dose received or administered is not necessarily the dose which is received at the organ which will be the target for a particular chemical, i.e the administered dose is not strictly proportional to the effective dose. Physiologically based pharmacokinetic models attempt to deal with the physiology of exposed individuals. They possess a high degree of predictive power for estimating the adverse effects of exposures to chemicals. However, no single model can be used for all chemicals and for any chemical there is frequently insufficient pharmacokinetic information available (Covello & Merkofer, 1993).

For carcinogenic effects there is a belief (particularly in the US risk assessment procedure – USEPA, 1989a) that the dose-response curve approaches linearity at low doses, i.e. there is no threshold below which no effects will result. Figure 10.3 indicates the different approaches, the simplest being the linear model. If nothing is known about a threshold, the dose-response function could be anywhere between zero and the straight line through the origin, as suggested by the curved line.

The multi-stage model is considered as dealing effectively with the linear approach, and specifically the linearised multi-stage model of carcinogenesis is used. This is based on two assumptions:

(i) At low doses the relationship between exposure and dose can be reasonably approximated to a straight line.
(ii) Any non-zero exposure entails a finite risk, such that this straight line passes through the origin and can be described in terms of its slope.

The slope (also called cancer potency factor or potency slope) of the dose-response function provides an indication of how powerful a chemical is in causing cancer, i.e. the carcinogenic toxicity. The cancer slope factor is the cancer risk (proportion affected) per unit of dose (i.e. risk per mg/kg/day).

The US approach to cancer risks has not found favour in the UK and Europe,

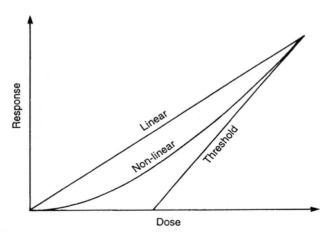

Figure 10.3. Possible extrapolations to low doses: dose-response functions

where the concept of a threshold dose is preferred. The UK Committee on Carcinogenicity in Food, Consumer Products and the Environment (COC) has argued that current models are not validated and large uncertainty results in the risk estimates. In the USA, it has been suggested (Hrudey & Krewski, 1995) that, depending on the slope factor, it should be possible to define a concentration at which the risks are so low that they can be disregarded.

10.3 EXPOSURE ASSESSMENT

10.3.1 Choice of scenarios

The risk assessment may be restricted to a limited range of critical source–pathway–target scenarios or cover an extensive range of plausible scenarios. The latter might include:

- Ingestion of drinking water.
- Inhalation of vapours while showering.
- Ingestion of fish and shellfish.
- Ingestion of homegrown meat products.
- Ingestion of homegrown dairy products.
- Ingestion of homegrown fruits.
- Ingestion of homegrown vegetables.
- Inhalation of vapours inside residence.
- Inhalation of vapours outside residence.
- Accidental ingestion of water while swimming.
- Dermal exposure while showering.
- Dermal exposure while swimming.
- Inhalation of particulates outside residence.
- Inhalation of particulates inside residence.
- Ingestion of soil by a child.
- Ingestion of soil by an adult.

Exposure scenarios are always site specific. For certain land uses specific exposure pathways may be dominant. For example, for residential land the potential direct ingestion of soil by children may be so dominant for some contaminants such as lead that intakes from this pathway may exceed those from all other pathways combined. For certain pathways, uncertainties may be high in the exposure assessment. For example, it has been argued that the uncertainties associated with dermal exposure may exceed those of all other pathways combined (Paustenbach, 1987; McKone, 1990). The uncertainty reflects a lack of quantitative information on key parameters such as the nature, frequency and duration of contact events as a function of type of activity and the bioavailability of soil-borne contaminants. The lack of direct measurements of dermal uptake from soil for all but a few chemicals force the assessor to the use of predictive models. Ferguson (1996) notes that two recent models (McKone & Howd, 1992) look promising, but as with other models have not been adequately validated. The latter problem is also relevant to the modelling of soil-vapour ingress into buildings, which may be a significant risk pathway from source to target (Ferguson, 1996).

All possible exposure pathways involve a number of assumptions and levels of uncertainty. Some are difficult to address using only normal site data without making some very general assumptions. For example, estimation of the potential dose from inhalation of respirable dust is not possible in the absence of information on the proportion of respirable dust in the soil and the concentration of the contaminant in this fraction.

10.3.2 Input data

Assuming that a "reasonable number" of measurements are available (see Chapter 5), there are a number of options for inputting data to exposure assumptions:

- To use the maximum observed concentration – this is necessary when only limited data are available.
- To use the 95% upper confidence limit (UCL) of the arithmetic mean – the USEPA's preferred statistic when sufficient data are available.
- To use the 95 percentile (i.e. the value below which 95% of actual values lie) or some other percentile.
- To use the mean of the observed concentrations.

The last can seldom be justified because it is not possible from any site investigation to know the true mean, which is the appropriate parameter (USEPA, 1992d) regardless of the pattern of daily exposure and regardless of the type of statistical distribution that might best describe the sampling data. The 95% upper confidence limit (UCL) is employed rather than the observed mean to allow for uncertainties due to limited sampling.

It is important to note that the method of calculating the UCL differs for a normally distributed and a log-normally distributed population: most contaminant profiles tend to be log-normally distributed (Gilbert, 1987; USEPA, 1992d).

The need for adequate data was discussed in Chapter 5 and is illustrated here by another case. The principal substances of concern on this approximate 2.5 ha site were polycyclic aromatic hydrocarbons (PAHs). During the initial survey eight surface samples were taken, of which only six were analysed for PAHs. A maximum value of 102 mg/kg total PAHs was obtained.

In a second survey, 56 surface samples were taken. The mean concentration was 73 mg/kg. The maximum observed concentration for total PAHs was 753 mg/kg. The data were approximately log-normally distributed. The calculated 95% UCL of the mean was 96 mg/kg. This value was used in subsequent estimates of risk. In a third survey a further 29 surface samples were taken. The mean concentration was 106 mg/kg. The highest observed value for total PAHs was 523 mg/kg. The 95% UCL of the mean was 229 mg/kg. When the data from the second and third surveys was combined a 95% UCL of 126 mg/kg was obtained.

As discussed in Chapter 5, the sampling frequency must be commensurate with the area of the site that is of concern. For example, on a modern housing development a small domestic garden might occupy no more than about 25–50 m^2. A child could thus be restricted in its play to this area, whereas in a larger garden it could roam more freely. The confidence that can be attached to a risk assessment can be no greater than the confidence that can be attached to having located the one garden, say within one hundred, where the risks are highest. If the probability of finding a critical concentration is only 20%, then there must be doubts about the validity of any estimates of potential risks.

In a case involving an existing garden of about 100 m^2 area, 10 samples were taken. Although the overall assessment was that risks were acceptable, the estimate in respect of risks from PAHs was dominated by a single very high concentration, which because of the sampling frequency represented at least 10% of the garden (the sample was one of six taken from the rear garden). It was conceivable that a small child might adopt the area represented by the sample as its 'favourite spot', in which case the risks would be much higher. So although the overall risks seemed acceptable, further sampling was carried out to determine how large the area of high concentration was. It was found to be less than 1 m in diameter and thus could be regarded as presenting negligible risk.

The USEPA Superfund manual (USEPA, 1989a) states, "averaging soil data over an area the size of a residential backyard (e.g. an eighth of an acre [about 500 m^2] may be the most appropriate for evaluating residential soil pathways". A more reasonable size range assumption in the UK would be 25–250 m^2 depending on the type of development, i.e. from small terraced housing through to large detached homes. The recent USEPA Soil Screening Levels (1996a) refer to a typical residential lot (half an acre, about 2000 m^2). New UK guidance on averaging areas (Department of the Environment, 1997b) relates to different land uses, for example 10 000 m^2 for playing fields, being slightly larger than a football pitch.

The data used in the assessment must also be appropriate. For example:

(i) If present risks to children playing on a site are to be assessed, the data must be from surface samples (say 0.0–0.05 m), rather than from samples taken from 1 m down.

(ii) If future risks to children are to be assessed on a site that is to be developed, then consideration must be given to concentrations to whatever depth there might be disturbance during site works, with an attendant possibility of material being brought to the surface.

(iii) If risks to gardeners are being considered, then information to depths of at least 0.5 m will be required.
(iv) If risks to remediation workers are to be assessed, then data for samples from the full depth of material to be disturbed must be considered.
 (v) If risks to construction workers are to be assessed, then data for samples for the full depth of disturbance must be considered (for example, at least 1 m for service trenches, up to several metres for foundations and several metres for deep sewers).

The targets at risk as well as their routes of exposure will differ for different site uses and even within site uses. Table 10.2 provides a matrix of example potential exposure routes from a contaminated site. The assessor must choose on which critical targets and exposure scenarios to focus the assessment and then proceed to make certain exposure assumptions about the target which will influence their intake or absorption of a chemical.

Table 10.2. Matrix of example potential exposure routes from a contaminated site (Petts & Eduljee, 1994, adapted from USEPA, 1989a)

Exposure route	Residential	Commercial/ industrial	Recreational
Ground water			
Ingestion	L	A	–
Surface water			
Ingestion	L	A	L, C
Dermal contact	L	A	L, C
Sediment			
Dermal contact	C	A	L, C
Air			
Inhalation of vapour			
Phase chemicals			
indoors	L	A	–
outdoors	L	A	L
Inhalation of particulates			
indoors	L	A	–
outdoors	L	A	L
Soil/dust			
Incidental ingestion	L, C	A	L, C
Dermal contact	L, C	A	L, C
Food			
Ingestion	L	–	L

Key:
L = lifetime exposure
C = exposure in children may be significantly greater than in adults
A = exposure to adults (highest exposure is likely to occur during occupational activities)
– = Exposure of this population via this route is not likely to occur

10.3.3   Exposure assumptions

Three types of dose can be considered:

- (i) the administered dose, being the amount ingested, inhaled or in contact with the skin;
- (ii) the intake dose, being the amount actually absorbed into the body; and
- (iii) the target dose, being the amount which actually reaches the target organ.

As introduced earlier, all three are difficult to calculate with any certainty, the last being particularly uncertain.

The calculation of intake of a chemical depends on:

- the concentrations of the chemicals in the affected media;
- the activity patterns of the potentially exposed populations, defining the types of exposure, their duration and frequency;
- the contact rate with the medium, e.g. rate of ingestion of soil or water.

These factors can be represented by a generic equation that can be adapted for various types of exposure (USEPA, 1989b):

$$I = (C \times CR \times EF \times ED)/(BW \times AT)$$

where I = intake or the amount of chemical taken into the body, C = chemical concentration in the affected medium, CR = contact rate with the affected medium per unit time or event, EF = exposure frequency, describing how often exposure occurs, ED = exposure duration, describing how long each exposure event occurs, BW = body weight, and AT = averaging time or the period over which exposure is averaged.

Asante-Duah (1996) includes the equations for target exposures by the different primary routes of exposure. An adaptation of the generic equation to deal with consideration of a chronic daily intake via ingestion would look like the following:

$$CDI = (CS \times IR \times CF \times FI \times EF \times ED)/(BW \times AT)\ \text{mg/kg/day}$$

where CS = measured concentration in soil (mg/kg); IR = ingestion rate of soil from all sources in mg/day; CF = conversion factor in kg/mg ($= 0.000001$); FI = fraction ingested from site as a fraction of the total from all sources (in range 0.0–1.0); EF = exposure frequency in days/year; ED = exposure duration in years; BW = body weight in kg; AT = averaging time in days (by convention for non-cancer risks $= ED \times 365$, and for lifetime cancer risks $= 70 \times 365$).

An adaptation to deal with consideration of a chronic absorbed dose due to dermal exposure would look like the following:

$$AD = (CS \times CF \times SA \times AF \times ABS \times EF \times ED)/(BW \times AT)\,\text{mg/kg/day}$$

where, CS = measured concentration in soil in mg/kg; CF = conversion factor in kg/mg (= 0.000001); SA = surface area of skin exposed in m^2; AF = adhesion factor (amount of soil adhering to skin) in mg/m^2; ABS = absorption factor–fraction of contaminant in soil adhering to skin that is absorbed in range 0.0–1.0; EF = exposure frequency in days/year; ED = exposure duration in years; BW = body weight in kg; AT = averaging time in days (by convention for non-cancer risks = $ED \times 365$, and for lifetime cancer risks = 70×365).

The equation parameters should reflect site or area-specific activity patterns; however, default or standard exposure scenarios can also be used. USEPA (1989a) uses the concept of the "reasonably maximum exposed individual" (RMEI). This is the individual subjected to the "reasonable maximum exposure" which is an exposure which is high yet still has a reasonable likelihood of occurring (essentially synonymous with the more value-laden "reasonable worst case"). Key features of the RME is that one would expect at least 90% of actual exposures to be lower (hence "maximum") and that it could occur (hence "reasonable"). However, there has been concern about use of the RMEI concept. Guidelines released in 1992 (USEPA, 1992c) stress that worst-case analysis should be used as a screening tool to determine whether exposure is significant, not as the basis for characterising the actual or plausible health risks.

USEPA suggests exposure intake values for the RMEI and for an average individual for general application (see Table 10.3, which also includes some figures for an average American adult). The idea is that intake variable values for a given exposure pathway should be selected so that the combination of all intake variables should result in an estimate of the reasonable maximum exposure by that pathway. Some intake variables may not be at their maximum values, but in combination with other variables will yield an RME.

Such standard characteristics need to be applied with caution. Even where sufficient information is available to compute an average exposure rate, this will hide potentially significant variations within populations. For example, to estimate ingestion of contaminants in drinking water, risk assessors often assume an average ingestion rate of 0.03 litres of water for each kg of body weight. However, formula-fed infants and young children have average intake rates as much as eight times those of adults relative to body weight (Cothern, Coniglio & Marcus, 1986; Covello & Merkhofer, 1993).

If the assessment is to be truly site specific, the assumed exposure and individual characteristics need to be related to the site in question; i.e. a RMEI for the site in question must be deduced. The characteristics of two alternative RMEIs used to assess potential risks from an informal recreational area are given in Table 10.4. The main risk was judged to be cancer arising from inhalation of

Table 10.3. USEPA "reasonably maximum" and "average American" exposed individual characteristics

Characteristic	Reasonable maximum exposed individual	Average American
Body weight (kg)	70	70
Event frequency (per year)	350	350
Exposure period (years)	30	9
Lifetime (years)	70	70
Water consumption:		
Water consumption (l/event)	2.0	1.4
Fraction of water contaminated	1.0	0.75
Inhalation of vapours during showering:		
Inhalation rate (m^3/hr)	0.6	0.6
Event duration – inhalation (hours/event)	0.2	0.12
Inhalation of particulates outside residence:		
Inhalation rate (m^3/hr)	1.67	1.4
Event duration (hours/event)	3.0	0.44

contaminated particulates; thus, the different age bands were chosen so that changes in body weight and inhalation rates with age and activity patterns could be taken into account.

While the assessor may use default values for baseline calculations, the more the intention is to try to produce a realistic assessment of risks rather than a worst-case and conservative estimate the more assessors have to use their own judgement. Table 10.5 compares the default assumptions taken from the USEPA literature with the assumptions made in a number of site-specific assessments in the UK. The cases include children playing in a contaminated garden, school children of different age ranges, children with access to a public open space, remediation workers, and gardeners directly employed to look after the landscaped areas of a commercial development.

Table 10.5 indicates a significant variation in the assumptions. It is important to recognise that there is not necessarily a "correct" answer in any particular case: the assumptions made must always depend on the objectives of the assessment (e.g. how "realistic" it is intended to be) and will always depend on professional judgement. However, when there is deviation from those set out in standard protocols it is essential to justify the deviations.

Exposure frequency requires judgement. The values in Table 10.5 were based on the following assumptions:

365 days for child in garden: A worst-case scenario – a "more realistic" 194 was also used allowing for holidays and inclement weather, holidays etc.

Table 10.4. Examples of site-specific RMEIs for parkland

Individual using site for recreation		Inhalation rate (m^3/hr)
0–5 years	Visits site with adult for 6 hours every third day. Walks, runs and plays. Does not rest.	1.1
6–9 years	Visits site with adult or alone for 6 hours every third day. Walks, runs and plays. Does not rest.	1.7
10–18 years	Visits site with adult or alone for 6 hours every third day. Walks, runs and plays. Does not rest.	2.7
19–38 years	Jogs on site for 1 hour every day apart from holidays.	3.9
39–70 years	Walks dog on site for 1 hour each morning and evening every day apart from holidays.	1.7
Individual growing up near site and then working on it:		
0–5 years	Visits site with adult for 6 hours every third day. Walks, runs and plays. Does not rest.	1.1
6–9 years	Visits site with adult for 6 hours every third day. Walks, runs and plays. Does not rest.	1.7
10–18 years	Visits site with adult for 6 hours every third day. Walks, runs and plays. Does not rest.	2.7
19–40 years	Works full time on site.	2.1
41–60 years	Works full time on site.	2.1
61–70 years	Walks dog on site for 1 hour each morning and evening every day apart from holidays.	1.7

198 days for public open space (POS):	5 days/week for 24 weeks (summer) + 3 days/wk for 26 weeks (winter) + 0 days/wk for 2 weeks (holiday)
131 days for schools:	(194 term days)*((119 days/yr rainfall > 1 mm)/365)
240 days for remediation worker:	(52 × 5-day weeks) − (4 weeks holidays etc.)
200 days for gardener:	(52 × 5-day weeks) − (40 days inclement weather) − (4 weeks' holidays etc.)

Table 10.5. Assumptions used in some site-specific risk assessments

Site	Target	Age	Exposure duration (years)	Exposure frequency (days/yr)	Bodyweight (kg)	Surface area (m²)	Ingestion rate (mg/day)	Fraction ingested	Absorption factor
EPA "default" values	Child		70 yrs lifetime 30 yrs national 90% at one residence 9 yrs national median at one residence	365	16 (1–6 years – national median)	about 0.70 total (3 < 6)	200 (1–6)	site specific depending on activity patterns	varies with chemical
	Adult				70 average	About 1.94 total adult male	100 (> 6)		
Existing garden	Child	4 (–14)*	10	365	16	0.2000	200	0.5	0.01
Public open space with play area for small children	Child	5 (–16)*	11	198	18.5	0.0500	50	0.5	0.01
Secondary school to be built on above site	Child	11 (–18)*	7	131	34	0.0773	50	0.5	0.01
Combined infants and primary school	Child	5 (–11)*	7	131	18.5	0.0500	50	0.5	0.01
Commercial development	Remediation worker	Adult	1	240	70	0.3120	100	1	0.01
	Gardener	Adult	10	200	60	0.3120	100	1	0.01

*e.g. from just 4 to just 14.

Several questions are pertinent to the above assumptions. For example, is it reasonable to conclude that just 1 mm of rain during a day would be sufficient to prevent access to the playing field?

Bodyweights for adults are averages. USEPA uses 70 kg whereas the World Health Organisation (WHO) uses 60 kg. Tabulations of bodyweights versus sex and age are available from a number of sources (e.g. Heimendenger, 1964; Hamill, Johnson & Roche, 1979; USEPA, 1991b).

The ingestion rates for the schools and public open space (POS) are low in comparison to the USEPA default suggestions. Ferguson & Marsh (1993) have reviewed the information available on ingestion rates. The figures suggested by various authors are highly variable and make widely different assumptions:

 (i) about the ages at which exposure occurs (e.g that a child below one will be insufficiently mobile to be exposed to soil);
 (ii) whether to aggregate ingestion of soil and indoor dust;
(iii) whether to assume ingestion occurs every day or only on a proportion of days during the year so that the average is proportionally lower (since the USEPA model has a separate variable for exposure duration it would seem more appropriate to take a figure for "every day"); and
(iv) how to allow for the "pica" child who deliberately and habitually ingests soil (geophagia) or frequently mouths contaminated toys etc.

Ferguson & Marsh (1993) do not recommend a figure, but seem to accept 40 mg/kg as reasonable up to age 6 years. However, they also state,

> we have used a computer model to simulate the combined effect of inadvertent soil ingestion at 40 mg/kg and occasional deliberate soil ingestion, over a few days or months, at 10 g/day. This shows that excess consumption at 10 g/day for only ten days could be enough to double the average daily dose (averaged over the first six years of childhood). If such behaviour persists for two or three months, the daily dose over the first six years could increase tenfold.

Given this statement, the USEPA default value of 200 mg/kg for a small child, or even the Australian default value of 100 mg/kg, seem reasonable.

10.4   EFFECTS ASSESSMENT

10.4.1   Basic approaches

Section 10.2.2 introduced two approaches to dealing with cancer and non-cancer risks: i.e. the threshold and no-threshold approaches. For non-cancer risks estimates of the hazard posed by a chemical are expressed as measures of the exposure to the chemical that could occur over a prolonged period without ill-effect (a "reference" dose exposure). Exposures below this reference level are assumed to be "safe". Risk estimates for exposure to chemicals that cause non-cancer effects are conducted by comparing the actual or potential exposures to the reference level for the chemical concerned. Different measurements of non-cancer toxic hazard are employed for different routes of exposure. In the USEPA system, oral or dermal exposures are generally compared to reference doses (RfDs):

> an estimate (with uncertainty spanning perhaps an order of magnitude or greater) of an exposure level for the human population, including sensitive subpopulations, that is likely to be without an appreciable risk of deleterious effects during a lifetime.

Inhalation exposures are compared to reference concentrations (RfCs). Reference doses are commonly expressed as mg/kg/day. Reference concentrations are expressed in mg/m^3 or similar units.

For the direct ingestion pathway, a *hazard quotient* is calculated as a ratio of the chronic daily intake to a reference dose expressed in the same units, i.e.:

$$\text{Hazard quotient} = \text{CDI/reference dose}$$

The reference dose might be a published acceptable daily intake (ADI) or tolerable daily intake (TDI), a generic reference dose developed for the purpose (e.g. USEPA RfD), or a value derived for site-specific use.

The *hazard index* for a contaminant is the summation of the hazard quotients for each exposure route taken into account in the assessment. The overall site hazard index is the summation of the hazard indices for each of the contaminants taken into account in the assessment.

For carcinogenic effects a no-threshold approach is generally taken in the USA, although as noted earlier there has been a tendency in the UK to use the threshold approach as for non-cancer risks. In the no-threshold approach the slope of the dose-response function provides an indication of how powerful a chemical is in causing cancer. Two estimates of this slope are employed in the USEPA procedure: slope factor expresses the slope in dose-related units, while unit risk expresses the slope in concentration-related units. Lifetime excess risk of developing cancer is calculated for the ingestion pathway using the following equation:

$$\text{Excess cancer risk} = CDI \times SF$$

where, CDI = the chronic daily intake (mg/kg/day), and SF = the slope factor $(mg/kg/day)^{-1}$.

The estimates of risk derived from such calculations are *hypothetical estimates* of *individual excess risk*. Individual excess risk is an estimate of the probability that an individual will *develop* cancer from the specified exposure (e.g. 1 in 1 million or 1×10^{-6}). The term *excess risk* indicates that the risk is *in addition to* the risk that the individual has of developing cancer from other causes.

Individual excess risk is:

(i) An estimate of the probability of *developing* cancer, *not* the direct probability of dying from cancer, although judgements about the latter can be made (e.g. there is a 50% chance of dying from cancer if it is contracted).
(ii) An estimate of risk *to an individual*. It is not an estimate of the number of cancers that would be expected in a population (except in the unlikely event that all members of the population had identical exposures and personal characteristics such as sex and body weight).
(iii) A *hypothetical* risk estimate, derived from information on toxic hazard and exposure, not from an observed incidence of cancer in a defined population.

As for the non-cancer risks, the excess risks for a single contaminant are summed for each exposure route, and the excess risks for individual contaminants are summed to give an overall site estimate of excess cancer risks.

10.4.2   Reference sources

Acceptable or tolerable daily intakes are available from a range of standard sources (e.g. JEFCA 1972; 1993). USEPA sources used in the Risk*Assistant model are:

● the Integrated Risk Information System (IRIS);
● the Health Effects Assessment Summary Tables (HEAST).

The RfD for a chemical is set for the most dose-sensitive of the various non-carcinogenic effects which a chemical may induce, by first determining a no observed effect level (NOEL) and then applying a safety factor to account for interspecies variability, sensitivity etc. For example, lead is associated with damage to blood, skeletal development and the central nervous system. The latter is by far the most sensitive to lead intake, and is the basis on which the RfD (and the acceptable daily intake in the UK) is derived.

The USEPA safety factor combines an uncertainty factor (UF) to deal with interspecies variability etc., and a modifying factor (MF) which reflects the confidence in the quality of data for predicting human risk. In addition, USEPA RfDs are accompanied by an overall statement of the agency's confidence in the RfD (high, medium, or low). Some examples of data from a database of RfDs are provided in Table 10.6.

Similar concepts underlie the setting of ecotoxicological criteria. Considerable data exist for measures of acute aquatic toxicity in the form of LD_{50}, LC_{50} and EC_{50} values: these represent the lethal dose and concentration required to kill 50% of the exposed organisms and an effect-specific concentration of a chemical that is required to cause a particular effect. Ecological standards have been derived by applying a safety factor (application factor) to such measures of acute

Table 10.6. Example output from Database (data source IRIS except where indicated)

CAS	Substance		Uncertainty factor	Modifying factor	Confidence
Ingestion		RfD (mg/kg/day)			
7440-36-0	Antimony	0.0004	1000	1	Low
7782-49-2	Selenium and compounds	0.005	3	1	High
74-90-8	Hydrogen cyanide	0.02	100	5	Medium
57-12-5	Cyanide	0.02	100	5	Medium
143-33-9	Sodium cyanide	0.04	100	5	Medium
151-50-8	Potassium cyanide	0.05	100	5	Medium
7440-39-3	Barium	0.07	3	1	Medium
Inhalation		RfC (mg/m^3)			
74-90-8	Hydrogen cyanide	0.0005*			
7440-39-3	Barium	0.003			

* = data source is HEAST
Blanks = no hazard information found

toxicity: typically a factor of 100 (USEPA, 1989b). This would be taken as the threshold or safe level of the pollutant in the organism. Smaller factors than 100 (e.g. 10) may be used where sufficient data are available for a large number of relevant species. Larger factors (e.g. 1000) may be used where data are available for only a limited number of species in a taxonomic group (e.g. where earthworms are considered to be representative of soil invertebrates in general). Sources of information on ecological dose-response data include the USEPA's AQUIRE (Aquatic Information Retrieval) database (USEPA, 1990d) and the Ecological Effects Database (USEPA, 1988f). Van Straalen and Denneman (1989) proposed a more rigorous method for deriving ecotoxicological standards by extrapolation from data on species that matched to the site being assessed.

An ecological hazard quotient/index can be derived in the same manner as the human health index. Suter (1993) provides an example of the application of the hazard quotient to the assessment of the impact of a waste site on an adjoining estuarine river and wetlands. Effects on herons and mammals were estimated using the method and existing toxicity data for individual chemicals. Herons were assumed to get 35% of their diet from contaminated fish and the expression of toxicity was the highest NOEL from a chronic test of any bird species, or if only acute data were available for a chemical the LC_{50} adjusted by 1/5 as an acute-chronic correction factor.

No single list of reference doses etc. can be published by an authority without revision and review. In the case of occupational exposure, it is accepted that occupational exposure standards (OESs) and maximum allowable concentrations (MACs) should be the subject of continuing review, and that at least annual updates are required (as in the case of the UK Health & Safety Executive's publication of OESs). There has been recent acknowledgement that all is not well with IRIS (Anon, 1996b). The USEPA has announced plans to introduce external peer review to IRIS as part of a pilot project to improve the quality of the database.

In general, choice of reference doses should correspond to the model that is being used: e.g. if a USEPA model is being used, USEPA RfDs should be used. Table 10.7 illustrates the effect that choosing a different data set can have on the estimates of risk. The table shows the different hazard quotients which would be derived using either the USEPA RfD or the Dutch maximum tolerable risk (MTR) dose for a site with a suite of analytes. Using the RfDs produces a site HI of 3.798, compared to 2.885 using the Dutch MTR. Further inspection of the table shows that assuming chromium is present in the more toxic hexavalent form rather than the less toxic trivalent form makes a large difference to the index – see the note to the table. It is relevant to note that views on chromium toxicity and cancer potency are changing (Anon, 1996b; James, 1996), and that neither Cr^{VI} or Cr^{III} is stable in the soil environment. Similarly, the type of phenol (generic) and cyanide present will affect the estimate. Results such as those in Table 10.7 may indicate to the risk assessor that there could be considerable benefit in additional, more detailed analytical studies.

Table 10.7. Hazard indices calculated using different reference values[1]

Analyte	Concentration (mg/kg)	USEPA reference doses (mg/kg/day)	Hazard quotient from RfDs[2]	Dutch maximum tolerable risk (MTR) dose (mg/kg/day)	Hazard quotient from MTR
Arsenic	400	0.0003	1.252	0.0021	0.179
Cadmium	150	0.0005	0.281	0.001	0.141
Chromium	10 000	0.005	1.879[3]	0.005	1.879
Copper	1 300	0.04	0.030	0.14	0.009
Lead	20 000	0.1	0.188	0.036	0.521
Mercury	20	0.0003	0.063	0.00061	0.031
Nickel	700	0.02	0.033	0.05	0.013
Zinc	3 000	0.3	0.009	1	0.003
Cyanide	5 000	0.02	0.047	0.05	0.094
Phenol[4]	1 000	0.6	0.016	0.06	0.016
Site indices for ingestion			3.798		2.885

[1]For remediation worker using assumptions listed in Table 10.5.
[2]Current at time of study.
[3]RfD for Cr^{VI}. RfD for Cr^{III} = 1 mg/kg/day giving a hazard quotient of 0.009 and reducing site hazard index to 1.930.
[4]For the compound phenol (C_6H_5OH), analysis is for total phenols.

The figures for lead also merit attention. Whilst it is necessary to include lead in the assessment of toxic hazards as shown, it is important to recognise that the primary concern is for the neurological effects associated with increased blood lead levels. Separate models to predict soil/blood lead responses have been developed and should be used where potential risks to children are of special concern (SEGH, 1993).

In relation to cancer potency factors the International Agency for Research on Cancer (IARC), based in France, has an ongoing programme that evaluates the available relevant epidemiological and experimental data on chemicals to which humans are known to be exposed.

The USEPA has qualitatively classified substances in terms of the strength of evidence available that they are human carcinogens, and with respect to the dose-response functions for different routes of exposure, but not with regard to the organ systems they affect or the nature of the tumours induced. A system of six alphanumeric categories is currently used to characterise the weight of evidence that a chemical may induce cancer in humans, as follows:

A = known human carcinogen
B1 = probable human carcinogen – limited human data

B2 = probable human carcinogen – inadequate or no human data
C = possible human carcinogen
D = not classifiable as human carcinogen
E = evidence that not carcinogenic in humans

However, proposals for an amended system were published in 1996 (USEPA, 1996c; Anon, 1996c). These include the replacement of the current alphanumeric classification with "weight of evidence narrative" and the introduction of just three "descriptors" for classifying human carcinogenic potential: known/likely; cannot be determined; and not likely. The Canadian classification system is similar to the US but instead of six alphanumeric categories it uses six groups. Those in Groups I and II, carcinogenic and probably carcinogenic, are considered to have no thresholds for effects. For those in Groups III–VI there is the assumption of a threshold dose and tolerable daily intakes are calculated (CCME, 1995).

Slope factors are not available for many suspected carcinogens. Those for polycyclic aromatic hydrocarbons (PAHs) are of particular importance because one of them, benzo(a)pyrene, has a high slope factor, and in the absence of other information, it is customary to treat the total PAH concentration as though it were all benzo(a)pyrene. Since not all of the PAHs normally determined are regarded as human carcinogens, this can lead to an overestimate of risks. The USEPA has produced separate guidance (USEPA, 1993) on the relative cancer potency of PAHs.

It is inappropriate to use the oral slope factor to evaluate the risks associated with dermal exposure to carcinogens such as benzo(a)pyrene, which cause cancer at the point of contact (USEPA, 1989c). These types of skin carcinogens and other locally active compounds must be assessed separately. In contrast, use of the oral slope factor is appropriate for chemicals such as arsenic which are believed to cause skin cancer through systemic rather than local action.

10.4.3 Additive and synergistic effects

There is no agreed method for the assessment of health risks from exposure to mixtures of chemicals: the toxic effects of two or more chemicals in combination may be additive (i.e. the sum of the individual toxicities), synergistic (i.e. more than the sum), or antagonistic (i.e. less than the sum). The USEPA methodology assumes additivity of risks from individual chemicals. For carcinogens this means that the overall carcinogenic risk is the sum of the risks from individual chemicals, while for exposure to mixtures of non-carcinogens the individual hazard indices are summed.

Substances differ markedly in their action in the body. For example, cadmium accumulates in the body, in particular in the kidneys and liver, and is only slowly secreted (the half-life of cadmium in the body is about 30 years). If the accumulated dose exceeds a critical value then damage to the kidneys may result. A variety of other harmful effects have also been linked to cadmium (Department of the Environment, 1980). In contrast, lead is more readily secreted from the body. The critical organ is the blood and if concentrations become too high in children it is considered that lasting damage may result (e.g reduced intelligence). Phenol causes burning and necrosis of the skin, and is readily absorbed through the skin being translocated to the liver where it can cause permanent injury.

Certainly, the overall site index provides a useful tool for comparing risks between sites and between zones on the same site. A more disaggregated approach may be justifiable in some instances when the acceptability of the potential risks arising from a site is being evaluated. However, permitting each of 10 compounds to reach a hazard quotient of 90% of the defined acceptable hazard index for the substance (giving a site index of 9 if the acceptable index for each was 1 or 1.8 if it was set at 0.2) would leave most assessors feeling very uncomfortable.

The importance of recognising synergistic effects (which are not generally allowed for in the production of generic guidelines) and the difficulty of allowing for such effects in practice are indicated by emerging information on the oestrogenic effects of pesticides. It has been found that a mix of two of the pesticides (dieldrin, toxaphene or endosulfan) is 160 to 1600 times more potent than the individual chemicals (Anon, 1996d; Arnold *et al.*, 1996). Similarly, a combination of two polychlorinated biphenyls (PCBs) has been shown to be five times as active as the individual PCBs. Suggestions are beginning to emerge that there may be no safe threshold for chemicals that have the potential to disrupt the hormone system (Anon, 1996d).

10.5 RISK EVALUATION

10.5.1 Judgements about acceptable risks

A judgement about the acceptability of a hazard quotient or index must be made both when the results of risk estimation are being evaluated and also when site-specific risk-based criteria are being derived. Determining whether a computed site hazard index for non-cancer risks is acceptable involves more than determining simply that it is less than one (i.e. the computed chronic daily intake (CDI) is less than the reference dose). For many substances, there will be exposures not related to the site (e.g. workplace, home, school) and these will differ for different human targets. No guidance has been generally available on what allowance to make for these non-site exposures which may be difficult to quantify. A complete estimation of risks would involve summing exposures by these non-site pathways with those from the site. However, the data on which to base this assessment may be difficult to come by and the resulting calculations full of uncertainties.

The new procedural risk assessment guidance being developed in the UK at the time of writing (Department of the Environment, 1997b) considers the concept of the tolerable daily soil intake, which is the tolerable daily intake minus the mean daily intake. The latter is a measure of the background intake of the contaminant from the ambient concentrations in food, water and air for the UK population. Derivation of the latter data has been part of the department's research programme on toxicological data (Department of the Environment, 1997c).

The practical option may be to assume that only a proportion of the TDI (or RfD) may come from the site. Setting this proportion at one half should be reasonably protective for many compounds, but it should be noted that for some compounds groups of the populations already approach or may exceed the TDI.

Thus, in general:

$$PDIS = TDI - MDI \text{ (or } EAI)$$

where: $PDIS$ = permitted or tolerable daily intake from site via all routes of exposure
TDI = tolerable daily intake
MDI or EAI = mean daily intake or estimated average intake for defined population group

If the MDI (EAI) equals or exceeds the TDI then the PDIS would be zero: i.e. there is no balance to allocate to contaminated soils. In these, circumstances the PDIS becomes a matter of pure judgement.

If the estimate of risks only includes some pathways (e.g. direct ingestion and dermal exposure) because other routes are regarded as likely to be negligible or because the data are not available on which to base the calculation, an allowance must also be made for exposure by non-computed pathways. To allocate the whole of the difference between the TDI and MDI to just one source (e.g. soil on a contaminated site) may not be considered to be health protective, bearing in mind that the receptor may be exposed to contaminated soil at more than one location. Deciding on the amount allocated to soil is part of an overall risk management decision where the cost of reducing risks from any one source must balance the benefits.

Deciding on what constitutes an acceptable excess cancer risk is no easier. When evaluating the risks, both the calculated estimate and the magnitude and type of population at risk should be taken into account. Account must also be taken of the confidence there is that the site investigation has been thorough and that the estimated risks are therefore reasonably reliable. For cancer risks it is necessary to set numerical risk targets that define acceptability. The only UK guidance which has considered acceptability is in relation to guidance on accident hazards (i.e. acute risks of immediate harm, not chronic cancer risks). The Health and Safety Executive suggested an acceptability band of $1 \times 10^{-5} - 1 \times 10^{-7}$ risk per annum of receiving a dangerous dose for an individual member of the public (HSE, 1992). The USEPA has set the range of acceptability of risk as 1×10^{-4} to 1×10^{-6} risk of death over a lifetime.

About one-third of all deaths in the UK are attributed to cancer (see annual mortality statistics published by the Office of Population Censuses and Surveys), so that even a risk of 1×10^{-3} (1 in 1000) might seem small if only a small population were to be exposed to the risk. However, in practice, it would be difficult to justify risks outside of the range 1×10^{-6} to 1×10^{-4}; those below 1×10^{-6} could be regarded as negligible, whereas those above 1×10^{-4} are generally regarded as unacceptable.

It seems legitimate to ask whether different standards should be applied to new developments compared to existing developments. For example, should the aim for a new school site be 1×10^{-6}, whereas on an existing school site could a value of between 1×10^{-5} and 1×10^{-4} be accepted? Arguments might relate in the former case to an objective not to produce an increase in risk to people, whereas in the latter case they could relate to the costs involved in the remediation of existing risks and who might have to pay for these. There can be no easy answer to such a question. In a particular case it may depend on the confidence which the assessor has in their ability to persuade the authorities that risks above 1×10^{-6} are unacceptable (requiring action) or the parents that risks below 1×10^{-4} are acceptable (requiring no action).

Full evaluation of risks requires not only consideration of probabilities or other expressions of the magnitude of risks (e.g. a hazard quotient) but also of the consequences of the risk being realised. This requires consideration of social and economic factors. For example, if the risk of a newly planted tree's dying is one in two, the consequences are aesthetic and monetary. One response might be simply to plant twice the number of trees originally planned. If, on the other hand, 1 child in the 1000 attending a local comprehensive school were to get skin burns due to contact with a chemical, not only would it be "unacceptable" to the child, but it would be likely to cause such concern to the children and their parents that it would be worth reducing the risks to well below this level, say to 1 in 100 000. How far one might go to reduce the calculated risks would in part depend on the confidence that one had in the calculations being reasonably realistic and in the underlying data on which they are based.

A further example of a probably unacceptable risk would be that of explosion due to the presence of methane. Whether the explosion occurred in a house or a school, its impact would be such that most assessors would want to reduce the risks to a very low level by adopting appropriate protective measures. The calculation of risks for this type of event differs from that for assessment of risk from contamination, relying largely on combining the probabilities of a series of events occurring: for example, that methane is present in the ground; that there is a pathway for migration to beneath the building; that there is a crack in the floor; that the gas will enter a confined space; that an explosive concentration will accumulate; and that there will be a source of ignition.

There has been a temptation among many risk assessors to make risk evaluations by comparison of the assessed risk with other risks to which people are exposed. The UK Health and Safety Executive has included comparative risk tables derived from mortality statistics in its tolerability discussion (1992). However, comparing an individual's risk of dying from exposure to a contaminated site with the risk of dying in a motor vehicle accident, or of cancer, or of being hit by lightning etc. has a number of serious problems, for example:

(i) The risk data are often averaged across an entire population, not the small group of people actually exposed to the risk (for example, many data on aircraft accidents).
(ii) The risk data may relate to "voluntary" risks which people take in order to derive some direct benefit, whereas exposure to a risk from a contaminated site might be viewed as an involuntary risk which should be lowered as far as possible (discussed further in Chapter 11).

The UK Health and Safety Executive has drawn a distinction between "acceptable" and "tolerable" risks (HSE, 1992). It prefers the latter concept, which refers to a willingness to live with a risk so as to secure certain benefits *and* in the confidence that the risk is being properly controlled. As in relation to any type of risk, the relevance of tolerability to the evaluation of contaminated sites will be site specific and vary depending on the purpose of the risk assessment being conducted.

10.5.2   Uncertainty analysis

As will have become evident, uncertainty in the calculated risk estimate can arise from a number of sources, for example:

(i) During the hazard identification stage, through incomplete source characterisation; failure to identify key contaminants as critical or indicator chemicals; failure to characterise the potential on-site population or surrounding population.
(ii) During the hazard assessment stage, through inappropriate sampling and a failure to identify a contaminant hot spot; failure to understand groundwater flows; a misuse of generic guidelines and a failure to understand the need for a site-specific risk assessment; inappropriate chemical analysis.
(iii) During the site-specific risk estimation stage, in the applicability of animal toxicity data to humans; in the incorrect choice of activity or exposure scenarios (e.g. using an average consumption pattern when more sensitive targets may consume twice the average); in extrapolations from high to low doses to derive a reference dose; in the determination of background intake (i.e. concentration of the contaminated soil to risk).

As introduced in Chapter 2, and elaborated in Chapters 5–7, uncertainty can arise at least as much from a combination of human error, ignorance, and investigation and assessment errors as from fundamental data and information deficiencies at the risk estimation stage.

As discussed in Chapter 2 and at the beginning of this chapter, the primary concern has been in the consistent use of worst-case assumptions throughout the risk assessment process. Maxim (1989) concluded that the "better safe than sorry" ethic in risk management provided the rationale for the conservative or worst-case approach. The precautionary principle might also engender such an approach. However, on the other hand, quantified estimates of risk provide an opportunity to derive site-specific risk understanding whereas the use of generic guidelines derived to protect the whole population might itself lead to an overly conservative (protective) approach.

The purpose of uncertainty analysis is to:

- determine the effect on the final risk estimates (e.g. from a particular substance) of using alternative parameter values;
- indicate the relative contribution of each scenario to risks from the contaminated medium; and
- indicate the proportion of the total risk from a chemical attributable to contamination of each medium.

The two common techniques for studying uncertainty in input parameters and exposure assumptions are probability analysis and sensitivity analysis (introduced in Chapter 2 and see also Figure 10.1). In probabilistic analysis uncertainty is propagated through the assessment by expressing the variability in the input parameters as probability distribution functions (normal, log-normal, triangular, uniform etc. distributions). The resulting risk estimate is also expressed as a probability distribution function, from which required values can be extracted: for example, the median risk, the 95 percentile etc.

When the available data are insufficient to express the parameters as distribution functions, it may still be possible to describe the potential ranges of values which these parameters might assume. In sensitivity analysis, the risk assessment is run with different combinations of minimum, average and maximum values. It is then possible to identify parameters which, when varied, have a significant effect on the final estimate, and those parameters to which the final result seems to be relatively insensitive.

The final analysis would usually (as a minimum) consist of a series of tables and narrative statements. For example, the estimates for a standard reasonable maximum exposed individual using USEPA default values might be compared with those for an average adult, or the risks arising to two site-defined individuals might be compared. For site-defined individuals, changes in the exposure pattern might be considered (e.g. number of days spent on the site each year) or changes in intakes (e.g. concentrations of particulates in inhaled air). Note that this analysis should result in some higher estimates of risk (i.e. more worst-case assumptions are tested) and some lower estimates of risk (i.e. some less "conservative" assumptions are tested) compared to the "main" calculation.

When linear models are being employed it is easy to see how variation in the assumptions will affect the calculated risks: doubling a particular parameter will double or halve the end result. When stochastic models are being used, variation of the input parameters is more difficult, since instead of substituting a single value a different distribution for the parameter would need to be supplied (presumably, however, the "new" range could consist of a single value). Such analysis is less necessary, however, since such models provide a distribution of risk estimates and, say, the 95 percentile value can be compared rapidly with other percentile values.

Such analyses are also beneficial when deciding on a remediation strategy. They may show that 80% of the estimated risks arise from one exposure route; eliminating this route may be sufficient to bring overall risks within an acceptable range. Probabilistic risk output can allow for interested parties to see the full range of variability and uncertainty, and the nature and extent of the professional judgements in a risk assessment. However, the output needs careful explanation just as a deterministic output. It is not as yet apparent that individuals find decision making easy when a range or distribution of risk estimates is presented to them.

10.6   RISK ASSESSMENT CASE STUDY

10.6.1   The site

The site contains a nursery school (starting age 4), and combined infants and primary school (ages 5–10). The schools, with their grassed playing field, occupy about 2 ha. The site is essentially flat. There are flowerbeds around the school itself. The main part of the infant/primary school was built about 20 years ago and has piled foundations. The nursery school, and some of the primary school children, are housed in temporary buildings raised about 0.5 m above the ground surface. The playing field is generally not in use at weekends or during the school holidays, but access is comparatively easy. The site is surrounded by houses built in the 1900s.

As a result of a screening process to identify potentially contaminated sites, the school was found to be built on a clay extraction pit filled with wastes of unknown origin over 70 years ago. Investigation revealed the presence of industrial wastes including "oily" materials. The site was sampled on an approximately 20 m grid (51 sampling locations) with samples being taken from the immediate surface layer (top 50 mm) and at 0.5 m intervals to about 5 m total depth. The fill depth is uneven, but generally varies between 3 and 4 metres. The groundwater table, at the time of the investigation, was at a depth of about 1 m.

10.6.2   Contaminants of concern

While a variety of contaminants arising from coal carbonisation and associated with the urban setting of the school were identified, a preliminary screening of the data resulted in a decision being made that the contaminants of primary concern were polycyclic aromatic hydrocarbons (PAHs). These were present in concentrations of up to 747 mg/kg in the immediate surface layers with an average concentration of 161 mg/kg. Concentrations of up to 3.5% were found at depths of 0.5 m and below. Some PAHs (e.g. benzo(a)pyrene) are proven human carcinogens; others are toxic. Benzo(a)pyrene is regarded as a potent carcinogen by ingestion and also by direct skin contact.

Comparison with the ICRCL trigger values (which were relevant at the time of the assessment) showed that the measured concentrations were in excess of both the threshold value for domestic gardens and other situations where there is intensive contact by children (50 mg/kg) and the action value (500 mg/kg). Therefore, it was decided to carry out a site-specific risk assessment.

It was also decided that other issues, such as the possible migration of contaminants off-site to affect the neighbouring properties and the possible presence of volatile organic compounds (VOCs) such as benzene, would have to be addressed, but these are not discussed here.

10.6.3 Objectives of the risk assessment

The immediate objective of the risk assessment was to determine the current risks to all those who might come into contact with the site, so that decisions could be made about:

(i) whether any immediate measures were required to mitigate the risks, including, if necessary, closing the school; and
(ii) whether any longer-term measures should be taken to deal with the potential problems that had been revealed.

Those identified as potentially at risk included the school children; teachers; ground maintenance staff; those maintaining underground services or carrying out building works that involved disturbance of the ground; and children and adults trespassing on the site at weekends and in the evening.

10.6.4 Exposure pathways

It was considered that exposure of the children might occur through ingestion, inhalation, absorption through the skin and absorption through cuts and grazes. Exposure would occur while playing outside and also through dust brought into the buildings. The teachers would be similarly exposed, but were considered less likely to have the intimate contact that a child might have: for example, while playing games. Those cutting the grass would be most likely to be exposed via inhalation.

Table 10.8 attempts to relate qualitatively the probability of exposure to the magnitude of the risk if exposure occurs.

Table 10.8. Qualitative risks (probability of exposure/risk if exposure occurs)

Exposure route	Child	Teacher	Grounds worker	Services worker	Comments
Ingestion of soil/dust	high/high	high/high	high/high	high/high	Exposure is certain. PAHs carcinogenic when ingested.
Ingestion of drinking water	high/nil	high/nil	low/nil	low/nil	Water not contaminated, therefore nil risk.
Inhalation of soil/dust	high/low	high/low	high/high	medium/low	PAHs carcinogenic by inhalation but no potency data available.
Inhalation of vapours from drinking water	high/nil	high/nil	nil/nil	nil/nil	Water not contaminated therefore nil risk.
Contact with soil/dust	high/high	high/high	high/high	high/high	Benzo(a)pyrene causes cancer through contact but no guidance data available.
Contact with drinking water	high/nil	high/nil	low/nil	low/nil	Water not contaminated therefore nil risk.
Contact with groundwater	low/high	low/high	low/high	high/high	Groundwater contaminated.
Ingestion of groundwater	low/high	low/high	low/high	low/high	Groundwater contaminated.

It was decided that the children and teachers would only be exposed to the immediate surface layers. The samples taken included some from goal mouths etc. where the bare earth had been exposed to depths of up to about 150 mm below the surrounding surface level. The results on these samples did not differ significantly from the generality of surface samples. Similarly, soil samples from the flower beds (taken to depths of about 400 mm) did not differ significantly from the generality of surface samples. However, those engaged in deep cultivation or excavating services could be exposed to much higher concentrations.

The water-supply pipes were believed to be in contact with the wastes. The water supply was therefore checked for the presence of PAHs and other contaminants. No evidence of contamination was found. Thus, it was decided that there were no risks arising from exposure to the drinking-water supply through ingestion, vaporisation of contaminants during showering or contact with the water. On the other hand, with water pipes laid at not less than 0.8 m depth, and the groundwater at only 1 m depth, workers could come into contact with contaminated groundwater.

10.6.5 Exposure assessment

As noted above, the maximum total PAHs concentration (the sum of 16 individually determined compounds) was 747 mg/kg. Analyses were available for 51 surface samples. Inspection of the data (see Figures 10.4 and 10.5) showed them to be approximately log-normally distributed. Statistical parameters are presented in Table 10.9.

The concentrations measured at 1 m were judged most relevant to the expo-

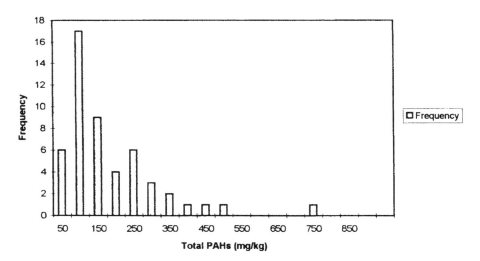

Figure 10.4. Distribution of results for surface soils: measured values

Table 10.9. Statistical parameters for surface samples and samples from 1.0 m

	Surface mg/kg	1.0 m mg/kg
Mean	161	5 644
Geometric mean	122	4 298
Median	113	4 008
Standard error	19	770
Standard deviation	134	5 496
Minimum	24	877
Maximum	747	35 309
95% upper confidence level	203	6 740

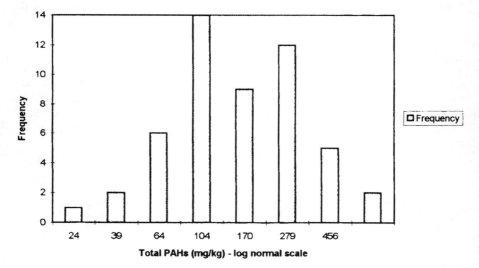

Figure 10.5. Distribution of results for surface soils: log-normal

sure of an excavation worker. Concentrations at this depth ranged up to 3.5%. The average concentration and the 95% UCL were 5644 and 6740 mg/kg respectively.

The average proportions of each of the 16 PAHs determined in the 51 samples are listed in Table 10.10 (column C).

PAHs have varying cancer potencies and the risk assessment must therefore be based on an assessment of the risks posed by individual compounds. In the USEPA procedure (USEPA, 1989a), compounds are judged relative to the slope factor for benzo(a)pyrene. The derived slope factors (7.3 × the potency listed in column A) are listed in Table 10.10 (column B).

Table 10.10. Average proportion of individual PAHs

Compound	A Potency relative to benzo(a)pyrene (order of magnitude)[1] (USEPA, 1993)	B Slope factor derived from benzo(a)pyrene ($=7.3*A$)	C Average proportion of total PAHs	D Contribution to effective slope factor for total PAHs ($=B*C$)
naphthalene	non-carcinogen		0.14	0
acenaphthalene	non-carcinogen		0.00	0
acenapthene	non-carcinogen		0.01	0
fluorene	non-carcinogen		0.01	0
phenanthrene	non-carcinogen		0.06	0
anthracene	non-carcinogen		0.08	0
fluoranthene	non-carcinogen		0.11	0
pyrene	non-carcinogen		0.05	0
benzo(alpha)anthracene	0.1	0.73	0.07	0.0511
chrysene	0.001	0.0073	0.06	0.0044
benzo(b)fluoranthene	0.1	0.73	0.11	0.0803
benzo(k)fluoranthene	0.001	0.0073	0.01	0.0007
benzo(a)pyrene	1	7.3	0.014	1.0220
indeno(1,2,3,c,d)pyrene	0.1	0.73	0.10	0.0730
dibenz(a,h)anthracene	1	7.3	0.02	0.1460
benzo(ghi)perylene	non-carcinogen		0.03	0
				Total $= 1.3775$

As it is usual to regard the risks from PAHs as additive, estimation of risks can be simplified by calculation of a weighted average slope factor, taking into account the potency of the individual compounds and the average proportion present in the samples (see Table 10.10, column D).

The calculated total weighted slope factor for all PAHs used in the calculations was 1.38 $(mg/kg/day)^{-1}$ compared to the slope factor of 7.3 $(mg/kg/day)^{-1}$ for benzo(a)pyrene. It is interesting to note that a subsequent check of the risk assessment referring to the Dutch maximum tolerable risk dose for benzo(a)pyrene taken from the RISC-HUMAN model appeared to result in a different weighted slope factor of 0.05 (mg/kg/day), indicating a lower potency and serving to emphasise earlier discussion about the importance of understanding the basis of data used.

Considering the exposure assumptions, reasonably maximum exposed individuals were judged to be:

- a child joining the nursery school on their 4th birthday and staying until just before their 11th birthday;
- a female teacher joining the school at age 22 and staying for 10 years;
- a young worker carrying out excavations on the site for 10 working days.

A female teacher was judged at greater risk because of a likely lower average bodyweight than a male teacher.

The other exposure-related assumptions are listed in Table 10.11.

It was assumed that exposure occurs to similar levels on each day which the school is open. For example, even if the child does not play outside because of poor weather, there will be exposure inside through dust and dirt trafficked into the school buildings.

10.6.6   Risk estimation

Only those risks arising from ingestion were taken into account. Although, for example, benzo(a)pyrene is carcinogenic by inhalation, no reference data could be found for use in an estimation of risks. It should be noted that risks due to direct contact effects are not covered. The detailed calculations are set out in Table 10.12. The exposure equation used is noted at the bottom of the table, i.e.:

$$\text{Intake} = (CS \times IR \times CF \times FI \times EF \times ED)/(BW \times AT)\ \text{mg/kg/day}$$

where CS = the measured concentration; IR = the ingestion rate; CF = conversion factor; FI = the fraction ingested from the site as a proportion of all the total from all sources; EF = the exposure frequency; ED = the exposure duration; BW = body weight; AT = the averaging time in days per lifetime (70 years). The risk is the intake $\times$ the slope factor.

287

Table 10.11. Assumptions

Assumption		Child	Female teacher	Excavation worker	Comments
Initial age	(years)	4	22	22	–
Bodyweight	(kg)	16.5 increasing	60	65	Data on average bodyweights for children taken from published source
Ingested rate	(mg/day)	100	100	100	US EPA suggested value
Fraction of soil ingested from site	(%)	50	50	50	This means that of the 100 mg soil ingested each day 50% comes from the site and 50% from elsewhere including the home
Exposure duration	(years)	7	10	1	–
Exposure frequency	(days/years)	194	200	10	Number of days of term time in year in question. Teachers assumed to attend an additional two days each term.
Averaging time (for cancer risks)	(years)	70	70	70	US EPA convention. Assumes risk is strictly linear and additive – in principle a lifetime dose taken in one year has the same effect as the same dose accumulated steadily over a lifetime
Soil concentration used in calculation		203 (95% UCL)	203 (95% UCL)	6740 (95% UCL) 10 248 (90th percentile)	95% UCL used by convention. 90th percentile value taken for worker since excavation could be in relative hot spot

By adding each of the calculated risks, Table 10.12 shows that the increased lifetime cancer risk of a child was 3×10^{-5}; of a teacher 2×10^{-5}; and of a worker 0.4×10^{-5}.

10.6.7   Risk evaluation

Readers can best judge for themselves the reasonableness of the assumptions made. The model used is linear and thus doubling a factor used in the calculation will either double or halve the calculated risks. For example, doubling the ingestion rate to 200 mg/day would double the calculated risk, while doubling the bodyweight halves the risk. Some of the assumptions are arbitrary, for example that 50% of the ingested soil comes from the site.

It is usual to take an average bodyweight for an age in such calculations. If it were assumed that the exposed child has the 90 percentile low bodyweight rather than an average bodyweight for their age, this would increase the calculated risks by 22%.

As discussed earlier in the chapter, there is no UK guidance on what constitutes an acceptable or tolerable risk in such circumstances. The question facing the assessor was therefore to decide whether any immediate actions were required and if so what these should be, and whether any longer-term action is required; and of course the urgency of such actions. To aid the deliberations the additional risks were re-expressed in relation to an additional term's exposure by a child of age 4. These were calculated to be 2×10^{-6}.

The education authority also had to decide how to report the findings to the staff, the parents and the children themselves, who would naturally want to know the reasons for any change in their routine or works at the school. Some suggested actions and comments are listed in Table 10.13. It was decided that risks to excavation workers could be adequately controlled by use of protective clothing and good hygiene.

10.7   DERIVATION OF SITE-SPECIFIC RISK-BASED REMEDIATION CRITERIA

The same processes used to estimate current and future potential risks from a site can be used *in reverse* to calculate site-specific criteria; i.e. an acceptable risk level (concentration at point of exposure) is set and then by back calculation the concentration in a particular medium that would give rise to that risk is determined by consideration of dilution, degradation, dispersion factors etc. along the relevant pathways for the site. The concept is depicted in Figure 10.6. This is also the basis for the derivation of generic risk-based criteria, but in this case generic assumptions are used rather than site-specific ones (although the generic assumptions might be relevant to a site-specific case – see Chapter 8).

Note that if risks arise from more than one exposure route and/or medium, there must first be an apportionment of the allowable risks before the requisite

Table 10.12. Calculation of additional lifetime risk of acquiring cancer

Age	CS mg/kg	IR mg/day	CF kg/mg	FI	EF days/yr	ED years	BW kg	AT days	Intake mg/kg/day	SF 1/(mg/kg/day)	Risk
Child											
4	203	100	0.000001	0.5	194	1	16.5	25 550	4.67E-06	1.38	6.45E-06
5	203	100	0.000001	0.5	194	1	18.6	25 550	4.14E-06	1.38	5.72E-06
6	203	100	0.000001	0.5	194	1	20.8	25 550	3.71E-06	1.38	5.11E-06
7	203	100	0.000001	0.5	194	1	23.4	25 550	3.29E-06	1.38	4.55E-06
8	203	100	0.000001	0.5	194	1	26.0	25 550	2.96E-06	1.38	4.09E-06
9	203	100	0.000001	0.5	194	1	28.1	25 550	2.74E-06	1.38	3.78E-06
10	203	100	0.000001	0.5	194	1	31.0	25 550	2.49E-06	1.38	3.43E-06
								Total additional risk	1.32E-05		3.31E-05
Teacher											
22	203	100	0.000001	0.5	200	10	60	25 550	1.32E-05	1.38	1.83E-05
Excavation worker											
22	6 740	100	0.000001	0.5	10	1	65	25 550	2.03E-06	1.38	2.8E-06
22	10 248	100	0.000001	0.5	10	1	65	25 550	3.09E-06	1.38	4.26E-06

Notes: Intake = (CS*IR*CF*FI*EF*ED)/(BW*AT)
Risk = sf*intake

Table 10.13. Some options for action

Timeframe	Action	Comments
Immediate	• Close school temporarily while mitigating work undertaken.	• Extremely difficult to find alternative accommodation at short notice. Alarming?
	• Open school for autumn term but aim to have new premises available for subsequent term. In the meantime try to limit exposure by taking some or all of the measures listed below.	• Parents and teachers would have to be reassured that the additional exposure involved would be acceptable and that there would be no further procrastination.
	• Stop use of grassed playing area. Keep children on hard paved area or inside at breaktimes. Supervise hand washing before eating and after any period using the play areas.	• Difficult control problem for teachers if children confined all day. Very hard to keep small children clean.
	• Restrict play to small hard-surfaced area – this could possibly be extended fairly quickly.	• Dust likely to be blown and trafficked on to area.
	• Maintain high standards of housekeeping in school and "deep clean" if dust sampling suggests this is necessary.	
	• Institute permit to work system for all works involving disturbance of the ground.	
Medium term – works to start within about three months	• Impose small depth of clean soil (say 150 mm) over grassed areas, extend hard play areas, create "artificial" play surfaces, excavate and replace soil in flower beds.	• Artificial play surfaces expensive if treated as "temporary" facility.
Long term – e.g. to start at end of the school year	• Close school and abandon site for school use.	• Have to find alternative site in urban area at short notice convenient to present school population etc. Moving school away from area would be resisted by local population.
	• Keep school and impose engineered cover system.	• Only small opportunity for works during summer holidays unless alternative temporary accommodation can be provided. School buildings remain on contaminated ground and monitoring and control will be required as long as the school remains open. School is in close urban area and there would be additional risks to children and general public from lorry movements.

Table 10.13 (*cont.*)

Timeframe	Action	Comments
	• Impose cover system, demolish present school and build new one.	• Expensive. Would require alternative accommodation for a longer period (6–12 months?) or the present school to remain open if the new school could be located on a different part of the site.
	• Carry out remediation involving demolition and excavation and replacement of contaminated materials.	• Expensive, time consuming, possibility of extensive environmental impacts on neighbourhood, risks from two sets of traffic (outward and inwards).

calculations can be made. In addition, background exposures must be taken into account.

Site-specific criteria can be set for a number of purposes. For example, they might be set:

• to define whether remediation is required or the suitability of different parts of a site for different uses;
• as remediation objectives or targets (e.g. the residual concentration of a contaminant in soil that is permitted following application of a process-based remediation method);
• as criteria for when certain actions should be taken (e.g. adoption of different levels of protection for remediation or construction workers);
• to define suitability for disposal of soils via particular routes.

Judgement with regard to whether remediation is required might be made on the basis of the data used in the risk assessment. If the data arise from a grid sampling pattern, then individual squares can be designated as passing or failing (any square for which one of the four corner samples exceeded the criterion would be classified as failing), since there would rarely be any justification for interpolation of concentrations between points. If this resulted in a requirement to remove too large a quantity of contaminated soil, then a further, more detailed sampling exercise would be required, perhaps concentrating only on critical contaminants. The cost of this additional investigation (and the time delays) would have to be balanced against the costs of possibly excavating a greater volume of soil than is strictly necessary. There might, in practice, be additional costs of supervision and compliance testing if the excavation exercise becomes too complex, and a loss of value of the site if potential purchasers and investors cannot be convinced that what is left, if resampled, would meet the remediation criteria.

Regardless of the purpose of the derived criteria, there must be advance agreement on how they are to be applied. For example, in a process-based

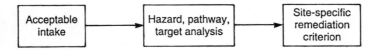

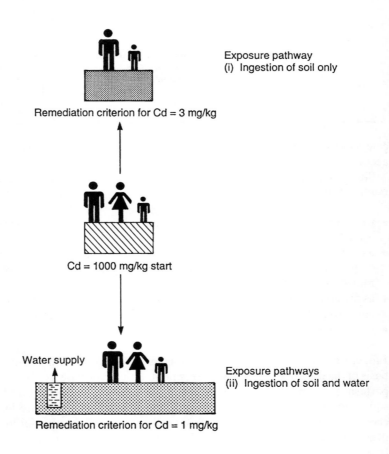

Figure 10.6. Derivation of risk-based remediation criteria (after WDA, 1993)

remediation treatment, what frequency of sampling is required? Does a single figure above a criterion represent failure or are some statistical parameters to be set: for example, that the 95% UCL of the mean should not exceed the criterion, or that no more than 10% of measured values should exceed the limit? In this connection, it should be remembered that the assessment has probably been done using some such derived statistic. Thus, the result of a back calculation must represent either a maximum permissible concentration in any sample or an equivalent parameter – which will differ depending on the purpose of the criteria. When making such calculations, account must be taken of the statistical distribution of concentrations in the medium under study. For example, concentrations in the product of a process-based treatment method might be expected to be normally distributed, whereas concentrations in material left in the ground are likely to be log-normally distributed (although see Skaarup and Pedersen (1996) for a case where both input and output were log-normally distributed).

There are strong arguments (Bowers, Shifrin & Murphy, 1996) for the most cost-effective approach being that the post-remediation *mean* concentration should equal the risk-based remediation target (contamination related objective – CRO). This is appropriate because the CRO is based on the assumption that over time an individual would have an exposure to the average concentration in the ground. The risk presented by the unremediated site is calculated (if USEPA protocols are followed) from the 95% UCL of the mean, since it is not possible to know the true mean. Thus, to allow analogously for variations in sampling etc., the 95% UCL of the mean of measured concentrations following remediation should not exceed the CRO. Inherent in this approach, as concentration distributions for each site will differ, is that the permissible 95% UCL for the same target average concentration will differ between sites (the corollary is that if generic criteria are employed, the average concentrations ensuring compliance will differ for different sites). In general, the more data that are available, the closer the 95% UCL will be to the true mean. Statistical methods are available for the calculation of *confidence response goals* (CRG) (Bowers, Shifrin & Murphy, 1996).

Whatever the purpose of the criteria, the fundamental question facing the assessor is, "What constitutes an acceptable risk?" This has been partly addressed above. It is not an easy question to answer.

If, for example, national authorities have produced screening values indicating negligible risks to human health based on careful consideration of the policy and practical issues (e.g. 1×10^{-6} excess cancer risk), then the assessor is likely to be "safe" in adopting the same limiting factor – indeed, to use another whether more or less stringent may be difficult to justify. However, if such useful indirect guidance is not available (some jurisdictions may conceivably provide direct guidance) assessors have to make their own judgements.

The question of the summation of risks is again an issue, as it is at the assessment stage. When responsible authorities set criteria they have so far always done so on a compound by compound basis. Thus, if there is a site on which 10 compounds are at the limit value for a 1×10^{-6} risk, then the site hazard index would be 1×10^{-5}. If this is unacceptable, then remediation criteria for individual compounds would have to be set at one-tenth of the negligible risk screening value (in this case at 1×10^{-7}). Thus, whatever the merits of summing risks at the assessment stage, there is little alternative when setting site-specific criteria to treat each contaminant separately, although some grouping of similar compounds might be sensible (e.g. a figure might be set for total PAHs based on benzo(a)pyrene).

An example of back calculation is provided in Table 10.14. It refers to the gardener cited in Table 10.5, i.e. the assumptions used there are used in the back calculation. The calculations are based on the following assumptions:

- that an excess individual cancer risk for a single compound arising from the site of 1×10^{-6} is negligible compared to cancer rates in the general population;
- that a hazard index for a single compound of 0.1 via the ingestion pathway represents an acceptable risk;
- that it is sufficient to take ingestion into account in the calculations and that this will be protective of other routes of exposure.

Table 10.14. Determination of some site-specific criteria relating to a gardener on a commercial/office site

		Cancer risks			
Analyte	Non-cancer risks RfD (mg/kg/day)	Concentration giving HQ = 0.1 via ingestion (mg/kg)	Slope factor	Concentration giving 1×10^{-6} risk for ingestion (mg/kg)	Clean-up criterion (mg/kg)
Arsenic	0.0003	33	1.75	4.4	4
Cadmium	0.0005	548			550
Chromium VI	0.005	5475			5 500**
Chromium III	1	10%			10 000
Copper	0.04	438			440
Lead	0.1	10 950			11 000
Mercury	0.00003	33			33
Nickel	0.02	2 190			2 200
Zinc	0.3	32 850			33 000
Cyanide	0.02	2 190			2 200
Phenol	0.6	65 700*			250*
Benzo(a)pyrene			7.3	1.05	1

*Assuming a free soil moisture content of 10% this could mean a concentration of 60% phenol in the soil moisture phase – such a concentration would be expected severely to damage the skin. The calculation is thus flawed and a criterion must be derived based on other considerations. A value of 250 mg/kg possibly giving 2500 mg/l in soil moisture should be below the level at which direct damage to skin will occur.

**Does not take into account possible effects on skin.

The values for arsenic (4 mg/kg) and benzo(a)pyrene are low, but not dispropor-tionate to those set as screening values by some jurisdictions. In practice, they led to a review in the risk evaluation part of the assessment of the assumptions used; it was decided that a more realistic scenario would be a contract gardener (rather than a retained one) with no more than 100 days' exposure per year and with contact with the site for no more than five years. These changes produce a fourfold reduction in the cancer risk and a decision that a "more realistic" arsenic criterion of 18 could be applied.

10.8 CONCLUSIONS

In Chapter 2, the Ontario view of site-specific quantified risk assessment was discussed: a view which cautions against its use based on an argument that generic guidelines derived by a quantified approach provide for protection, whilst remediations resulting from site-specific assessments may be overly cau-tious. However, in the UK at least, the lack of comprehensive guidelines for the range of contaminants relevant to sites and to the risks of concern (particularly ecorisks) will continue to force site-specific assessments for some sites, not least where exposure potential is not adequately covered by the generic guidelines. The derivation of site-specific remediation values which may provide for a less restrictive remediation than may be suggested by reference to generic guidelines is also likely to continue to encourage the quantified approach.

However, this chapter has tried, while explaining the general approach, also to provide a clear warning as to the complexity and uncertainty of site-specific, quantified risk assessment. The ready availability of commercial risk assessment packages has the potential to engender an apparent technical robustness and ease of application that belie the uncertainties and skill requirements. Even within large environmental consultancy organisations, experience of conducting such assessments is still relatively limited. There is no doubt that a team which undertakes such an assessment must have toxicological skills within it. However, this is not to downplay the fundamental issue which has dominated discussion in the book: i.e. that an assessment which is based on inadequate hazard identifica-tion, and inappropriate site investigation relative to the source–pathway–target chain(s) of concern, will be flawed even before the uncertainties inherent in the assessment process and the judgements about exposure and toxicity are added.

Judgements about the acceptability of an estimated risk can be made by an explicit discussion of the inputs to the assessment and the assumptions which underlie it within such an assessment. Indeed, contrary to a view that the most difficult issue is how to communicate the risk estimate, the most important requirement is a transparent approach to the assessment, as discussed in the next chapter.

11

Risk Assessment and Contaminated Land: Conclusions

11.1 RECOGNITION OF THE BENEFITS OF RISK ASSESSMENT

In the UK, redevelopment has been the primary reason for the assessment of sites, with all of the attendant limitations of time, money and problem-solving focus which the development process can entail. The assessment of contaminated land has often primarily involved a consideration of feasibility of development, and of competing and intuitive assessments. The assessment itself has been based on intrusive site investigation to determine soil contaminant levels which could be compared directly with available (limited) generic guidelines. Characterisation of sites relative to their previous land uses and current soil contaminant levels has frequently been a mechanistic activity undertaken with little regard to the nature and range of potential risks presented by a site. Indeed, in the context of redevelopment consideration of risks has been limited. Unlike some other countries the UK has not (until recently) taken a proactive approach to the identification of sites which may present a risk. Therefore, the need to develop structured risk assessment approaches to decision making has been less pressing.

The 1990s have begun to see a change in emphasis, although at the time of writing only a limited change in practice. Risk management has become the preferred approach to a range of environmental problems, including contaminated land. Risk assessment has become the most favoured tool to assist management decisions. Although risk assessment has, for over 15 years, dominated the management of technological and major safety risks where the effects may be acute and immediate, the UK has been relatively slow to embrace its value in assessing the risks of chronic releases of chemicals in the environment. However, outstanding problems of application and uncertainty have now been recognised as manageable and the benefits of risk assessment are promoted (Petts, 1993; 1994b; WDA, 1993; Harris, Herbert & Smith, 1995; Department of the Environment, 1995a; HSE, 1996), i.e.:

(i) It is a systematic, objective and transparent process which requires the assessor to understand all of the potential linkages and interactions be-

tween the contaminant source and sensitive targets and to evaluate the most critical of these so that appropriate courses of action are considered.
(ii) It assists in separating complex environmental problems into manageable components which focus on the most critical risks.
(iii) It can direct limited financial resources to dealing with the sources of the most significant potential environmental (including health) risks.
(iv) It provides a framework for considering long-term legal and financial liabilities.
(v) It can be used to demonstrate that generic guidelines (where available) provide sufficient protection for the targets at risk at a particular site.
(vi) It can allow for site-specific decisions to be made about acceptable risks and risk reduction requirements which reflect local environmental, economic and social conditions and priorities; these decisions may be that either higher or lower risks are acceptable than might be suggested in generic guidelines.
(vii) It can facilitate communication between site assessors and decision makers by use of a common language and recognised methods for describing the nature and extent of uncertainties about environmental consequences.
(viii) It can facilitate communication between site assessors, decision makers and other interested parties by providing explicit data that are amenable to review by others.

The potential benefits of the risk management approach to contaminated sites have been obscured by a number of perceptions and prejudices about the complexity, uncertainty and costs of the process, which have most often been based on a lack of full understanding of risk assessment. The most frequently encountered, and most basic, perception is that risk assessment always requires a quantitative estimate of risk. There is no doubt that this reflects experience of the use of risk assessment in relation to safety hazards in the UK and to health risks in the USA.

In practice, risk assessment can be qualitative, semi-quantitative or quantitative. As promoted throughout this book, risk assessment is fundamentally a process of making judgements about risks. If risk assessment can be understood as such, then it is more likely to be used. The judgement about risk can be complete and sufficient at the hazard identification stage (Chapter 4) if no contaminative uses of the site are apparent, or if no pathways and targets can be identified which could be affected by contaminants on a site. A qualitative evaluation or judgement about risks can also be made if measured contaminant levels are below generic guidelines relevant to the contaminants, pathways and targets of concern at the specific site (if those guidelines have been derived by a risk assessment process). Chapter 9 discussed semi-quantified approaches which allow for multiple sites to be compared, and for data related to a single site to be assessed, on a common basis and potentially more explicitly than may be apparent in narrative approaches. Chapter 10 discussed quantitative risk assessment, which is characterised by the estimate of risk to actual and potential targets either on a site-specific basis or as a means of deriving generic guidelines. Such a quantitative estimate is demanding of data and skill and is unlikely to be necessary for the majority of contaminated sites, usually being limited in application to the most risky (or high hazard) sites, which have passed through a screening against generic guidelines.

Based on a view that risk assessment has to result in quantified output, the second most common argument against the usefulness and effectiveness of risk assessment is that data and scientific understanding are often insufficient. While such insufficiency may rule out the use of some data-intensive risk assessment methods, it does not necessarily follow that other systematic tools designed to understand and document uncertainty (for example, as discussed in Chapter 10) will not be helpful in the decision process.

Linked to the above has been the third misperception of the value of risk assessment (not specific to the issue of contaminated land): i.e. that it is merely a tool to aid decisions, conducted separately from the decision process and, hence, required to be as objective and quantified as possible. When this output has been seen to meet social and political questioning, technical and expert concerns have been aroused. While the value of objectivity is supported throughout this book, a strong message supports the view of risk assessment as a process rather than merely a tool or computerised method. Risk assessment as a process provides for a blend of robust scientific and technical analysis, effective communication and the involvement of interested parties so that priorities, concerns and interests are considered during the assessment so as to optimise the potential for a consensus about the levels and acceptability of the risks. It is only recently that such a view of risk assessment is beginning to be discussed officially among the technical community (e.g. NAS, 1996), although it has been supported for years in discussion among the social and decision sciences (e.g. Wynne, 1987; O'Riordan, 1991; Renn, 1992).

Finally, a significant problem which this book has stressed and sought to counter is a view that site investigation is separate from risk assessment, i.e. that the latter only starts once the data relative to a site have been gathered. Site investigation is an inherent component of the risk assessment of contaminated land and has to be conducted in full recognition of the requirements to inform the risk judgement.

11.2   EVALUATING THE STATE OF THE ART

As a conclusion, a number of criteria common to formal analysis (Covello & Merkhofer, 1993) are adopted as a means of identifying problems in the state of the art of risk assessment for contaminated land and in order to summarise requirements. These criteria can be considered in two parts:

(i) criteria relating to the "technical" application of risk assessment, i.e. completeness and accuracy; and

(ii) criteria relating to the use of risk assessment within the management process, i.e. acceptability and credibility.

These criteria link together and overlap.

11.2.1 Completeness

Completeness has two dimensions: (i) whether the assessment can account for all potential elements of the risk problem, i.e. the source–pathway–target chain, and (ii) whether the conduct of the assessment itself leads to omissions. The complexity of the environment and of linkages between human, ecosystem and man-made components means that no assessment can provide a complete understanding of all the possible risks. However, other omissions can be managed.

Potential risk problems are sometimes deliberately limited in the assessment (for various political, economic and scientific reasons): for example, the tendency to focus on public health impacts to the detriment of consideration of ecological effects, or to focus on the explosive and flammable risks associated with gases to the possible detriment of chronic health effects (to humans and flora). At the hazard identification stage serious omissions in the assessment can arise from a failure to understand the limitations of data sources such as maps in providing a full picture of previous land uses. Important potential hazards can be missed (particularly areas of waste disposal).

Generic guidelines which do not provide for a comprehensive listing of potential contaminants of concern can lead to a tendency to focus in sample analysis on those contaminants for which a figure which suggests an acceptable concentration exists, rather than on those contaminants which may be present on a site and/or which may present a risk to different targets. As generic guidelines which cover all potential contaminants will take many years (possibly decades) to derive, the onus is on the site assessor to understand the plausible and critical contaminants of concern.

Unanticipated changes in the environment may also affect the assessment, particularly where site investigation data are incomplete (for example, gas data collected over too limited a period to determine actual gassing conditions – Chapter 7), or where assumptions about future conditions are deliberately limited (for example, the failure to consider the potential for flooding of a site and wash-off of contaminants, or an assumption in the assessment that control over the use of the land to limit exposure will be ensured through planning conditions).

Whether important omissions will be made is a function partly of the diligence and foresight of the assessor and partly of the capabilities and appropriateness of the risk assessment methods (including site investigation) being used. Ensuring that the assessment is open to review provides the best means of identifying serious omissions. Within environmental impact assessment a fundamentally important step which aims to ensure that all potentially significant environmental impacts are considered is the *scoping* stage, which provides a focus for the assessment by identifying the key issues of concern and ensuring that they are subject to an appropriate level of assessment. This stage is optimised through consultation and discussion with any relevant parties (regulatory, non-statutory, public). In relation to the risk assessment of contaminated sites there is also potential for such scoping to ensure that all potential risks are identified and the risk assessment focused on those agreed to be plausible and critical.

11.2.2 Accuracy

Accuracy relates to whether the assessment produces results that correctly describe and convey risks (Covello & Merkhofer, 1993). All risk estimates are "best estimates". Risk assessment is itself an exercise in probability evaluation: for example, no investigation can explore more than a fraction of the volume of materials which lie on and below the site surface, and scientific uncertainty prevents us from being able to determine absolute concentrations of contaminants which indicate "safe" or "unsafe" conditions. There has been a tendency to argue that all risk assessments are always deficient in information. However, lack of information does not mean that the results are inaccurate if the risk assessment has correctly accounted for the sources of uncertainty and explicitly reported their effects.

A lack of accuracy arises from human error and from scientific uncertainty. It is evident that the primary source of uncertainty lies in assessments which have either failed to identify targets which could be at risk or conversely assuming a pathway of exposure which could not exist.

Lack of accuracy arising from human error has dogged site investigation: the design of sampling strategies which miss hot spots of contaminants; the taking of samples with inappropriate or contaminated equipment; and gas monitoring which only measures concentrations and ignores flows are just three examples of problems discussed in Chapters 5–7. Site investigation errors are due to a lack of training and skills and also to a failure to devote adequate resources to the work.

Sources of inaccuracy in the assessment of data include inaccurate sample analysis and data processing; the use of parameters in ranking tools which are inadequately justified and hence open to abuse (Chapter 9); the use of inappropriate assumptions in the consideration of exposure of different targets (discussed in Chapter 10); trying to fit models to sparse data; the use of surrogate data (including generic guidelines and standards) which are not relevant to the characteristics of the exposure of the sensitive targets of concern (Chapter 8); relying on experts who do not represent a full range of knowledge relevant to the risks of concern; and utilising inappropriate models. Scientific uncertainty is inherent in the use of simplified models of complex processes; in the aggregation of risk estimates in a way which simplifies synergistic or antagonistic effects; in the application of safety factors to deal with the problems of dose-response data derived from tests on laboratory animals etc.

There are several important means of dealing with uncertainty:

- Providing adequate training in the risk-based approach for site investigators, assessors and regulators and the appointment of competent staff to undertake assessments. It is important to note at this point that simply buying a computer program that can perform a site-specific assessment or do a Monte Carlo simulation does not qualify anyone to perform quantified risk assessments.
- Taking an iterative approach to the risk assessment, i.e. revisiting assumptions; discussing intuitive assumptions with experts; testing conceptual models of a site; and phasing investigations.
- Applying sensitivity analysis to risk models and estimates.
- Fully documenting all scenarios examined and omitted, the data collection methods and the assessment assumptions used etc. so that the assessment can be open to review by all relevant parties.

Risk assessment of contaminated sites is primarily an "art" with abundant opportunities for personal interpretation and bias. The latter can be detrimental to the outcome, but also provide the underpinning professional judgements that are inherent to an effective process. Risk assessment cannot be reduced to standardised instructions and guidelines. While the latter are important in minimising omissions and inconsistent and inaccurate approaches, the complexity of risk situations and the different contexts in which risk assessments are applied (discussed in Chapter 3 and in Table 11.1) mean that overprescriptive procedural guidelines might lead to assessments that either fail to use all relevant information or which disregard significant evidence.

11.2.3 Acceptability

The above discussion has focused on technical or scientific acceptability of the risk assessment. However, the risk assessment has to be acceptable to the decision makers, the client, the regulatory authority, the public etc. The definition of "acceptability" includes practical relevance, robustness and usability, as well as the extent to which the assessment is credible and trustworthy.

Credibility is a function of the extent to which:

(i) the assessment or assessor is trusted to have acted in the interests of a particular party, including the extent to which judgements and choices made in the assessment reflect the values important to that particular party;
(ii) the assessor/expert is viewed as technically competent; and
(iii) the basic elements of the decision-making system (e.g. its openness, accessibility etc.) accord with people's expectations (Covello, 1992; Petts, 1996).

The extent to which conservatism is built into the assessment (through the use of worst-case scenarios); the choice of which adverse health or environmental consequences to assess (for example cancer as opposed to other health effects, or the concentration on public health as opposed to adverse effects on animals, plants or ecosystems); and the choice of exposure assumptions, are examples of value judgements and/or policy choices which are inherent within risk assessments but which may not be acceptable to all those who are affected by the assessment. The extent to which these assumptions are transparent will affect acceptability. However, the extent to which different interested parties have been involved in the assessment process and the assumptions and choice of methods have been discussed with them (not least regulatory bodies) will be a fundamental determinant of acceptability.

Table 11.1 identifies three different decision contexts requiring risk assessment and the potentially interested stakeholders (those with a "stake" in the decision) and their likely concerns and interests. The listing is an oversimplification, but serves to emphasise conflicting interests even within individual stakeholders which place pressures on the risk assessment process. For example, some interests (e.g. those of developers) could lead to pressures to underestimate potential risks and to reduce the costs of site investigation. Other interests (e.g. those of site residents) may emphasise the importance of worst-case scenarios being included in the risk assessment.

It is communication within the risk assessment and of the risk judgement within the risk management process which is fundamentally important to acceptability.

11.3   COMMUNICATION

Risk assessment has often been discussed in terms of the production of a judgement about the risk by the assessor which is then communicated to the interested parties. While most risk assessors regard communication as an important component of risk management, it has primarily been viewed as a one-way activity (i.e. from the assessor to the relevant interested party). However, the general development of understanding of risk communication over a period from the early 1980s to the current time (Table 11.2) provides clear evidence that communication as a two-way process is essential.

Table 11.2 summarises the development over time of risk communication understanding in relation to a wide range of risks: from an initial emphasis on trying to quantify risks and communicate the numbers, through to an understanding of the need to involve people in decision making (including the assessment process). Many technical experts still believe that in the drive to the use of risk assessment to manage contaminated sites it is important to learn how to communicate the risk output (e.g. a 1×10^{-6} increased cancer risk). To an extent this is important, as recognised in item 8 of Table 11.2. However, the favoured approach of using risk comparisons (for example, a 1×10^{-6} risk is about the same as the risk of dying from being hit by lightning) is known to fail to deal with many of the qualitative dimensions of people's concerns about risks (Covello;

Table 11.1. Stakeholder interests in relation to contaminated land

Decision context	Key stakeholders	Key concerns/interests
Development of land	Developer	Optimisation of development potential. Reduction of delays to development or difficulties obtaining funding if contamination risk cannot be reduced. Rapid assessment of risk but minimisation of long-term liabilities.
	Funding institution(s)	Minimisation of risk liabilities. Confidence in risk assessment.
	Local planning authority	Optimisation of reuse of land. Impact of development, including risks.
	Potential residents	Minimisation of risks to residents. Confidence in risk assessment.
Potential problem site	Landowner/polluter	Risk liabilities including arising from Environment Act 1995/loss of land value.
	Landowner/occupier where not the polluter	Risk to property and health. Blight impacts/loss of land value. Disruption. Immediate assessment of risk.
	Statutory authorities	Adverse health and environmental impacts. Costs of remediation. Need to assess risks quickly. Confidence in risk assessment. Public concerns.
	Funding institutions	Risk liabilities.
	Local community	Blight impacts on land values. Loss of amenity. Concern over health impacts.
Operational site	Landowner/occupier	Minimisation of risk (including compliance) liabilities. Prioritisation of risks to reduce compliance liabilities and optimise remediation expenditure.
	Potential purchaser	Identification of unacceptable risks to gain reduction in purchase price or remove site from purchase options.
	Funding institutions	Minimisation of risk liabilities.

Table 11.2. Development stages in risk communication (Petts, 1996, adapted from Fischhoff, 1995)

Development stage	Explanation
1 All we have to do is get the numbers right.	Focus on technical assessment of risk.
2 All we have to do is let them have the numbers.	Start of risk communication – getting the message across.
3 All we have to do is explain the numbers.	Communication is hampered by uncertainties in the assessment and evidence that the public does not rely solely on the numbers.
4 All we have to do is show them that they have accepted the same risks in the past.	Reliance on risk comparisons in the mistaken belief that people perceive risks similarly.
5 All we have to do is show them that it is a good deal.	View that people make cost–benefit trade-offs. Start to see move away from discussion of physical risks.
6 All we have to do is treat them nicely.	Emergence of good neighbour approach as trust becomes an issue.
7 All we have to do is make them partners.	Recognition of need to involve people in decision as lack of trust in decision makers is problem in risk acceptance.
8 We must do all of the above.	Recognition of full complexity of requirements for effective communication.

1989, 1991). Although the risk communication literature has not entirely discounted the usefulness of risk comparisons, it emphasises the importance of choice of comparative data targeted to the specific audience taking into account their needs, concerns and level of knowledge (Petts, 1996).

The development stages of risk communication reflect the increasing understanding that people's concerns about risks are the result of a complex mix of sociological, cultural and psychological influences (see Krimsky & Golding, 1992 and Royal Society, 1992 for reviews). A broad range of qualitative characteristics underpin perceptions of risk which relate not only to the potential harm (delayed effects, effects on children, catastrophy potential etc.), but also to the potential for control and the extent to which institutions can be trusted to manage the risks. Stigmatisation has been seen as an important component of concerns about contaminated sites (Slovic, 1989). The term can be translated broadly to mean that an area in which people live is perceived negatively as a result of some environmental problem. In relation to contaminated sites stigma is related to blight (Petts, 1994a). Table 11.3 identifies the broad dimensions of blight in relation to the identification of potentially contaminated land.

Public concern, anxiety and stress in relation to the identification of contaminated sites have been documented in studies in the Netherlands, USA and UK (De Boer, 1986; Finsterbusch, 1989; Van der Pligt & De Boer, 1991; Denner, Perry & Sterrit, 1991; Petts, 1994a). The resulting demands on risk assessors and

Table 11.3. Possible blight scenarios in relation to the identification of contaminated land (Petts, 1994a)

Planning blight	Financial blight	Personal blight
• Need to revoke or discontinue planning permission – with compensation implications	• Delays in selling property or land	• Stress in property/landowners as a result of financial impacts
• Restrictions on development in vicinity	• Loss of land value	• Stress in property/landowners as a result of uncertainty
• Need to redesignate land use to a lower-value use	• Costs/liabilities of remediation on existing properties	• Adverse health impacts as a result of stress
• Loss of amenity	• Delay in obtaining permission, or loss of development potential	• Loss of confidence in authorities/responsible parties
• Difficulty in attracting developers	• Costs of legal assistance/advice	• Perceived loss of amenity
• Loss of potential developers to neighbouring areas	• Costs in terms of time spent dealing with the issue	• Disruption to personal plans
	• Loss to individuals of new business opportunities	• Disruption to daily life

decision makers in relation to information, justification of assessment methods and assumptions, and even on site investigation teams who may have to deal directly with questions, can be significant.

The important requirements of communication in terms of response to the concerns of different parties are:

• timing;
• communication about hazards, assessment assumptions and uncertainties;
• direct methods of communication and involvement of interested parties in decisions;
• planned communication responses.

The components of these will differ in different circumstances. For example, a local authority which has identified a potentially gassing closed landfill and is taking action to assess it will have to provide information to any residents on or close to the site at the start as well as during the assessment process, and will need to liaise directly with individuals and groups of residents. Compare this situation to that of a developer who is considering a vacant site for redevelopment. This consideration may be confidential and little communication may be necessary or possible until the assessment is complete and the development proposals are finalised. The communication will then have to present the full assessment as part of a planning application. However, if the assessment phase involved activities which may affect local landowners (e.g. requirement for access to land to install boreholes) or lead to potential risks to the environment, then communication will be required during the process.

At the time of writing there is little experience in the UK of the use of quantitative assessments in the public domain in relation to contaminated sites. However, we would not expect the response to such assessments to be any different to that experienced in relation to other environmental risk areas where the quantified output of risk assessments are discussed in public, i.e. in themselves they will not provide for public concerns. A small study of the relative effectiveness of narrative presentation of radon risks compared to technical risk information (the former being more effective than the latter) suggests the need for better understanding of the relative effectiveness of different communication methods (Golding, Krimsky & Plough, 1992).

In environmental risk management there is much discussion of the value of consensus-building approaches (see Renn, Webler & Wiedemann, 1995 for a review). Consensus building seeks to improve the quality of public involvement in decisions. Claus (1995) describes a problem area of soil contamination in a residential area of Varresbecker Bach, Wuppertal, Germany. Here the interested parties involved residents and property owners. The latter were involved in the choice of the site assessment team and a group was formed to receive and discuss the risk assessment information as it was generated. Such consensus approaches are most relevant to decision making in relation to problem sites where concerns, fears and divergent interests are evident. The adoption of such approaches to public involvement presents a challenge to traditional public participation requirements and processes of decision making, but also to the skills and general understanding of the techniques of communication among officials, site assessors etc.

11.4   EFFECTIVENESS OF RISK ASSESSMENT

Risk assessment has to inform risk management decisions. In relation to a particular site the latter could be:

(i) no action required as no potential risk;
(ii) ongoing monitoring of the site required as the risk appears to be low and does not present a priority for action but the conditions may change;
(iii) some remedial action is required to deal with certain potential risks, either to remove the hazard or to cut the potential pathway from the hazard to the target or to remove the target (rarely justified on grounds of significance of risk or economically);
(iv) the determination of remediation criteria or contamination-related objectives which the remedial action should achieve, and the identification of site-specific constraints.

The risk assessment must not only provide sufficient information to understand the degree of risk, but also allow relevant interested parties to understand the extent to which the assessment is complete and accurate and reflects the concerns of different interests. The information will also inform the choice of remedial action, dependent on the objectives to be achieved and the treatment options (process and/or engineering based) which may be appropriate to risk reduction.

The risk assessment of contaminated land is dependent on:

- skilled investigation and assessment teams;
- the identification of plausible and critical source–pathway–target scenarios;
- the design of an investigation strategy relevant to the scenarios of concern;
- the collection of robust and representative data;
- the choice of appropriate assessment models;
- the choice of appropriate generic guidelines relevant to the scenarios of concern;
- full understanding of the nature and extent of uncertainty in the assessment;
- an iterative approach to assessment which ensures that information and data are revisited or further data collected as the assessment develops;
- effective communication within the assessment process to ensure that all relevant concerns and interests are considered.

While the current practice has significant limitations and shortcomings, the risk assessment approach to contaminated land still provides the best means of addressing environmental risks and optimising appropriate risk management decisions. It can help set priorities and develop standards; reduce the range of uncertainty in decision making; and prescribe levels of risk that should remain after the application of risk reduction measures. Risk assessment provides an empirical foundation for balancing costs, benefits and risks. It should provide for transparency in decision making. However, to achieve each of these risk assessment has to be based on a clear understanding of the source–pathway–target relationship, and is completely dependent on the quality of the data collected to inform judgements. Risk assessment can provide no more than a best estimate of the potential risk, but the site assessor can optimise this achievement by a full understanding of the data requirements to inform this estimate and of the means to achieve these data.

References

ACE (Association of Environmental Consultancies) (1993) *Code of Practice (Contaminated Land)*. AEC, Hertford, UK.

ACS (American Chemical Society) (1992) *Proceedings of the Eighth Annual Waste Testing and Quality Assurance Symposium*. ACS, Arlington, Virginia, USA.

Advisory Committee on Business and the Environment (1993) *Third Progress Report to and Response from the Secretary of State for the Environment and the President of the Board of Trade*. Department of the Environment, London.

AEC (Association of Environmental Consultancies) (1993) *Draft Code of Practice (Contaminated Land)*. AEC, Hertford, UK.

AERIS (1991) *Aid for Evaluating the Redevelopment of Industrial Sites: AERIS Model Version 3.0): Technical Manual and Users Guide*. AERIS Software Inc., Richmond Hill, Ontario, Canada.

AGS (Association of Geotechnical Specialists) (1992a) *Safety Awareness on Investigation Sites*. AGS, Beckenham, Kent, UK.

AGS (Association of Geotechnical Specialists) (1992b) *Safety Manual for Investigation Sites*. AGS, Beckenham, Kent, UK.

AGS (Association of Geotechnical Specialists) (1996) *Guide to the Model Document Report: GeoEnvironmental Site Assessment*. AGS, Beckenham, Kent, UK.

Aller, L., Bennett, T., Lehr, J., Petty, R. & Hackett, G. (1987) *DRASTIC: a Standardised System for Evaluating Groundwater Pollution Potential using Hydrogeological Settings*. EPA/600/2-87/035, USEPA, Cincinnatti, USA.

Alloway, B.J. (ed.) (1990) *Heavy Metals in Soils*. Blackie & Son, Glasgow.

Alloway, B.J. & Ayres, D.C. (1993) *Chemical Principles of Environmental Pollution*. Blackie Academic & Professional, Glasgow.

Andelman, J.B. & Underhill, D.W. (eds) (1987) *Health Effects from Hazardous Waste Sites*. Lewis Publishers, Michigan, USA.

Anderson, E., Chrostowski, P. & Vreeland, J. (1990) Risk assessment issues associated with cleaning-up inactive hazardous waste sites. In *Integrating Insurance and Risk Management for Hazardous Wastes*. H. Kunreuther & M.V. Rajeev (eds), pp 15–40. Kluwer Academic Publishers, Mass, USA.

Anon (1986) *Report into the Non-Statutory Public Inquiry into the Gas Explosion at Loscoe, Derbyshire, 24 March 1986*. 2 volumes, Derbyshire County Council, Matlock, UK.

Anon (1994) *Sustainable Development: the UK Strategy*. HMSO, London.

Anon (1995) Proceedings International Symposium and Trade Fair on the Clean-up of

Manufactured Gas Plants, Prague 1995. *Land Contamination and Reclamation*, **3(4)**, 1/1–14/1 (whole issue).

Anon (1996a) Birth defects, other disorders linked to superfund site exposure, ATSDR says. *Environmental Science and Technology*, **30(10)**, 429A–30A.

Anon (1996b) Pilot program will update IRIS health risk database for 11 chemicals. *Environmental Science and Technology*, **30(6)**, 240A–41A.

Anon (1996c) Proposed cancer risk guidelines open door to use of new data. *Environmental Science and Technology*, **30(6)**, 238A–39A.

Anon (1996d) Pesticide duos are more dangerous. *Chemistry and Industry*, **17 June**, 436.

ANZECC (1992) *Australian and New Zealand Guidelines for the Assessment and Management of Contaminated Sites.* Australian and New Zealand Environment and Conservation Council/National Health and Medical Research Council, Canberra, Australia.

Argyraki, A., Ramsey, M.H. & Thompson, M. (1997) Measurement uncertainty in contaminated land investigations. In *Current Trends in Contaminated Land Research*, pp 10–25. Society of the Chemical Industry, London (in press).

Arnold, S.F., Klotz, D.M., Collins, B.M., Vonier, P.M., Guillette, L.J. & McLachlan, J.A. (1996) Synergistic activation of estrogen receptor with combinations of environmental chemicals. *Science,* **272 (7 June)**, 1489–92.

Asante-Duah, D.K. (1996) *Managing Contaminated Sites: Problem Diagnosis and Development of Site Restoration.* John Wiley & Sons, Chichester, UK.

ASL (Analytical Services Laboratories), Wastewater Technology Centre & Grace Dearborn Inc. (1995) *Enhanced Bioremediation and Near-real-time Monitoring of Contaminated Soils.* Report for The British Columbia Ministry of Environment, Lands and Parks, and Environment Canada. British Columbia Ministry of Environment, Vancouver, Canada.

ASTM (American Society for Testing and Materials) (1990) *Standard Practice for the Design and Installation of Groundwater Monitoring Wells in Aquifers.* ASTM D5092 [revised annually]. ASTM, Philadelphia, USA.

ASTM (American Society for Testing and Materials) (1993a) *Standard Practice for Environmental Site Assessments: Transaction Screen Process.* Standard E.50.02.1. ASTM, Philadelphia, USA.

ASTM (American Society for Testing and Materials) (1993b) *Standard Practice for Environmental Site Assessments: Phase I, Environmental Site Assessment Process.* Standard E.50.02.2 ASTM, Philadelphia, USA.

ASTM (American Society for Testing and Materials) (1995) *Standard Guide for Riskbased Corrective Action Applied at Petroleum-Release Sites.* E1739-95. ASTM, West Conshohocken, PA, USA.

Barlaz, M.A., Ham, R.K. & Schaefer, D.M. (1990) Methane production from municipal refuse: a review of enhancement techniques and microbial dynamics. *Critical Reviews in Environmental Control*, **19**, 557–84.

Barnes, D.G. & Dourson, M. (1988) Reference Dose (RfD): description and use in health risk assessments. *Regulatory Toxicology and Pharmacology*, **8**, 471–86.

Barnthouse, L.W. (1992) The role of models in ecological risk assessment: a 1990's perspective. *Environmental Toxicology and Chemistry,* **11**, 1751–60.

Barry, D.L. (1985) Former iron and steelmaking plants. In *Contaminated Land: Reclamation and Treatment.* M.A. Smith (ed.), pp 311–40. Plenum, London.

Barry, D.L. (1987) Hazards from methane and carbon dioxide. In *Reclaiming Derelict Land.* T. Cairney (ed.), pp 181–99. Blackie & Son, Glasgow, UK.

Barry, D.L. (1991) Hazards in land recycling. In *Recycling Derelict Land.* G. Fleming (ed.), pp 28–63. Thomas Telford, London.

Bartell, S.M., Gardner, R.H. & O'Neill, R.V. (1992) *Ecological Risk Estimation*. Lewis Publishers, Chelsea, MI, USA.

Bauer, L.D., Gardner, W.H. & Gardner, W.R. (1972) *Soil Physics*. 4th edn. John Wiley & Sons, New York.

BDA (British Drilling Association) (1981) *Code of Safe Drilling Practice, Part 1: Surface Drilling*. BDA Brentwood, UK.

BDA (British Drilling Association) (1991) *Guidelines for the Drilling of Landfill, Contaminated Land and Adjacent Areas*. BDA, Brentwood, UK.

Begley, R. (1996) Risk-based remediation guidelines take hold. *Environmental Science and Technology*, **30(10)**, 438A–44A.

Berliner, B. & Spuehler, J. (1990) Insurability issues associated with managing existing hazardous waste sites. In *Integrating Insurance and Risk Management for Hazardous Wastes*. H. Kunreuther & R. Gowda (eds), pp 131–68. Kluwer Academic Publishers, Norwell, Mass, USA.

Blacker, S. & Goodman, D. (1994) Risk-based decision-making case study: application at a Superfund site. *Environmental Science & Technology*, **28(11)**, 471A–77A

Boulding, J.R. (1994) *Description and Sampling of Contaminated Soils: A Field Guide*. Lewis Publishers, Boca Raton, Florida, USA.

Bowen, H.J.M. (1979) *Environmental Chemistry of the Elements*. Academic Press, London.

Bowers, T.S., Shifrin, N.S. & Murphy, B.L. (1996) Statistical approach to meeting soil cleanup goals. *Environmental Science and Technology*, **30(5)**, 1437–44.

Bradbury, J.A. (1994) Risk communication in environmental restoration programs. *Risk Analysis*, **14(3)**, 357–63.

Brick, C. & Moore, J. (1996) Diel variation of trace metals in the Upper Clark Fork River Montana. *Environmental Science and Technology*, **30(6)**, 1953–60.

Bridges, E.M. (1987) *Surveying Derelict Land*. Clarendon Press, Oxford.

British Medical Association (1991) *Hazardous Waste and Human Health*. Oxford University Press, Oxford, UK.

Brown, M. (1979) *Laying Waste: the Poisoning of America by Toxic Chemicals*. Washington Square Press, New York.

BSI (British Standards Institution) (1981a) *Code of Practice for Site Investigations*. BS 5930. BSI, London.

BSI (British Standards Institution) (1981b) *Water Quality: Part 6: Sampling: Section 6.1 Guidance on the Design of Sampling Programmes*. BS 6068: Part 6: Section 6.1: 1981 (confirmed 1990) [ISO 5667-1:1980]. BSI, London.

BSI (British Standards Institution) (1987) *Water Quality: Part 6 Sampling: Section 6.4 Guidance on Sampling from Lakes, Natural and Man Made*. BS 6068: Part 6: Section 6.4:1987 [ISO 5667-4:1987]. BSI, London.

BSI (British Standards Institution) (1988) *DD175:1988 Draft for Development Code of Practice for the Identification of Contaminated Land and its Investigation*. BSI, London.

BSI (British Standards Institution) (1991a) *Water Quality: Part 6 Sampling: Section 6.2 Guidance on Sampling Techniques*. BS 6068: Part 6: Section 6.2: 1991 [ISO 5667-2:1991] . BSI, London.

BSI (British Standards Institution) (1991b) *Water Quality: Part 6 Sampling: Section 6.6 Guidance on Sampling of Rivers and Streams*. BS 6068: Part 6: Section 6.6: 1991 [ISO 5667-6:1991]. BSI, London.

BSI (British Standards Institution) (1993a) *BS 7750: Environmental Management Systems*. BSI, London.

BSI (British Standards Institution) (1993b) *Water Quality: Part 6 Sampling: Section 6.11 Guidance on Sampling of Groundwaters*. BS 6068: Part 6: Section 6.11: 1993 [ISO 5667-11:1993]. BSI, London.

BSI (British Standards Institution) (1995) *Soil Quality Part 3 Chemical Methods: Section*

3.9 Extraction of Trace Elements Soluble in Aqua Regia. BS 7755: Section 3.9: 1995 (ISO 11466:1995). BSI, London.

BSI (British Standards Institution) (1996a) *Water Quality: Part 6 Sampling: Section 6.3 Guidance on the Preservation and Handling of Samples.* BS 6068: Part 6: Section 6.3: 1996 [ISO 5667-3:1985]. BSI, London.

BSI (British Standards Institution) (1996b) *Water Quality: Part 6 Sampling: Section 6.12 Guidance on Sampling of Bottom Sediments.* BS 6068: Part 6: Section 6.12: 1996 [ISO 5667-12:1995]. BSI, London.

BSI/ISO (British Standards Institution/International Organisation for Standardisation) (1993) *ISO 10381-6: 1993 Soil Quality Sampling Part 6: Guidance on the Collection, Handling and Storage of Soil for Assessment of Aerobic Microbial Processes in the Laboratory.* BS 7755: Section 2.6. BSI, London.

BSI/ISO (British Standards Institution/International Organisation for Standardisation) (1995a) *Draft BS7755: Part 2: Section X: Sampling – Sampling Techniques [ISO DIS 10381-2].* BSI, London.

BSI/ISO (British Standards Institution/International Organisation for Standardisation) (1995b) *Draft BS7755: Part 2: Section X: Sampling – Guidance on Safety [ISO DIS 10381-3].* BSI, London.

BSI/ISO (British Standards Institution/International Organisation for Standardisation) (1995c) *Draft BS7755: Part 2: Section X: Soil Quality – Sampling – Design of Sampling Programmes [ISO DIS 10381-1].* BSI, London.

BSI/ISO (British Standards Institution/International Organisation for Standardisation) (1995d) *Draft BS7755: Part 2: Section X: Soil Quality – Sampling – Guidance on the Procedures for Investigation of Natural, Near Natural and Cultivated Sites [ISO DIS 10381-4].* BSI, London.

Bullock, P. & Gregory, P.J. (eds) (1991) *Soils in the Urban Environment.* Blackwell Scientific Publications, Oxford, UK.

Burmaster, D.E. & Anderson, P.D. (1994) Principles of good practice for the use of Monte Carlo techniques in human health and ecological risk assessments. *Risk Analysis*, **14(4)**, 477–81.

Butler, B.E. (1996) Consultation with national experts: managing contaminated land. *UNEP Industry & Environment*, **19(2)**, 52–6.

Cairney, T. (1993) *Contaminated Land: Problems and Solutions.* Blackie Academic, London.

Cairney, T. (1995) *The Re-Use of Contaminated Land: A Handbook of Risk Assessment.* John Wiley & Sons, Chichester, UK.

Cairney, T., Clucas, R.C. & Hobson, D.M. (1990) Evaluating subterranean fire risks on reclaimed sites. In *Proceedings of the Third International Symposium on the Reclamation, Treatment and Utilisation of Coal Mining Wastes*, pp 237–43. Balkerina, the Netherlands.

Calabrese, E.J. & Baldwin, L.A. (1993) *Performing Ecological Risk Assessments.* Lewis Publishers, Chelsea, Michigan, USA.

Calabrese, E.J. & Kostecki, P.T. (1992) *Risk Assessment and Environmental Fate Methodologies.* Lewis Publishers, Boca Raton, Florida, USA.

Cameron, L. (1991) *Guide to Site and Soil Description for Hazardous Waste Site Characterization: Volume 1, Metals.* EPA 600/4-91/029. Environmental Monitoring Systems Laboratory, Las Vegas, USA.

Canter, L.W. (1991) Environmental impact assessment for hazardous waste landfills. *Journal of Urban Planning & Development*, **117(2)**, 59–76.

Card, G.B. (1993) *Protecting Development from Methane.* CIRIA Report CP/8. Construction Industry Research and Information Association, London.

CCME (Council of Canadian Ministers of the Environment) (1991) *Interim Canadian*

Environmental Quality Criteria for Contaminated Sites. Report CCME EPC-CS34. CCME, Winnipeg, Canada.

CCME (Council of Canadian Ministers of the Environment) (1992) *National Classification System for Contaminated Sites.* Environmental Protection Committee Report No. CCME/EPC-C539E, March. CCME, Ottawa, Canada.

CCME (Council of Canadian Ministers of the Environment) (1993a) *Guidance Manual on Sampling, Analysis and Data Management, Volume I: Main Report.* Report CCME EPC-NCS-62E. CCME, Winnipeg, Canada.

CCME (Council of Canadian Ministers of the Environment) (1993b) *Guidance Manual on Sampling, Analysis and Data Management, Volume II: Analytical Method Summaries.* CCME, Winnipeg, Canada.

CCME (Council of Canadian Ministers of the Environment) (1994) *Subsurface Assessment Handbook for Contaminated Sites.* Report CCME EPC-NCSRP-48E. CCME, Winnipeg, Canada.

CCME (Council of Canadian Ministers of the Environment) (1995) *A Protocol for the Derivation of Environmental and Human Health Soil Quality Guidelines.* Draft. CCME, Winnipeg, Canada.

CCME (Council of Canadian Ministers of the Environment) (1996a) *A Framework for Ecological Risk Assessment: General Guidance.* Report CCME-EPC-111. CCME, Winnipeg, Canada.

CCME (Council of Canadian Ministers of the Environment) (1996b) *Guidance Manual for Developing Site-Specific Soil Quality Remediation Objectives for Contaminated Sites in Canada.* CCME, Winnipeg, Canada.

CEC (Commission of the European Communities) (1980) Directive on the Protection of Groundwater against Pollution Caused by Certain Dangerous Substances. 80/68/EEC. *Official Journal of the European Communities,* No. L20, 26.1.80, 43–48.

CEC (Commission of the European Communities) (1993a) *Green Paper on Remedying Environmental Damage.* COM(93) 47 Final, 14 May. CEC, Brussels.

CEC (Commission of the European Communities) (1993b) Council Regulation No. 1836/93 allowing for voluntary participation by companies in a Community eco-management and audit scheme. *Official Journal of the European Communities,* No. L168, 10.7.93, 1–18

Clark, L. (1988) *The Field Guide to Water Wells and Boreholes.* Geological Society, London.

Claus, F. (1995) The Varresbecker Bach Participatory Process: the model of citizens initiatives. In *Fairness and Competence in Citizen Participation: Evaluating Models for Environmental Discourse.* O. Renn, T. Webler & P. Wiedemann (eds), pp 189–202. Kluwer Academic Publishers, Dordrecht, the Netherlands.

Close, M.E. (1993) Assessment of pesticide contamination of groundwater in New Zealand. *Journal of Marine & Freshwater Research,* 27, 257–66.

Costes, J.M., Texier, J., Zmirou, D. & Lambrozo, J. (1995) A prioritisation system for former gasworks sites based on the sensitivity of the environment. In *Contaminated Soil '95.* W.J. van der Brink, R. Bosman & F. Arendt (eds), pp 605–6. Kluwer Academic Publishers, Dordrecht, the Netherlands.

Cothern, C.R., Coniglio, W.A. & Marcus, W.L. (1986) Estimating risk to human health. *Environmental Science and Technology,* 20(2), 111–16.

Council of Europe (1972) *European Soil Charter.* B(72)63, May. Council of Europe, Luxembourg.

Council of Europe (1992) *Recommendation on Soil Protection.* R(92)8, May. Council of Europe, Luxembourg.

County Surveyors Society (1982) *Gas Generation from Landfill Sites.* Special Activity Group No. 7, Report. County Surveyors Society, London.

Covello, V.T. (1989) Informing people about risks from chemicals, radiation and other toxic substances. In *Prospects and Problems in Risk Communication.* W. Leiss (ed.), pp 1–50. University of Waterloo Press, Canada.

Covello, V.T. (1991) Risk comparisons and risk communication: issues and problems in comparing health and environmental risks. In *Communicating Risks to the Public: International Perspectives.* R.E. Kasperson & P.J.M. Stallen (eds), pp 79–124. Kluwer Academic Publishers, Dordrecht, the Netherlands.

Covello, V.T. (1992) Risk communication: a new and emerging area of communication research. In *Risk Assessment.* Proceeedings of a Conference 6–9 October 1992, pp 474–89. Health & Safety Executive, London.

Covello, V.T. & Merkhofer, M.W. (1993) *Risk Assessment Methods: Approaches for Assessing Health and Environmental Risks.* Plenum Press, New York, USA.

Covello, V.T. & Mumpower, J. (1985) Risk analysis and risk management: an historical perspective. *Risk Analysis,* 5, 103–20.

Cripps, J.C., Bell, F.G. & Culshaw, M.G. (eds) (1986) *Groundwater in Engineering Geology.* Geological Society, London.

Crowhurst, D. (1987) *Measurement of Gas Emissions from Contaminated Land.* Building Research Establishment, Fire Research Station, Borehamwood, Herts, UK.

Crowhurst, D. & Manchester, S.J. (1993) *The Measurement of Methane and other Gases from the Ground.* Report 131. Construction Industry Research and Information Association, London.

Cullen, A.C. (1994) Measures of compounding conservatism in probabilistic risk assessment. *Risk Analysis,* 14(4), 389–93.

Dalton, M.G., Hunstman, E. & Bradbury, K. (1991) Acquisition and interpretation of water-level data. In *Groundwater Monitoring.* D.M. Nielsen (ed.), pp 367–95. Lewis Publishers, Chelsea, Michigan, USA.

Davies, B.E. (ed.) (1980) *Applied Soil Trace Elements.* John Wiley & Sons, London.

De Boer, J. (1986) Community response to soil contamination – risk and uncertainty. In *Contaminated Soils.* J.W. Assink & W.J. van den Brink (eds), pp 211–19. Kluwer Academic Publishers, Dordrecht, the Netherlands.

Deloraine, A. *et al.* (1995) Case control assessment of the short-term health effects of an industrial toxic waste landfill. *Environmental Research,* 68, 124–32.

Denner, J.M., Perry, R. & Sterrit, R.M. (1991) *Hazardous Wastes and the Public: United Kingdom.* Working paper No WP/91/52/EN. European Foundation for the Improvement of Living and Working Conditions, Dublin.

Department of the Environment (1980) *Cadmium in the Environment and its Significance to Man.* HMSO, London.

Department of the Environment (1987a) *Problems Arising from the Redevelopment of Gas Works and Similar Sites.* 2nd edn. HMSO, London.

Department of the Environment (1989) *Waste Management Paper No 27: Landfill Gas.* 2nd edn. HMSO, London.

Department of the Environment (1990) *Appraisal of Hazards Related to Gas Producing Landfills.* Report No. CWM/016/90. Department of the Environment, London.

Department of the Environment (1991) *Public Registers of Land which May be Contaminated: A Consultation Paper.* May. Department of the Environment, London.

Department of the Environment (1992a) *Approved Document C: Site Preparation and Resistance to Moisture.* HMSO, London.

Department of the Environment (1992b) *Transport Pathways of Substances in Environmental Media: a Review of Available Models.* DoE Report No. DoE/HMIP/RR/92/030. Department of the Environment, London.

Department of the Environment (1993) *Waste Management Paper 26A: Landfill Completion.* HMSO, London.

Department of the Environment (1994a) *Framework for Contaminated Land.* Department of the Environment, London.

Department of the Environment (1994b) *Sampling Strategies for Contaminated Land.* CLR Report No. 4. Department of the Environment, London.

Department of the Environment (1994c) *A Framework for Assessing the Impact of Contaminated Land on Groundwater and Surface Water, Volumes One & Two.* CLR Report No. 1. Department of the Environment, London.

Department of the Environment (1994d) *Digest of Environmental Protection and Water Statistics.* No. 16. HMSO, London.

Department of the Environment (1994e) *Guidance on Preliminary Site Inspection of Contaminated Land.* CLR Report No. 2. Department of the Environment, London.

Department of the Environment (1994f) *Documentary Research on Industrial Sites.* CLR Report No. 3. Department of the Environment, London.

Department of the Environment (1995a) *A Guide to Risk Assessment and Risk Management for Environmental Protection.* HMSO, London.

Department of the Environment (1995b) *Prioritisation and Categorisation Procedure for Sites which May be Contaminated.* CLR Report No. 6. Department of the Environment, London.

Department of the Environment (1995c) *Digest of Environmental Statistics.* No. 17. HMSO, London.

Department of the Environment (1995/96) *Industry Profiles* (various titles). Department of the Environment, London.

Department of the Environment (1997a) *Standard Procedures for the Identification, Assessment Treatment and Monitoring of Contaminated Land.* Expected publication 1997 as CLR Report No. 11. Department of the Environment, London.

Department of the Environment (1997b) *Risk Assessment Procedure for Contaminated Land.* Expected publication 1997 as CLR Report. Department of the Environment, London.

Department of the Environment (1997c) *Contaminants in Soils: Collation of Toxicological Data and Intake Values for Humans.* Expected publication 1997 as CLR Report No. 9. Department of the Environment, London.

Department of the Environment (1997d) *The Contaminated Land Exposure Assessment Model (CLEA): Technical Basis and Algorithms.* Expected publication 1997 as CLR Report No. 10. Department of the Environment, London.

Department of the Environment (1997e) *Guideline Values for Contamination in Soils.* CLR Report No. 10, GV Series. Department of the Environment, London.

Devinny, J.S., Everett, L.G., Lu, J.C.S. & Stollar, R.L. (1990) *Subsurface Migration of Hazardous Wastes.* Van Nostrand Reinhold, New York, USA.

Droppo, J.G., Buck, J.W., Strenge, D.L. & Hoopes, B.L. (1993) Risk computation for environmental restoration activities. *Journal of Hazardous Materials*, **35**, 341–52.

Emberton, J.R. & Parker, A. (1987) The problems associated with building on landfill sites. *Waste Management & Research*, 5, 473–82.

Environment Canada (1995) *Draft: Canadian Soil Quality Guidelines for Arsenic.* Environment Canada, Hull, Quebec, Canada.

Erb, T.L., Philipson, W.R., Teng, W.L. & Liang, T. (1985) The analysis of landfills with historic airphotos. In *The Surveillant Science: Remote Sensing of the Environment.* R.K. Holz (ed.), pp 90–4. John Wiley & Sons, New York, USA.

Evans, O.D. & Thompson, G.M. (1986) Field and interpretative techniques for delineating subsurface petroleum hydrocarbon spills using soil gas analysis. In *Proceedings of Conference on Petroleum Hydrocarbons and Organic Chemicals in Groundwater: Prevention, Detection and Restoration.* National Water Well Association, Dublin.

Farland, W. & Dourson, M. (1993) Noncancer health endpoints: approaches to quantitative risk assessment. In *Comparative Environmental Risk Assessment*. C.R. Cothern (ed.), pp 87–106. Lewis Publishers, Boca Raton, Florida, USA.

Ferguson, C. (1992) The statistical basis for spatial sampling of contaminated land. *Ground Engineering*, **June**, 34–8.

Ferguson, C. (1996) Assessing human health risks from exposure to contaminated land: a review of recent research. *Land Contamination & Reclamation*, **4(3)**, 159–70.

Ferguson, C. & Abbachi, A. (1993) Incorporating expert judgement into statistical sampling designs for contaminated sites. *Land Contamination & Reclamation*, **1(3)**, 135–42.

Ferguson, C.C. & Denner, J.M. (1993a) Soil remediation guidelines in the UK: a new risk-based approach. In Proceedings of Conference *Developing Cleanup Standards for Contaminated Soil, Sediment and Groundwater: How Clean is Clean?* pp 205–12. Water Environment Federation, Alexandria, Virginia, USA.

Ferguson, C.C. & Denner, J.M. (1993b) Soil guideline values in the UK: new risk-based approach. In *Contaminated Soil '93*. F. Arendt, G.J. Annokkée W.J. van den Brink (eds), pp 365–72. Kluwer Academic Publishers, Dordrecht, the Netherlands.

Ferguson, C.C. & Denner, J.M. (1994) Developing guideline (trigger) values for contaminants in soil: underlying risk analysis and management concepts. *Land Contamination & Reclamation*, 2(3), 117–23.

Ferguson, C.C. & Denner, J.M. (1995) UK action (or intervention) values for contaminants in soil for protection of human health. In *Contaminated Soil '95*. W.J. van den Brink, R. Bosman & F. Arendt (eds), pp 1199–200. Kluwer Academic Publishers, Dordrecht, the Netherlands.

Ferguson C. & Marsh J. (1993) Assessing human health risks from ingestion of contaminated soil. *Land Contamination and Reclamation*, **1(4)**, 177–85.

Ferguson, C.C., Krylov, V.V. & McGrath, P.T. (1995) Contamination of indoor air by toxic soil and vapours: a screening risk assessment model. *Building and Environment*, **30**, 375–83.

Finsterbusch, K. (1989) Community responses to exposures to hazardous wastes. In *Pyschosocial Effects of Hazardous Toxic Waste Disposal on Communities*. D.L. Peck (ed.), pp 57–80. Charles C. Thomas, Springfield, Illinois, USA.

Fischhoff, B. (1995) Risk perception and communication unplugged: twenty years of process. *Risk Analysis*, **15(2)**, 137–45.

Foote, T.W. (1993) Contaminated sites: a Canadian perspective. In *Demonstration of Remedial Action Technologies for Contaminated Land and Groundwater*. Vol 2 – Part 1, pp 4–16, EPA/600/R-93/012b. USEPA, Cincinnati, USA.

Franzius, V. & Grimski, D. (1995) Recent developments in contaminated land remediation in the Federal Republic of Germany: current programmes and future research. *Land Contamination & Reclamation*, **3(1)**, 47–54.

Freeze, R.A. & Cherry, J.A. (1979) *Groundwater*. Prentice-Hall, Englewood Cliffs, New Jersey, USA.

Garetz, W.V. (1993) Current concerns regarding implementation of risk-based management: how real are they? In *Comparative Environmental Risk Assessment*. C.R. Cothern (ed.), pp 11–31. Lewis Publishers, Boca Raton, Florida, USA.

Gaskin, J.E. (1988) *Quality Assurance Guidelines and Principles for the Handling and Management of Water Quality Data*. Water Quality Branch, Inland Waters and Lands, Conservation and Protection, Environment Canada, Ottawa, Canada.

Gasser, U.G., Walker, W.J., Dahlgren, R.A., Borch, R.S. & Burau, R.G. (1996) Lead release from smelter and mine waste impacted materials under simulated gastric conditions and relationship to speciation. *Environmental Science and Technology*, **30(3)**, 761–9.

Gendebien, A., Pauwels, M., Constant, M., Ledrut-Damanet, M-J., Nyns, E-J., Willumsen, H.-C., Butson, J., Farby, R. & Ferrero, G-L. (1992) *Landfill Gas from Environment to Energy.* EUR 14017/1 EN. Commission of the European Communities, Luxembourg.

Gerrard, S. & Kemp, R. (1993) The risk management of UK landfill sites: catching will-o-the-wisp. *Project Appraisal,* **June**, 66–76.

Gilbert, R.O. (1987) *Statistical Methods for Environmental Pollution Monitoring.* Van Nostrand Reinhold, New York, USA.

GLC (Greater London Council) (1976) Some guidelines on the re-use of industrially contaminated land. *London Environmental Bulletin,* **No. 98 (2nd Series),** 1/1–1/8.

Golding, D., Krimsky, S. & Plough, A. (1992) Evaluating risk communication: narrative vs technical presentation of information about radon. *Risk Analysis,* **12(1),** 27–35.

Goldsborough, D.G. & Smit, P.J. (1995) RISC: computer models for soil investigation, risk analysis and urgency estimation. In *Contaminated Soil '95.* W.J. van den Brink, R. Bosman & F. Arendt (eds), pp 581–8. Kluwer Academic Publishers, Dordrecht, the Netherlands.

Grisham, J.W. (ed) (1986) *Health Effects of the Disposal of Waste Chemicals.* Pergamon Press, New York.

Groh, H. & Pahl, A. (1993) Developing a software tool to support investigation strategies to tackle problems in the field of contaminated sites. In *Contaminated Soil '93.* F. Arendt, G.J. Annokkee, R. Bosman & W.J. van den Brink (eds), pp 665–72. Kluwer Academic Publishers, Dordrecht, the Netherlands.

Grolimund, D., Borkovec, M., Barmettler, K. & Sticher, H. (1996) Colloid facilitated transport of strongly sorbing contaminants in naturally porous media: a laboratory study. *Environmental Science & Technology,* **30(10),** 3118–23.

Hall, D.H., Jefferis, M., Gronow, J. & Harris, R.C. (1995) Probabilistic risk assessment for landfill regulation. In *Contaminated Soil '95.* W.J. van den Brink, R. Bosman & F. Arendt (eds), pp 617–19. Kluwer Academic Publishers, Dordrecht, the Netherlands.

Hamill, P.V.V., Johnson, C.L. & Roche, A.F. (1979) Physical growth: National Center for Health Statistics percentiles. *American Journal of Clinical Nutrition,* **32 (March),** 607–29.

Harget, M.J. & Miller, F.J. (1996) Hazard and risk ranking criteria for closed landfill sites. *Land Contamination & Reclamation,* **4(2),** 77–84.

Harris, M.R., Herbert, S.M. & Smith, M.A. (1995) *Remedial Treatment for Contaminated Land, Volume III: Site Investigation and Assessment.* SP103. Construction Industry Research and Information Association, London.

Harris, M.R., Herbert, S.M. & Smith, M.A. (1996) *Remedial Treatment for Contaminated Land, Volume VI: Sites, Volume VI: Containment and Hydraulic Measures.* SP106. Construction Industry Research and Information Association, London.

Harris, M.R., Herbert, S.M. & Smith, M.A. (1997) *Remedial Treatment for Contaminated Land, Volume XII: Policy and Legislation.* SP112. Construction Industry Research and Information Association, London.

Heimendenger, J. (1964) Die ergebnisse von körpermessungen an 5000 Basler Kindern von 2–18 jahren. *Helvetica Paediatrica Acta* **19 (suppl. 13),** whole volume 193 pp.

Hemel, A., Van Mulder, J., Ritsema, H. & Schiphuis, J.J. (1992) *Urgentiebepalingssysteem voor orienterend bodemonderzoek.* Van Hall Institute, Groningen, the Netherlands.

Hertzman, C., Ostry, A. & Teschke, K. (1989) Environmental risk analysis: a case study. *Canadian Journal of Public Health,* **80,** 8–15.

HMIP (Her Majesty's Inspectorate of Pollution) (1995) *Procedure for Operator and Pollution Risk Appraisal.* Consultation Document. HMIP, London.

Hobson, D.A. (1993) Rational site investigations. In *Contaminated Land: Problems and Solutions.* T. Cairney (ed.), pp 29–67. Blackie Academic, London.

Holdgate, M.W. (1979) *A Perspective of Environmental Pollution.* Cambridge University

Press, Cambridge.

Holtkamp, A.B. & Gravesteyn, L.J.J. (1993) Large-scale voluntary clean-up operation for contaminated sites in the Netherlands now on its way. In *Contaminated Soil '93*. F. Arendt, G.J. Annokkée & W.J. van den Brink (eds), pp 27–34. Kluwer Academic Publishers, Dordrecht, the Netherlands.

Hooker, P.J. & Bannon, M.P. (1993) *Methane: its Occurrence and Hazards in Construction*. CIRIA Special Publication. Construction Industry Research and Information Association, London.

House of Commons Select Committee on the Environment (1990) *Contaminated Land*. First Report, 3 Volumes. HMSO, London.

House of Commons Select Committee on the Environment (1996) *Contaminated Land*. Session 1996/1997. HMSO, London.

Howsan, P. & Thakoordin, M. (1996) Groundwater quality monitoring: using flow-through cells. *Journal of the Chartered Institute of Water and Environmental Management*, **10**, 407–10.

Hrudey, S.E. & Krewski, D. (1995) Is there a safe level of exposure to a carcinogen? *Environmental Science and Technology*, **29(8)**, 370A–75A.

HSE (Health and Safety Executive) (1987) *Disposal of Explosives Waste and the Decontamination of Explosives Plant, HS(G)36*. HMSO, London.

HSE (Health and Safety Executive) (1989) *Avoiding Danger from Underground Services, HS(G)47*. HMSO, London.

HSE (Health and Safety Executive) (1991) *Protection of Personnel and the General Public during Development of Contaminated Land, HS(G)66*. HMSO, London.

HSE (Health and Safety Executive) (1992) *The Tolerability of Risk from Nuclear Power Stations*. HMSO, London.

HSE (Health and Safety Executive) (1995a) *Managing Construction for Health and Safety: Construction (Design and Management) Regulations 1994*. HSE, London.

HSE (Health and Safety Executive) (1996) *The Use of Risk Assessment in Government Departments*. HSE Books, Sudbury.

Hunnes, E.G. (1995) Sampling strategies – an important choice in site investigation. In *Contaminated Soil '95*. W.J. van den Brink, R. Bosman & F. Arendt (eds), pp 135–44. Kluwer Academic Publishers, Dordrecht, the Netherlands.

Hushon, J.M. (ed.) (1990a) *Expert Systems for Environmental Applications*. American Chemical Society, Washington, DC, USA.

Hushon, J.M. (1990b) The defense priority model for Department of Defence remedial site ranking. In *Expert Systems for Environmental Applications*. J.M. Hushon (ed.), American Chemical Society Symposium Series 431, pp 206–16. American Chemical Society, Washington, DC, USA.

HWC (Health and Welfare Canada) (1989) *Guidelines for Canadian Drinking Water Quality*. 4th edn. Canadian Government Publishing Centre, Ottawa, Canada.

ICE (Institution of Civil Engineers) (1991) *Inadequate Site Investigations*. Thomas Telford, London.

ICE (Institution of Civil Engineers) (1993) *Guidance for the Safe Site Investigation in Construction, Volume 4: Investigation by Drilling of Landfills and Contaminated Land*. Thomas Telford, London.

ICRCL (Interdepartmental Committee on the Redevelopment of Contaminated Land) (1979) *Notes on the Redevelopment of Gasworks Sites*. ICRCL, 18/79. Department of the Environment, London.

ICRCL (Interdepartmental Committee on the Redevelopment of Contaminated Land) (1987) *Guidance on the Assessment and Redevelopment of Contaminated Land*. ICRCL 59/83, 2nd edn. Department of the Environment, London.

ICRCL (Interdepartmental Committee on the Redevelopment of Contaminated Land)

(1990a) *Notes on the Restoration and Aftercare of Metalliferous Mining Sites.* ICRCL, 70/90. Department of the Environment, London.

ICRCL (Interdepartmental Committee on the Redevelopment of Contaminated Land) (1990b) *Notes on the Redevelopment of Landfill Sites.* ICRCL, 17/78, 8th edn. Department of the Environment, London.

IEA (Institute of Environmental Assessment) (1995) *Guidelines for Baseline Ecological Assessment.* E & FN Spon, London.

ISO (International Standards Organisation) (1989a) *EN 45002: General Requirements for the Technical Competence of Testing Laboratories.* Available as BS7502 from British Standards Institution, London.

ISO (International Standards Organisation) (1989b) *EN 45001: General Criteria for the Operation of Testing Laboratories.* Available as BS7501 from British Standards Institution, London.

ISO (International Standards Organisation) (1993) *ISO 10381-6:1993 Soil quality: Sampling, Part 6: Guidance on the Collection, Handling, and Storage of Soil for the Assessment of Aerobic Microbial Processes in the Laboratory.* Available in the UK as BS 7755: Section 2.6:1994 from British Standards Institution, London.

ISO (International Organisation for Standardisation) (1996a) *Soil Quality – Sampling – Design of Sampling Programmes.* [CD 10381-51]. Available from the Secretary EH4, British Standards Institution, London.

ISO (International Organisation for Standardisation) (1996b) Technical Committee 190 (Soil Quality), Sub-Committee 7 (Soil and Site Assessment), proposed international standard on *Characterisation of Soil in Relation to Protection of Groundwater.* Current version available from DIN, Berlin.

IWM (Institute of Wastes Management) (1990) *Monitoring of Landfill Gas.* IWM, Northampton, UK.

James, B.R. (1996) The challenge of remediating chromium-contaminated soil. *Environmental Science and Technology*, **30(6)**, 248A–51A.

JEFCA (Joint FAO/WHO Expert Committee on Food Additives) (1972) *Evaluation of Certain Food Additives and Contaminants: Mercury, Lead and Cadmium.* WHO Technical Report Series No. 505, 1972 and corrigendum. World Health Organisation, Geneva.

JEFCA (Joint FAO/WHO Expert Committee on Food Additives) (1993) *Forty-first Meeting, Geneva, 9–18 February 1993, Summary and Conclusions.* Document PCS/93.8. World Health Organisation, Genva.

Jefferis, S.A. (1993) In-ground barriers. In *Contaminated Land: Problems and Solutions.* T.Cairney (ed.), pp 111–40. Blackie Academic & Professional, London.

Jenni, K.E., Merkhofer, M.W. & Williams, C. (1995) The rise and fall of a risk-based priority system: lessons from DOE's Environmental Restoration Priority System. *Risk Analysis*, **15(3)**, 397–410.

Jessiman, B., Richardson, G.M., Clark, C. & Halber, B. (1992) A quantitative evaluation of ten approaches to setting site-specific clean-up objectives. *Journal of Soil Contamination,* **1(1),** 39–59.

Jones, K., Alcock, R.E., Johnson, D.L., Northcott, G.L., Semple, K.T. & Woolgar, P.J. (1996) Organic chemicals in contaminated land: analysis, significance and research priorities. *Land Contamination and Reclamation,* **4(2),** 189–98.

Jordfald, G. (1991) The national clean-up plan for contaminated industrial and mining sites. In *Proceedings of Environment Northern Seas 1991, Seminar on Clean-Up of Polluted Industrial and Mining Sites.* Volume 3, August 1991. Stavanger Forum, Stavanger, Norway.

Jorgenson, S.E. (ed.) (1984) *Modelling the Fate and Effects of Toxic Substances in the Environment.* Elsevier, Amsterdam, the Netherlands.

Journel, A.G. & Huijbregts, C.J. (1978) *Mining Statistics.* Academic Press, London.

Kasamas, H. (1991) *Contaminated Sites – The Situation in Austria.* Paper presented at the Fifth Annual NATO CCMS Conference on the Demonstration of Remedial Action Technologies for Contaminated Land and Groundwater, November 1991, Washington DC.

Kasamas, H. (1995) Contaminated sites programme in Austria. In *Contaminated Soil '95.* W.J. van der Brink, R. Bosman & F. Arendt (eds), pp 1659–62. Kluwer Academic Publishers, Dordrecht, the Netherlands.

Keenan, R.E., Finley, B.L. & Price, P.S. (1994) Exposure assessment: then, now and quantum leaps in the future. *Risk Analysis,* **14(3)**, 225–31.

Keith, L.H. (1991a) *Quality Control and GC Advisor.* Lewis Publishers, Boca Raton, Florida, USA.

Keith, L.H. (1991b) *Environmental Sampling and Analysis: A Practical Guide.* Lewis Publishers, Chelsea, Michigan, USA.

Kelly, R.T. (1980) Site investigation and material problems. In *Proceedings of a Conference on the Reclamation of Contaminated Land,* pp B2/1–B2/14. Society of Chemical Industry, London.

Klauenberg, B.J. & Vermulen, E.K. (1994) Role for risk communication in closing military waste sites. *Risk Analysis,* **14(3)**, 351–6.

Kocher, D.C. & Hoffman, F.O. (1996) Comment on "An approach for balancing health and ecological risks at hazardous waste sites". *Risk Analysis,* **16(3)**, 295–7.

Kovalick, W.W. & Kingscott, J.W. (1995) Progress in clean-up and technological developments in US Superfund program. In *Contaminated Soil '95.* W.J. van der Brink, R. Bosman & F. Arendt (eds), pp 29–38. Kluwer Academic Publishers, Dordrecht, the Netherlands.

Krimsky, S. & Golding, D. (eds) (1992) *Social Theories of Risk.* Praeger, Connecticut, USA.

Leach, B.A. & Goodger, H.K. (1991) *Building on Derelict Land.* Special Publications 78. Construction Industry Research and Information Association, London.

Levine, A. (1981) *Love Canal: Science, Politics and People.* Lexington, Massachusetts, USA.

Liikala, T.L., Olsen, K.B., Teel, S.S. & Lanigan, D.C. (1996) Volatile organic compounds: collection and preservation methods. *Environmental Science & Technology,* **30(12)**, 3441–7.

Litz, H. & Blume, H.P. (1992) System for predicting the vulnerability of soils to organic chemicals. In *Proceedings of International Symposium on Environmental Contamination in Central and Eastern Europe,* pp 835–42. Florida State University, Tallahassee, Florida, USA.

Lord, J.A. (1991) Recycling landfill and chemical waste sites. In *Proceedings of Containment of Pollution and Redevelopment of Closed Landfill Sites.* Paper 6.1. Leamington Spa, UK.

Lovell, J. (1993) Environmental samples and carefully controlled shipping to prevent degradation. *Pollution Prevention,* **June,** 59–61.

LPC (Loss Prevention Council) (1992) *Pollutant Industries.* LPC, London.

Mackay, D. (1991) *Multimedia Environmental Models: the Fugacity Approach.* Lewis Publishers, Michigan, USA.

Maney, J.P. & Dallas, A. (1991) The importance of measurement integrity. *Environmental Laboratory,* **3(5)**, 20–5 and 52.

Marsh, G.M. & Caplan, R.J. (1987) Evaluating health effects of exposure at hazardous waste sites: a review of the state-of-the-art, with recommendations for future research. In *Health Effects from Hazardous Waste Sites.* J.B. Andelman & D.W. Underhill (eds), pp 3–80. Lewis Publishers, Michigan, USA.

Martin, G.R., Smoot, J.L. & White, K.D. (1992) A comparison of surface-grab and cross sectionally integrated stream-water-quality sampling methods. *Water Environment Research*, **64**, 866–76.

Marvan, I. & Herbert, D.J. (1996) *Evaluation of Six Near-real-time Analytical Methods.* Paper to NATO/CCMS International meeting on remediation of contaminated soils and groundwater, Adelaide, Australia.

Maughan, J.T. (1993) *Ecological Assessment of Hazardous Waste Sites.* Van Nostrand Reinhold, New York.

Maxim, L.D. (1989) Problems associated with the use of conservative assumptions in exposure and risk analysis. In *The Risk Assessment of Environmental and Human Health Hazards.* D.J. Paustenbach (ed.), pp 526–60, John Wiley & Sons, New York.

McCann, D.M. (1994) Geophysical methods for the assessment of landfills and waste disposal sites. *Land Contamination & Reclamation,* **2(2)**, 73–83.

McDonald, A. (1996) Dealing with genotoxic carcinogens: a UK approach. In *Environmental Impact of Chemicals: Assessment and Control.* M.D. Quint, D. Taylor & R. Purchase (eds). Royal Society of Chemistry, Cambridge, UK.

McFarland, R. (1992) Simple qualitative risk assessment techniques for the prioritisation of chemically contaminated sites in New South Wales, Australia. In *Risk Assessment,* Proceedings of a Conference, 6–9 October 1992, London, pp 456–64. Health & Safety Executive, London.

McKone, T.E. (1990) Dermal uptake of organic chemicals from a soil matrix. *Risk Analysis*, **10**, 407.

McKone, T.E. & Howd, R.A. (1992) Estimating dermal uptake of non-ionic organic chemicals from water and soil: I. Unified fugacity-based models for risk assessments. *Risk Analysis*, **12(4)**, 543–58.

MHSPE (Ministry of Housing, Spatial Planning and Environment) (1994) *Circular on Intervention Values for Soil Remediation.* Government Printing Office, The Hague, the Netherlands.

Ministry of Housing and Local Government (1963) *New Life for Dead Lands: Derelict Acres Reclaimed.* HMSO, London.

Montgomery, R.E., Remeta, D.P. & Gruenfeld, M. (1985) Rapid on-site methods of chemical analysis. In *Contaminated Land: Reclamation and Treatment.* M.A. Smith (ed.), pp 257–310. Plenum, London.

Moore, D.R.J. & Elliott, B.J. (1996) Should uncertainty be quantified in human and ecological risk assessments used for decision-making? *Human and Ecological Risk Assessment,* **2(1)**, 11–24.

Morgan, H. (ed) (1988) The Shipham Report: an investigation into cadmium contamination and its implications for human health. *The Science of the Total Environment,* **75**, 1–143.

Morgan, H. & Simms, D.L. (1988) Discussion and conclusions – the Shipham Report. *The Science of the Total Environment*, **75**, 135–43.

Morgan, M.G. & Henrion, M. (1990) *Uncertainty: a Guide to Dealing with Uncertainty in Quantitative Risk and Policy Analysis.* Cambridge University Press, New York.

Moskowitz, P.D., Pardi, R., Fthenakis, V.M., Holtzman, S., Sun, L.C. & Irla, B. (1996) An evaluation of three representative multi-media models used to support clean-up decision-making at hazardous, mixed, and radioactive waste sites. *Risk Analysis,* **16(2)**, 279–86.

MVROM (Ministerie van Volkshuisvesting, Ruimteljke Ordening en Milieubeheer) (1993a) *Protocol voor het Nader Onderzoek deel 1.* Sdu Uitgevij Koninginnegracht, Den Haag, Netherlands.

MVROM (Ministerie van Volkshuisvesting, Ruimteljke Ordening en Milieubeheer) (1993b) *Protocol voor het Oriënterend Onderzoek.* Sdu Uitgevij Koninginnegracht, Den Haag, Netherlands.

Myers, K., Vogt, T. & Wales, J. (1994) Hazard ranking criteria for contaminated sites. *Land Contamination & Reclamation*, **2(1)**, 13–18.

NAS (National Academy of Sciences) (1983) *Risk Assessment in the Federal Government: Managing the Process.* National Academy Press, Washington, DC, USA.

NAS (National Academy of Sciences) (1996) *Understanding Risk: Informing Decisions in a Democratic Society.* National Academy Press, Washington, DC, USA.

Nature Conservancy Council (1990) *Handbook for Phase I Habitat Survey.* Nature Conservancy Council, Peterborough, UK.

New York Department of Health (1978) *Love Canal: Public Health Time Bomb.* New York Department of Health, Albany, USA.

Nielsen, D.M. (ed.) (1991) *Practical Handbook of Groundwater Monitoring.* Lewis Publishers, Chelsea, Michigan, USA.

NNI (Nederlands Normalisatie-Institute) (1991) *NVN 5740: Soil: Investigation Strategy for Exploratory Survey.* NNI, Delft, the Netherlands.

NRA (National Rivers Authority) (1992) *Policy and Practice for the Protection of Groundwater.* National Rivers Authority, Bristol, UK.

NRA (National Rivers Authority) (1994a) *Contaminated Land and the Water Environment.* Water Quality Series No. 15. National Rivers Authority, Bristol, UK.

NRA (National Rivers Authority) (1994b) *Abandoned Mines and the Water Environment.* Water Quality Series No. 14. National Rivers Authority, Bristol, UK.

NRA (National Rivers Authority) (1995a) *Guide to Groundwater Protection Zones in England and Wales.* HMSO, London.

NRA (National Rivers Authority) (1995b) *Guide to Groundwater Vulnerability Mapping in England and Wales.* HMSO, London.

O'Brien, A.A., Steeds, J.E. & Law, G.A. (1992) Case study: investigation of Long Cross and Barracks Lane landfill sites. In *Proceedings of a Conference on Planning and Engineering of Landfills*, pp 31–4. Midlands Geotechnical Society, Birmingham, UK.

OECD (Organisation for Economic Cooperation and Development) (1989) *Compendium of Environmental Exposure Assessment Methods for Chemicals.* Environment Monograph 27. OECD, Paris, France.

Olsen, K.B., Wang, J., Setiadji, R. & Lu, J. (1994) Field screening of chromium, cadmium, zinc, copper and lead in sediments by stripping analysis. *Environmental Science and Technology*, **28(12)**, 2074–9.

OMEE (Ontario Ministry of Environment and Energy) (1994) *Draft – Rationale for the Development and Application of Generic Soil, Groundwater and Sediment Criteria for Clean-up of Contaminated Sites.* OMEE, Toronto, Canada.

OMEE (Ontario Ministry of Environment and Energy) (1996) *Guidelines for the Clean-up of Contaminated Sites in Ontario.* OMEE, Toronto, Canada.

O'Riordan, T. (1991) The new environmentalism and sustainable development. *The Science of the Total Environment*, **105**, 5–15.

Orr, W.E., Wood, A.M., Beaver, J.J., Ireland, R.J. & Beagley, D.P. (1991) Abbeystead outfall works: background to repairs and modifications and lessons learned. *Journal of Institution of Water and Environmental Management*, **5th February**, 7–22.

Palsma, A.J. & Diependaal, M.J. (eds) (1993) *Sustainable Soil Use: a TNO Concept.* TNO Environmental and Energy Research, Delft, the Netherlands.

Paustenbach, D.J. (1987) Assessing the potential environmental and human health risks of contaminated soil. *Comments on Toxicology*, **1**, 185–220.

Paustenbach, D.J. (ed.) (1989) *The Risk Assessment of Environmental and Human Health Hazards: A Textbook of Case Studies.* John Wiley & Sons, New York, USA.

Paustenbach, D.J., Rinehart, W.E. & Sheehan, P.J. (1991) The health hazards posed by chromium contaminated soils in residential and industrial areas: conclusions of an expert panel. *Regulatory Toxicology & Pharmacology*, **13**, 195–222.

Paustenbach, D.J., Shu, H.P. & Murray, F.J. (1986) A critical analysis of risk assessment of TCDD contaminated soil. *Regulatory Toxicology & Pharmacology*, **13**, 195–222.

Petts, J. (1993) Risk assessment for contaminated sites. In *Proceedings of a Conference on Site Investigations for Contaminated Sites*. Paper 1. IBC Technical Services, London.

Petts, J. (1994a) Contaminated sites: blight, public concerns and communication. *Land Contamination & Reclamation*, **2(4)**, 171–81.

Petts, J. (1994b) Dealing with contaminated land in a risk management framework. In *Proceedings of a Conference on Contaminated Land Policy, Risk Management and Technology*. IBC Technical Services, London.

Petts, J. (1996) Risk communication: research findings and needs. *Land Contamination & Reclamation*, **4(3)**, 171–8.

Petts, J. & Eduljee, G. (1994) *Environmental Impact Assessment for Waste Treatment and Disposal Facilities*. John Wiley & Sons, Chichester, UK.

Phillips, P., Denman, T. & Barker, S. (1997) Silent, but deadly. *Chemistry in Britain*, **January**, 35–8.

Pritchard, E. (ed.) (1995) *Quality in the Analytical Laboratory*. John Wiley & Sons, Chichester, UK.

Ramsay, L. (1995) Quality trade-offs in site investigations: advantages and disadvantages of using test methods instead of laboratory methods. In *Contaminated Soil '95*. W.J. van den Brink, R. Bosman & F. Arendt (eds), pp 113–24. Kluwer Academic Publishers, Dordrecht, the Netherlands.

Raybould, J.G., Rowan, S. & Barry, D.L. (1996) *Methane Investigation Strategies*. CIRIA Report 150. Construction Industry Research and Information Association, London.

RCEP (Royal Commission on Environmental Pollution) (1984) *Tackling Pollution – Experience and Prospects*. 10th Report. HMSO, London.

RCEP (Royal Commission on Environmental Pollution) (1996) *Sustainable Use of Soil*. 19th Report. HMSO, London.

Rees, J.F. & Grainger, J.M. (1982) Rubbish dump or fermenter? *Process Biochemistry*, Nov/Dec, 41–4.

Reichard, E., Cranor, C., Raucher, R. & Zaponni, G. (1990) *Groundwater Contamination Risk Assessment: A Guide to Understanding and Managing Uncertainties*. International Association of Hydrological Sciences, Institute of Hydrology, Wallingford, UK.

Renn, O. (1992) Risk communication: towards a rational discourse with the public. *Journal of Hazardous Materials*, **29**, 184–206.

Renn, O., Webler, T. & Wiedemann, P. (eds) (1995) *Fairness and Competence in Citizen Participation: Models for Environmental Discourse*. Kluwer Academic Publishers, Dordrecht, the Netherlands.

Renner, R. (1996) Ecological risk assessment struggles to define itself. *Environmental Science and Technology*, **30(4)**, 172a–4a.

Reynolds, J.M. & McCann, D.M. (1992) Geophysical methods for the assessment of landfill and waste disposal sites. In *Proceedings of 2nd International Conference on Polluted and Marginal Land*. M.C. Forde (ed.). Engineering Technical Press, Edinburgh.

Richardson, G.M. (1996) Deterministic versus probabilistic risk assessment: strengths and weaknesses in a regulatory context. *Human and Ecological Risk Assessment*, **2(1)**, 44–54.

Rix, I. (1994) Testing proficiency in soils analysis. *Land Contamination & Reclamation*, **2(1)**, 4–6.

Robitaille, G.E. (1992) Quantitative in situ soil gas sampling. In *Proceedings of the Eighth Annual Waste Testing and Quality Assurance Symposium*, pp 2–14. American Chemical Society, Washington, DC, USA.

Rogers, K.R. & Gerlach, C.L. (1996) Environmental biosensors: a status report. *Environmental Science & Technology,* **30(11),** 486A–519A.

Ross, S.M. (1989) *Soil Processes.* Routledge, London.

Ross, S.M. (ed.) (1994) *Toxic Metals in Soil-Plant Systems.* John Wiley & Sons, Chichester, UK.

Royal Society (1992) *Risk: Analysis, Perception and Management.* Report of a Royal Society Study Group. The Royal Society, London.

Ruby, M.V., Davis, A., Schoof, R., Eberle, S. & Sellstone, C.M. (1996) Estimation of lead and arsenic bioavailability using a physiologically based extraction test. *Environmental Science and Technology*, **30(2)**, 422–30.

Ruckelshaus, W.D. (1983) Science, risk and public policy. *Science,* **221**, 1026-8.

Schmidt, J.W. (1991) Contaminated land: a Canadian viewpoint. In *Proceedings of a Conference on Contaminated Land Policy, Regulation and Technology.* IBC Technical Services, London.

SEGH (Society for Environmental Geochemistry and Health) (1993) *Lead in Soil: Recommended Guidelines.* Science Reviews, Northwood, London.

SEPA (Swedish Environmental Protection Agency) (1997) *Development of Generic Guideline Values: Model and Data Used for Generic Guideline Values for Contaminated Soils in Sweden.* Report 4639. SEPA, Stockholm.

Sheppard, S.C., Gaudet, C., Sheppard, M.I., Cureton, P.M. & Wong, M.P (1992) The development of assessment and remediation guidelines for contaminated soils: a review of the science. *Canadian Journal of Soil Science*, **72**, 359–94.

Shook, G. & Grantham, C. (1993) A decision analysis technique for ranking sources of groundwater pollution. *Journal of Environmental Management*, **37(3)**, 201–6.

Shrivastava, P. (1993) The greening of business. In *Business and the Environment.* D. Smith (ed.), pp 27–39. Paul Chapman Publishing, London.

Siegrist, R.L. (1989) *International Review of Approaches for Establishing Cleanup Goals for Hazardous Waste Contaminated Land.* Institute of Georesources and Pollution Research, Norway.

Siegrist, R.L. & Jenssen, P.D. (1990) Evaluation of sampling method effect on volatile organic compound measurements in contaminated soils. *Environmental Science and Technology* **24**, 1387–92.

Simmons, S.M. (ed.) (1990) *Hazardous Wastes Measurement.* Lewis Publishers, Chelsea, Michigan, USA.

Skaarup, J. & Pedersen, M. (1996) *Rehabilitation of a Site Contaminated by Tar Substances using a New On-Site Technique.* Agency of Environmental Protection, Copenhagen, Denmark.

Slovic, P. (1989) A risk communication perspective on an integrated waste management strategy. In *Insurance and Risk Management for Hazardous Wastes.* H. Kunreuther & R. Gowda (eds), pp 195–216. Kluwer Academic Dordrecht. The Netherlands.

Smith, A.M., Sciortino, S., Goeden, H. & Wright, C.C. (1996) Consideration of background exposures in the management of hazardous waste sites: a new approach to risk assessment. *Risk Analysis,* **16(5),** 619–26.

Smith, M.A. (ed.) (1985) *Contaminated Land: Reclamation and Treatment.* NATO Challenges of Modern Society Volume 8. Plenum Press, New York.

Smith, M.A. (1991) Data analysis and interpretation. In *Recycling Derelict Land.* G. Fleming (ed.), pp 88–114. Thomas Telford, London.

Smith, M.A. (1992) Safety aspects of waste disposal to landfill. In *Proceedings of a Conference on Planning and Engineering of Landfills*, pp 9–21. Midlands Geotechnical Society, Birmingham, UK.

Smith, M.A. (1993a) Landfill gases. In *Contaminated Land: Problems and Solutions.* T. Cairney (ed.), pp 160–90. Blackie Academic & Professional, London.

Smith, M.A. (1993b) Experiences of the development and application of guidelines for contaminated sites in the United Kingdom. In Proceedings of Conference *Developing Cleanup Standards for Contaminated Soil, Sediment and Groundwater: How Clean is Clean?* pp 195–204. Water Environment Federation, Alexandria, Virginia, USA.

Smith, M.A. & Ellis, A.C. (1986) An investigation into methods used to assess gasworks sites for reclamation. *Reclamation and Revegetation Research*, **4**, 183–209.

Soczo, E.R., Meeder, T.A. & Versluijs, C.W. (1993) Ten years of soil clean-up in the Netherlands. In *Demonstration of Remedial Action Technologies for Contaminated Land and Groundwater*. Vol 2 – Part 1, pp 38–43. EPA/600/R-93/012b. USEPA, Cincinnati, USA.

Staff, M.G., Sizer, K.E. & Newson, S.R. (1991) The potential for surface emissions of methane from abandoned mine workings. In *Proceedings of Symposium on Methane – Facing the Problems*, Paper 1.1, March. Nottingham University, UK.

Steeds, J.E., Shepherd, E. & Barry, D.L (1996) *A Guide to Safe Working Practices for Contaminated Sites.* SP 119. Construction Industry Research and Information Association, London.

Stock, H.D. (1995) On-site versus off-site analysis – a matter of balance between heterogeneity and accuracy. In *Contaminated Soil '95*, W.J. van den Brink, R. Bosman & F. Arendt (eds), pp 107–12. Kluwer Academic Publishers, Dordrecht, the Netherlands.

Suter, G.W. (1993) *Ecological Risk Assessment.* Lewis Publishers, BOCA Raton, Florida, USA.

Tarr, J.A. & Jacobson, C. (1987) Environmental risk in historical perspective. In *The Social and Cultural Construction of Risk*. B.B. Johnson & V.T. Covello (eds), pp 317–44. D. Rheidel Publishing Co., Dordrecht, the Netherlands.

Taylor, L. (1993) Laboratory analysis techniques surveyed. *Pollution Prevention*, **3(5)**, 56–8.

Travis, C.C. & Cook, S.C. (1989) *Hazardous Waste Incineration and Human Health.* CRC Press, Boca Raton, Florida, USA.

Travis, C.C. & MacInnis, J.M. (1992) Vapour extraction of organics from the subsurface soils: is it effective? *Environmental Science and Technology*, **26(10)**, 1885–7.

Turnball, P. (1996) Guidance on environments known to be or suspected of being contaminated with anthrax spores. *Land Contamination and Reclamation*, **4(1)**, 37–45.

Tyler, R., Odoemelam, O., Shulec, P. and Marchlik, M. (1991) Uncertainty analysis: an essential component of risk assessment and risk management. In *Municipal Waste Incineration Risk Assessment.* C.C. Travis (ed.) pp 241–9 Plenum Press, New York.

USDHHS (US Department of Health and Human Services) (1985) *Occupational Safety and Health Guidance Manual for Hazardous Waste Site Activities.* National Institute for Occupational Safety and Health, Cincinnati, USA.

USEPA (United States Environmental Protection Agency) (1976) Interim procedures and guidelines for health risks and economic impact assessments of suspected carcinogens. *Federal Register*, **41**, 21–402.

USEPA (United States Environmental Protection Agency) (1979) *Environmental Assessment: Short-term Tests for Carcinogens and other Genotoxic Agents.* EPA/625/9-79/003. USEPA Health Effects Research Laboratory, Research Triangle Park, North Carolina, USA.

USEPA (United States Environmental Protection Agency) (1982) Appendix A – Uncontrolled hazardous waste site scoring system: a users' manual. *Federal Register,* **37(137)**, 31219–43.

USEPA (United States Environmental Protection Agency) (1986) *Reclamation and Redevelopment of Contaminated Land: Volume 1 – United States Case Studies.* EPA/600/2-86/066. Office of Research & Development, Cincinnati, USA.

USEPA (United States Environmental Protection Agency) (1988a) *Superfund Exposure Assessment Manual.* EPA/540/1-88/001. Office of Remedial Response, Washington, DC, USA.

USEPA (United States Environmental Protection Agency) (1988b) *US Production of Manufactured Gases: Assessment of Past Disposal Practices.* EPA/600/2-38/012 (NTIS PB88-165790). USEPA, Cincinnati, USA.

USEPA (United States Environmental Protection Agency) (1988c) *Standard Operating Safety Guides.* USEPA, Washington DC, USA.

USEPA (United States Environmental Protection Agency) (1988d) *Community Relations in Superfund: A Handbook.* EPA/540/G-88/002. Office of Emergency and Remedial Response, Washington, DC, USA.

USEPA (United States Environmental Protection Agency) (1988e) *Selection Criteria for Mathematical Models Used in Exposure Assessments: Groundwater Models.* EPA/600/8-88/075. Office of Health & Environmental Assessment, Washington, DC, USA.

USEPA (United States Environmental Protection Agency) (1988f) *Ecological Effects Database.* Office of Toxic Substances, Washington, DC, USA.

USEPA (United States Environmental Protection Agency) (1989a) *Risk Assessment Guidance for Superfund Volume I: Human Health Evaluation Manual (Part A).* EPA/540/1-89/002. Office of Emergency & Remedial Response, Washington, DC, USA.

USEPA (United States Environmental Protection Agency) (1989b) *Risk Assessment Guidance for Superfund Volume II: Environmental Evaluation Manual.* EPA/540/1-89/001. Office of Remedial Response, Washington, DC, USA.

USEPA (United States Environmental Protection Agency) (1989c) *Exposure Factors Handbook.* EPA/600/8-89/043. Office of Remedial Response, Washington, DC, USA.

USEPA (United States Environmental Protection Agency) (1989d) *Ecological Assessment of Hazardous Waste Sites: A Field and Laboratory Reference Document.* EPA/600/3-89/013. Office of Remedial Response, Washington, DC, USA.

USEPA (United States Environmental Protection Agency) (1990a) *Reducing Risk: Setting Priorities and Strategies for Environmental Protection.* SAB-EC-90-021. Science Advisory Board, Washington, DC, USA.

USEPA (United States Environmental Protection Agency) (1990b) Hazard ranking system – final rule. *Federal Register,* **55**, 51532–667.

USEPA (United States Environmental Protection Agency) (1990c) *The Revised Hazard Ranking: An Improved Tool for Screening Superfund Sites.* 9320.7-01 FS Nov. Office of Solid Waste and Emergency Response, Washington, DC, USA.

USEPA (United States Environmental Protection Agency) (1990d) *AQUIRE.* USEPA Research Laboratory, Duluth, Minnesota, USA.

USEPA (United States Environmental Protection Agency) (1991a) *Summary Report on Issues in Ecological Risk Assessment.* EPA/625/3-91/018. USEPA, Washington, DC, USA.

USEPA (United States Environmental Protection Agency) (1991b) *Risk Assessment Guidance for Superfund (Part A – Baseline Risk Assessment) Supplemental Guidance/Standard Exposure Factors.* USEPA OSWER Directive 9285.6-03. Office of Remedial Response, Washington, DC, USA.

USEPA (United States Environmental Protection Agency) (1991c) *Description and Sampling of Contaminated Soils: A Field Pocket Guide.* EPA/625/12-91/002. Center for Environmental Research Information, Cincinnati, USA.

USEPA (United States Environmental Protection Agency) (1991d) *Handbook of Suggested Practices for the Design and Installation of Groundwater Monitoring Wells.* EPA/600/4-89/034. USEPA, Washington, DC, USA.

USEPA (United States Environmental Protection Agency) (1991e) *Handbook: Groundwater Volume II*. EPA/625/6-90/016b. USEPA, Washington, DC, USA.

USEPA (United States Environmental Protection Agency) (1992a) *Reclamation and Redevelopment of Contaminated Land: Volume II European Case Studies*. EPA/600/R-92/031. Office of Research and Development, Cincinnati, USA.

USEPA (United States Environmental Protection Agency) (1992b) *Framework for Ecological Risk Assessment*. EPA/630/R-92/001. Risk Assessment Forum, Washington, DC, USA.

USEPA (United States Environmental Protection Agency) (1992c) Final guidelines for exposure assessment. *Federal Register,* **57**, 22888–938.

USEPA (United States Environmental Protection Agency) (1992d) *Supplemental Guidance to RAGS: Calculating the Concentration Term*. USEPA, Washington, DC, USA.

USEPA (United States Environmental Protection Agency) (1993) *Provisional Guidance for Quantitative Risk Assessment of Polycyclic Aromatic Hydrocarbons*. EPA/600/R-93/089. USEPA, Washington, DC, USA.

USEPA (United States Environmental Protection Agency) (1995a) *HNU-Hanby Immunoassay Test Kit: Innovative Technology Evaluation Report*. EPA/540/R-95/515. USEPA, Washington, DC, USA.

USEPA (United States Environmental Protection Agency) (1995b) *Field Analytical Screening Program: PCP Method: Innovative Technology Evaluation Report*. EPA/540/R-95/528. USEPA, Washington, DC, USA.

USEPA (United States Environmental Protection Agency) (1995c) *Clor-N-Soil PCB Test Kit: Innovative Technology Evaluation Report*. EPA/540/R-95/518. USEPA, Washington, DC, USA.

USEPA (United States Environmental Protection Agency) (1995d) *Rapid Optical Screen Tool (ROST™): Innovative Technology Evaluation Report*. EPA/540/R-95/519. USEPA, Washington, DC, USA.

USEPA (United States Environmental Protection Agency) (1995e) *Site Characterization Analysis Penetrometer System (SCAPS): Innovative Technology Evaluation Report*. EPA/540/R-95/520. USEPA, Washington, DC, USA.

USEPA (United States Environmental Protection Agency) (1995f) *Human Exposure Assessment: A Guide to Risk Ranking, Risk Reduction and Research Planning*. EPA/SAB/IAQC-95/005. USEPA, Washington, DC, USA.

USEPA (United States Environmental Protection Agency) (1996a) *Soil Screening Guidance: Technical Background Document*. EPA/540/R-95/128. USEPA, Washington, DC, USA.

USEPA (United States Environmental Protection Agency) (1996b) *Soil Screening Guidance: Users Guide*. EPA/540/R-96/018. USEPA, Washington, DC, USA.

USEPA (United States Environmental Protection Agency) (1996c) *Proposed Guidelines for Carcinogen Risk Assessment*. EPA/600/P-92/003c. USEPA, Washington, DC, USA.

Van den Berg, R., Denneman, C.A.J. & Roels, J.M. (1993) Risk assessment of contaminated soil: proposals for adjusted, toxicologically based Dutch soil clean-up criteria. In *Contaminated Soil 1993*. F. Arendt, G.J. Annokkée & W.J. van den Brink (eds), pp 349–64. Kluwer Academic Publishers, Dordrecht, the Netherlands.

Van der Pligt, J. & De Boer, J. (1991) Contaminated soil: public reactions, policy decisions and risk communication. In *Communicating Risks to the Public*. R.E. Kasperson & P.J.M. Stallen (eds), pp 127–44. Kluwer Academic Publishers, Dordrecht, the Netherlands.

Van Dyke, E. (1995) The contaminated sites policy in Flanders. In *Contaminated Soil '95*, W.J. van der Brink, R.Bosman & F. Arendt (eds), pp 39–48. Kluwer Academic Publishers, Dordrecht, the Netherlands.

Van Emon, J.M. & Gerlach, C.L. (1995) A status report on field-portable immunoassay.

Environmental Science and Technology, **29(7)**, 312a–17a.

Van Straalen, N.M. & Denneman, C.A.J. (1989) Ecotoxicological evaluation of soil quality criteria. *Ecotoxicology and Environmental Safety,* **18**, 231–51.

Veerkamp, W. & ten Berge, W. (1992) *Hazard Assessment of Chemical Contaminants in Soil,* Technical Report No. 40. ECETOC, Brussels.

Viellenave, J.H. & Hickey, J.C. (1991) Use of high resolution passive soil gas analysis to characterize sites contaminated with unknowns, complex mixtures, and semivolatile organic compounds. *Hazardous Materials Control,* **4(4)**, 42–9.

Voice, T.C. & Kolb, B. (1993) Static and dynamic headspace analysis of volatile organic compounds in soils. *Environmental Science and Technology,* **27(4)**, 709–13.

Wajzer, M.R. & Glover, J.M. (1995) Detection of buried metal tanks and drums using ground probing radar. *Land Contamination & Reclamation,* **3(3)**, 167–72.

Wales, J., Myers, K. & Vogt, T. (1993) Hazard ranking criteria for contaminated sites. In *Contaminated Soil '93.* F. Arendt, G.J. Annokkée & W.J. van den Brink (eds), pp 579–80. Kluwer Academic Publishers, Dordrecht, the Netherlands.

Ward, R.S., Williams, G.M. & Hill, C.C. (1993) Changes in landfill gas composition during migration. In *Proceedings of Discharge Your Obligations,* pp 381–92. Institute of Wastes Management, Northampton, UK.

Warwickshire Environmental Protection Council (1995) *Landfill Gas from Closed Sites in Coventry & Warwickshire: An Approach to Risk Management.* Warwickshire Environmental Protection Council, Nuneaton, UK.

WDA (Welsh Development Agency) (1993) *The WDA Manual on the Remediation of Contaminated Land.* Welsh Development Agency, Cardiff, UK.

Welinder, A.S. (1993) Property value and remediation: the Danish approach. In *Proceedings of the Fourth International KfK/TNO Conference on Contaminated Soil,* pp 63–7. Kluwer Academic Publishers, Dordrecht, the Netherlands.

Welsh Office (1988) *Survey of Contaminated Land in Wales.* Welsh Office, Cardiff, UK.

West, O.R., Siegrist, R.L., Mitchell, T.J. & Jenkins, R.A. (1995) Measurement error and spatial variability effects on characterization of volatile organics in the subsurface. *Environmental Science and Technology,* **29(3)**, 647–56.

Wild, A. (1993) *Soils and the Environment.* Cambridge University Press, Cambridge, UK.

Wildavsky, A. (1995) *But Is It True? A Citizen's Guide to Environmental Health and Safety Issues.* Harvard University Press, Cambridge, Mass, USA.

Wilkinson, J.G. (1979) *Industrial Wood Preservation.* The Rentokil Library. Associated Press, London.

Williams, G.M. & Aitkenhead, N. (1989) The gas explosion at Loscoe, Derbyshire. In *Proceedings of Symposium on Methane – Facing the Problems.* Paper 3–6, March. Nottingham University, UK.

WRC (Water Research Centre) (1992) *Methodology for Monitoring and Sampling Groundwater.* WRC, Medmenham, UK.

WRC/NRA (Water Research Centre/National Rivers Authority) (1993) *Pollution Potential of Contaminated Sites: a Review.* WRC, Medmenham, UK.

Wynne, B. (ed.) (1987) *Risk Management and Hazardous Waste: Implementation and the Dialectics of Credibility.* Springer Verlag, Berlin.

Young, P. (1992) *Representative Groundwater Sampling and an Overview of Surface Water Sampling, Sample Handling and Storage.* Course notes: Site Investigation, Centre for Hazard and Risk Management, November, Loughborough University, UK.

Young, P.J. & Parker, A. (1983) The identification and possible environmental impact of trace gases and vapours in landfill gas. *Waste Management & Research,* **1(3)**, 213–26.

Index